Advances in Kinetic Theory and Continuum Mechanics

R. Gatignol and Soubbaramayer (Eds.)

Advances in Kinetic Theory and Continuum Mechanics

Proceedings of a Symposium
Held in Honor of Professor Henri Cabannes
at the University Pierre et Marie Curie,
Paris, France, on 6 July 1990

With 63 Figures

Springer-Verlag
Berlin Heidelberg New York
London Paris Tokyo
Hong Kong Barcelona
Budapest

Professor Renée Gatignol
Laboratoire de Modélisation en Mécanique,
Université Pierre et Marie Curie, associé au C. N. R. S.
4, place Jussieu, F-75252 Paris Cedex 05, France

Professor Soubbaramayer
SPEA, CEN-Saclay, F-91191 Gif-sur-Yvette Cedex, France

ISBN 978-3-642-50237-8 ISBN 978-3-642-50235-4 (eBook)
DOI 10.1007/978-3-642-50235-4

Softcover reprint of the hardcover 1st edition 1991

57/3140-543210 – Printed on acid-free paper

Preface

This volume contains the proceedings of the symposium held on Friday 6 July 1990 at the University Pierre et Marie Curie (Paris VI), France, in honor of Professor Henri Cabannes on the occasion of his retirement. There were about one hundred participants from nine countries: Canada, France, Germany, Italy, Japan, Norway, Portugal, the Netherlands, and the USA. Many of his past students or his colleagues were among the participants.

The twenty-six papers in this volume are written versions submitted by the authors and cover almost all the fields in which Professor Cabannes has actively worked for more than forty-five years. The papers are presented in four chapters: classical kinetic theory and fluid dynamics, discrete kinetic theory, applied fluid mechanics, and continuum mechanics.

The editors would like to take this opportunity to thank the generous sponsors of the symposium: the University Pierre et Marie Curie, Commissariat à l'Energie Atomique (especially Academician R. Dautray and Dr. N. Camarcat) and Direction des Recherches et Etudes Techniques (especially Professor P. Lallemand). Many thanks are also due to all the participants for making the symposium a success. Finally, we thank Professor W. Beiglböck and his team at Springer-Verlag for producing this volume.

Paris,
January 1991

R. Gatignol
Soubbaramayer

Chers Collègues et amis,

Je suis très heureux et très honoré d'être invité à ouvrir cette journée toute entière consacrée à célébrer Henri Cabannes, son œuvre et sa personne, particulièrement remarquables l'une et l'autre à bien des titres. Henri Cabannes, c'est la vivacité personnifiée, celui qui fait vite et bien. Au volant de sa voiture, comme au déjeuner du restaurant de l'Université. Mais surtout vivacité d'esprit déjà étonnante pour ses camarades et ses professeurs lorsqu'il préparait les concours, encore plus surprenante aujourd'hui où elle fait toujours l'admiration de ses élèves et de ses jeunes collègues. Vivacité alliée à une grande virtuosité, hier dans les méthodes analytiques, aujourd'hui dans les méthodes numériques ; dynamisme qui aborde des problèmes nouveaux avec résolution, souci du résultat à atteindre ; et avec les succès que nous connaissons tous dans le domaine de la dynamique des gaz, de la magnétodynamique des fluides, des gaz raréfiés, de la théorie des cordes, de la théorie des chocs et des ondes de choc.

Henri Cabannes, c'est aussi le courage allié à la perspicacité. Homme de devoir, lorsqu'il a jugé où se situe le bien, il se lance dans l'action avec lucidité et détermination. Les plus anciens se rappellent sa conduite exemplaire pendant la guerre. Mais ces qualités, nous les avons tous découvertes dans l'action quotidienne. Henri Cabannes, c'est encore la fidélité dans toute sa vie envers ses collègues et ses élèves et aussi envers le laboratoire. Il y aurait encore tant de choses à évoquer : son amour des voyages, sa curiosité pour les nouveautés techniques, son inaltérable jeunesse d'esprit, ce talent, ce don qu'il est si difficile de conserver intact avec l'âge.

Je m'arrête là pour deux raisons. D'abord parce que le temps m'est compté et quand il s'agit de Henri Cabannes il faut faire attention au temps ; c'est pourquoi, contrairement à mon habitude, j'ai écrit cette brève allocution.

Et surtout aussi parce que Henri Cabannes est un homme discret qui a suivi depuis toujours le conseil donné par Jean Bernard dans le titre d'un chapitre de l'un de ses derniers ouvrages : "Il ne faut pas faire d'exhibitionnisme avec son cœur."

Il me reste à présent à remercier tout d'abord le laboratoire, l'ancien Laboratoire de Mécanique Théorique et l'actuel Laboratoire de Modélisation en Mécanique et en particulier tous ceux de ses membres qui ont préparé cette journée. Merci à tous ceux, personnalités et organismes, qui ont généreusement facilité son organisation et très spécialement à la DRET (Direction des Recherches et Etudes Techniques), au CEA (Commissariat à l'Energie Atomique) de Saclay et au Département des Applications Militaires. Merci surtout à vous tous, collègues et amis étrangers, qui avez fait le voyage à Paris et consacré cette journée à notre colloque : votre présence, vos communications, votre participation et l'ouvrage qui en gardera la mémoire, constituent pour Henri Cabannes le plus beau des cadeaux. Vite, vite, vite il faut m'arrêter : Cher Henri Cabannes, je suis sûr d'être l'interprète de tous, en te disant : "dans toutes nos paroles et toutes les actions de cette journée, dans toutes nos équations et nos transparents, dans tous nos gestes, nos poignées de mains et nos sourires, vois le témoignage de notre estime, de notre amitié et de notre admiration". Et maintenant, place au colloque !

P. Germain
Secrétaire Perpétuel
de l'Académie des Sciences

Contents

Part I

Classical Kinetic Theory and Fluid Dynamics

Trend to Equilibrium in a Gas According to the Boltzmann Equation

C. Cercignani

Dipartimento di Matematica, Politecnico di Milano,
Piazza Leonardo da Vinci 32, I-20133 Milano, Italy

Abstract : A proof that a gas in a container kept at constant and uniform temperature reaches a Maxwellian state is given. The cases of specularly reflecting walls and velocity reversing walls, previously considered by Desvillettes, are singular, in the sense that the Maxwellian is not uniquely determined by the boundary conditions.

1. INTRODUCTION.

Discussions of the trend to an equilibrium state (Maxwellian distribution) in kinetic theory are as old as the theory itself; beginning with Boltzmann's H-theorem. This theorem leads one to believe that the solution of the Boltzmann equation describing the evolution of the gas in a domain bounded by solid walls kept at a constant temperature, should tend to a Maxwellian distribution. Doubts on the possibility of proving this result in a mathematically rigorous way were cast by Truesdell and Muncaster [1], who claimed that the H-theorem "is not a sufficient condition for the strict trend to equilibrium" ; this is of course correct, if one disregards boundary conditions (which might be incompatible with the trend to equilibrium, when the temperature of the boundary is not uniform), or there are no boundaries. If, on the other hand, the boundary conditions are taken into account and are compatible with equilibrium, then the popular belief that the solution should tend to a Maxwellian is well justified [2].

Recently, the theory of the solutions of the Boltzmann equation has received a rigorous basis, thanks to the major contribution of Di Perna and Lions [3], which allows an accurate discussion of the trend to equilibrium along the lines advocated in Ref. [2].

2. BASIC EQUATIONS.

We shall start by recalling a few facts concerning the Boltzmann equation and the boundary conditions.

The Boltzmann equation (in the absence of body forces) reads as follows ([4],[5]):

$$\frac{\partial f}{\partial t} + \underline{\xi}.\frac{\partial f}{\partial \underline{x}} = \mathcal{A} \int_{R^3} \int_{\mathcal{B}} (f'f'_* - ff_*)\, B(\underline{V}, |\underline{V}.\underline{n}|)\, d\underline{\xi}_*\, d\underline{n} \tag{2.1}$$

where $f(\underline{x},\underline{\xi},t)$ is proportional to the probability density of finding a molecule at position $\underline{x}$, with velocity $\underline{\xi}$ at time t, $\underline{\xi}_*$ an integration variable (the velocity of the partner molecule in a collision), $\underline{n}$ the unit vector of the straight line joining the centers of the molecules at the moment of closest approach (ranging over the unit sphere $\mathcal{B}$), $\underline{\xi}'$ and $\underline{\xi}'_*$ the velocities that two molecules must have in order to have velocities $\underline{\xi}$ and $\underline{\xi}_*$ at the end of the collision. $\underline{V}$ is the relative velocity $\underline{\xi} - \underline{\xi}_*$, $\mathcal{A}$ a constant and $B(\underline{V}, |\underline{V}.\underline{n}|)$ a strictly positive kernel proportional to the differential cross section.

If we assume ([4],[5]) that the molecules do not interact among themselves when they are interacting with the wall $\partial\Omega$, we have the following condition :

$$|\underline{\xi}.\underline{n}|\, f^+(\underline{x},\underline{\xi},t) = \int_{\underline{\xi}'.\underline{n}<0} K(\underline{\xi},\underline{\xi}';\underline{x})\, f^-(\underline{x},\underline{\xi}',t)\, |\underline{\xi}'.\underline{n}|\, d\underline{\xi}'$$

$$(\underline{x} \in \partial\Omega \; ; \quad \underline{\xi}.\underline{n} > 0) \tag{2.2}$$

where $f^{\pm}$ are the restrictions to $\underline{\xi}.\underline{n} > 0$ and $\underline{\xi}.\underline{n} < 0$ of the traces of f on $\partial\Omega$, and K is a nonnegative kernel (which can be a distribution) such that :

$$\int_{\underline{\xi}.\underline{n}<0} K(\underline{\xi},\underline{\xi}';\underline{x})\, d\underline{\xi} = 1 \tag{2.3}$$

$$|\underline{\xi}.\underline{n}|\, M_w^+(\underline{x},\underline{\xi}) = \int_{\underline{\xi}'.\underline{n}<0} K(\underline{\xi},\underline{\xi}';\underline{x})\, M_w^-(\underline{x},\underline{\xi}')\, |\underline{\xi}'.\underline{n}|\, d\underline{\xi}'.$$

$$(x \in \partial\Omega \; ; \quad \underline{\xi}.\underline{n} > 0) \tag{2.4}$$

where M_w is a Maxwellian :

$$M_w = A\, e^{-\beta|\underline{\xi} - v|^2} \tag{2.5}$$

with the local temperature $(= 1/\beta)$ and the bulk velocity $\underline{v}$ of the wall. Eq. (2.3) ensures that no molecule goes through the wall, and Eq. (2.4) that a Maxwellian with the temperature and bulk motion of the wall automatically satisfies the boundary conditions. Two particular cases are the specular reflection and the inverse (or "bounce-back") reflection ; in these two cases, K is a Dirac delta ($\delta(\underline{\xi}' - \underline{\xi} - 2\underline{n}(\underline{n}.\underline{\xi})$ and $\delta(\underline{\xi}' + \underline{\xi})$, respectively), and Eq. (2.4) is satisfied for any Maxwellian with tangential, respectively zero, bulk velocity.

An important property of the boundary conditions under consideration is the following result which was stated by Darrozès and Guiraud [6] and proved by the Author ([7],[3]) :

Lemma : If Eqs. (2.2)-(2.4) hold, then :

$$\mathcal{J}_n = \int \underline{\xi}.\underline{n}\ f \log f\ d\underline{\xi} \leq -\beta_w \int \underline{\xi}.\underline{n}\ |\underline{\xi}|^2\ f\ d\underline{\xi} \qquad (\underline{x} \in \partial\Omega) \qquad (2.6)$$

where β_w is the inverse of the temperature of the wall (appearing in the Maxwellian M_w). Equality holds if, and only if f coincides with M_w(the wall Maxwellian) on $\partial\Omega$ (unless the kernel in Eq. (2.2) is a delta function). We remark that we freely use the trace of f on $\partial\Omega$, since it is known to exist ([8]).

3. TREND TO EQUILIBRIUM.

Let us consider an application of the so-called Boltzmann inequality ([1],[4],[5])to the Boltzmann equation :

$$\frac{\partial f}{\partial t} + \underline{\xi}.\frac{\partial f}{\partial \underline{x}} = \mathcal{A}\, Q(f,f) \qquad (3.1)$$

If we multiply both sides of this equation by log f and integrate with respect to $\underline{\xi}$, we obtain (directly if f is sufficiently smooth, with suitable tricks and use of distributional derivatives in the other cases ([8]) :

$$\frac{\partial \mathcal{H}}{\partial t} + \frac{\partial}{\partial \underline{x}}.\mathcal{J} = \mathcal{S} \qquad (3.2)$$

where

$$\mathcal{H} = \int_{R^3} f \log f\ d\underline{\xi} \qquad (3.3)$$

$$\mathcal{J} = \int_{R^3} \underline{\xi}\ f \log f\ d\underline{\xi} \qquad (3.4)$$

$$\mathcal{S} = \mathcal{A}\int_{R^3} \log f\, Q(f,f)\, d\underline{\xi} \tag{3.5}$$

Eq. (3.2) differs from a balance equation because the right side, generally speaking, does not vanish. The Boltzmann inequality ([1],[4],[5]) implies :

$$\mathcal{S} \leqslant 0 \quad \text{and} \quad \mathcal{S} = 0 \quad \text{if} \quad f \quad \text{is a Maxwellian} \tag{3.6}$$

Because of this inequality, Eq. (3.2) plays an important role in the theory of the Boltzmann equation. We illustrate the role of Eq. (3.2) in the case of space homogeneous solutions. In this case, the various quantities do not depend on x , and Eq. (3.2) reduces to :

$$\frac{\partial \mathcal{H}}{\partial t} = \mathcal{S} \leqslant 0 \tag{3.7}$$

This implies the so-called H-theorem (for the space homogeneous case): $\mathcal{H}$ is a decreasing quantity, unless f is a Maxwellian (in which case the time derivative of $\mathcal{H}$ is zero). We recall that in this case the mass, momentum and energy densities are constant in time ; we can thus build a Maxwellian M which has, at any time, the same density ρ , bulk velocity $\underline{v}$, and internal energy per unit mass e as any solution f corresponding to given initial data. Since $\mathcal{H}$ decreases unless f is a Maxwellian (i.e. $f = M$), it is tempting to conclude that f tends to M when $t \to \infty$. The temptation is strengthened when we realize that $\mathcal{H}$ is bounded from below by $\mathcal{H}_M$, the value taken by the functional $\mathcal{H}$ when $f = M$. In fact, $\mathcal{H}$ is decreasing, its derivative is nonpositive unless it takes the value $\mathcal{H}_M$; one feels that $\mathcal{H}$ tends to $\mathcal{H}_M$! This conclusion is, however, unwarranted, without a more detailed consideration of the source term $\mathcal{S}$ in Eq. (3.7). This analysis was carried out by Arkeryd [9]. Here, we restrict ourselves to remarking that, if $\mathcal{H}$ tends to $\mathcal{H}_M$, then it is easy ([4]) to conclude that f tends to M , thanks to the elementary inequality :

$$f \log f - f \log M + M - f \geqslant c\, g\left(\frac{|f - M|}{M}\right) |f - M| \tag{3.8}$$

where c is a constant (independent of f), and

$$g(z) = \begin{cases} z & \text{if} \quad 0 \leqslant z \leqslant 1 \\ 1 & \text{if} \quad z \geqslant 1 \end{cases} \tag{3.9}$$

Integrating both sides of Eq. (3.8) gives :

$$\mathcal{H} - \mathcal{H}_M \geqslant c \left[\int_{L_t} |f - M| \, d\xi + \int_{S_t} \frac{|f - M|^2}{M} \, d\xi \right] \tag{3.10}$$

where L_t and S_t denote the sets (depending on t) where $|f - M|$ is larger (resp. smaller) than M . Since $\mathcal{H}$ is assumed to tend to $\mathcal{H}_M$, it follows that both integrals tend to zero when $t \to \infty$. The fact that the second integral tends to zero implies, by Schwarz's inequality, that :

$$\int_{S_t} |f - M| \, d\xi \to 0 \tag{3.11}$$

Then

$$\int_{R^3} |f - M| \, d\xi = \int_{L_t} |f - M| \, d\xi + \int_{S_t} |f - M| \, d\xi \tag{3.12}$$

also tends to zero, and f tends strongly to M in L^1 .

If the state of the gas is not space homogeneous, the situation becomes more complicated. In this case, it is convenient to introduce the quantity

$$H = \int_\Omega \mathcal{H} \, dx \tag{3.13}$$

where Ω is the space domain occupied by the gas (assumed here to be time-independent). Then, Eq. (3.2) implies :

$$\frac{dH}{dt} \leqslant \int_{\partial\Omega} \underline{\mathcal{J}} \cdot \underline{n} \, d\sigma \tag{3.14}$$

where $\underline{n}$ is the inward normal, and $d\sigma$ the measure on $\partial\Omega$. Clearly, several situations may arise. Among the most typical ones, we quote :

1. Ω is a box with periodicity boundary conditions (flat torus). Then, there is no boundary, $dH/dt \leqslant 0$, and one can repeat about H what was said about $\mathcal{H}$ in the space homogeneous case. In particular, there is a natural (space homogeneous) Maxwellian associated with the total mass, momentum and energy (which are of course conserved). The proof that f tends to this Maxwellian is now more complicated.

2. Ω is a compact domain with specular reflection. In this case, the boundary term also disappears because the integrand of $\underline{\mathcal{J}} \cdot \underline{n}$ is odd on $\partial\Omega$, and the situation is similar to that in case 1. There might seem to be a difficulty for the choice of the natural Maxwellian, because momentum is not conserved. In the case of a box, a simple argument shows that the total momentum must vanish when $t \to \infty$. If $\partial\Omega$ is not a surface of revolution (in which case the gas might rotate as a solid body), the same conclusion should hold. Thus, if rotationally invariant domains are excluded, then M is a non - drifting Maxwellian with constant density and tempera-

ture. A simpler case is that of the reverse reflection ; in this case, the rotationally invariant domains do not constitute an exception.

3. Ω is the entire space. Then, the asymptotic behavior of the initial values at space infinity is of paramount importance. If the gas is initially more concentrated at finite distances from the origin, one physically expects and can mathematically prove that the gas excapes through infinity and the asymptotic state is a vacuum.

4. Ω is a compact domain, but the boundary conditions on $\partial\Omega$ are different from specular reflection. Then, the asymptotic state might be completely different from a Maxwellian, unless the temperature is constant along the boundary.

Case 1 was dealt with a rigorously way by Arkeryd [10], who used the tools of nonstandard analysis. Desvillettes [11] showed that one can dispense with the latter, and treated in detail case 2 Case 3 , in general, does not lead to equilibrium ; if the behavior at infinity is extremely bad, the solution may even cease to exist after a finite time. Particular cases are known ([1],[12],[13]), but a general discussion is lacking. Case 4 (in the case of a boundary at uniform temperature) is discussed below.

The aim of the remaining part of this section is in fact to discuss the trend to equilibrium, following the approach of Desvillettes [11], which is based on a remark by Di Perna and Lions [14]. The main result is the following :

Theorem : Let $f(\underline{x},\underline{\xi},t)$ be the solution of the Boltzmann equation (3.1), with initial data $f_o(\underline{x},\underline{\xi})$ such that :

$$f_o \geqslant 0\ ,\quad \int_\Omega \int_{R^3} f_o(\underline{x},\underline{\xi})(1 + |\underline{\xi}|^2 + |\log f_o(\underline{x},\underline{\xi})|)\, d\underline{x}\, d\underline{\xi} < +\infty \qquad (3.15)$$

Let f also satisfy the boundary condition (2.2), where the kernel is such that Eqs (2.3) and (2.4), with M_w a constant and uniform Maxwellian. Then, for every sequence t_n going to infinity, there exist a subsequence t_{n_k} and a local Maxwellian $M(\underline{x},\underline{\xi},t)$ such that $f_{n_k}(\underline{x},\underline{\xi},t) = f(\underline{x},\underline{\xi},t_{n_k}+t)$ converges weakly in$L^1(\Omega \times R^3 \times [0\ ,T])$ to $M(\underline{x},\underline{\xi},t)$ for any $T > 0$. Moreover, M satisfies the free transport equation :

$$\frac{\partial M}{\partial t} + \underline{\xi} \cdot \frac{\partial M}{\partial \underline{x}} = 0\ . \qquad (3.16)$$

and the boundary condition (2.2).

<u>Proof</u> : The proofs of Arkeryd [10] and Desvillette [11] can be applied to this case, provided the key inequality (2.6) (with β_w constant) is taken

into account. In fact, in this case, Hamdache [8] proved that conservation of energy and the H-theorem imply that :

$$\int_0^T \int_{R^3} \int_\Omega \int_{\mathcal{B}} \int_{R^3} [f(\underline{x},\underline{\xi}',t)\, f(\underline{x},\underline{\xi}'_*,t) - f(\underline{x},\underline{\xi},t)\, f(\underline{x},\underline{\xi}_*,t)]$$

$$\{\log[f(\underline{x},\underline{\xi}',t)\, f(\underline{x},\underline{\xi}'_*,t)] - \log[f(\underline{x},\underline{\xi},t)\, f(\underline{x},\underline{\xi}_*,t)]\}$$

$$B(\underline{V},\underline{V}.\underline{n})\, d\underline{\xi}_*\, d\underline{n}\, d\underline{x}\, d\underline{\xi}\, dt$$

$$+ \sup_t \int_\Omega \int_{R^3} f(\underline{x},\underline{\xi},t)(1+|\underline{\xi}|^2+|\log f(\underline{x},\underline{\xi},t)|)\, d\underline{x}\, d\underline{\xi} < +\infty \qquad (3.17)$$

Thus, $f_n(\underline{x},\underline{\xi},t) = f(\underline{x},\underline{\xi},t+t_n)$ is weakly compact in $L^1(\Omega \times R^3 \times [0,T])$ for any sequence t_n of nonnegative numbers and any $T > 0$. If $t_n \to \infty$, then there exist a subsequence t_{n_k} and a function $M(\underline{x},\underline{\xi},t)$ in $L^1(\Omega \times R^3 \times [0,T])$ such that f_{n_k} converges weakly to M in $L^1(\Omega \times R^3 \times [0,T])$ for any $T > 0$. In order to prove that M is a Maxwellian, we remark that, since the first integral in Eq. (3.17) is finite, then :

$$\int_{t_{n_k}}^{T+t_{n_k}} \int_{R^3} \int_\Omega \int_{\mathcal{B}} \int_{R^3} [f(\underline{x},\underline{\xi}',t)\, f(\underline{x},\underline{\xi}'_*,t) - f(\underline{x},\underline{\xi},t)\, f(\underline{x},\underline{\xi}_*,t)]$$

$$\{\log[f(\underline{x},\underline{\xi}',t)\, f(\underline{x},\underline{\xi}'_*,t)] - \log[f(\underline{x},\underline{\xi},t)\, f(\underline{x},\underline{\xi}_*,t)]\}$$

$$B(\underline{V},\underline{V}.\underline{n})\, d\underline{\xi}_*\, d\underline{n}\, d\underline{x}\, d\underline{\xi}\, dt$$

tends to 0 when k tends to infinity, and thus :

$$\int_0^T \int_{R^3} \int_\Omega \int_{\mathcal{B}} \int_{R^3} [f_{n_k}(\underline{x},\underline{\xi}',t)\, f_{n_k}(\underline{x},\underline{\xi}'_*,t) - f_{n_k}(\underline{x},\underline{\xi},t)\, f_{n_k}(\underline{x},\underline{\xi}_*,t)]$$

$$\{\log[f_{n_k}(\underline{x},\underline{\xi}',t)\, f_{n_k}(\underline{x},\underline{\xi}'_*,t)] - \log[f_{n_k}(\underline{x},\underline{\xi},t)\, f_{n_k}(\underline{x},\underline{\xi}_*,t)]\}$$

$$B(\underline{V},\underline{V}.\underline{n})\, d\underline{\xi}_*\, d\underline{n}\, d\underline{x}\, d\underline{\xi}\, dt \quad \to 0 \quad (k \to \infty) \qquad (3.18)$$

But, as Di Perna and Lions [14] first showed for R^3 and Hamdache [8] for an open set $\Omega \in R^3$, for all smooth nonnegative functions Φ , ψ with compact support :

$$\int_{R^3} \int_{\mathcal{B}} \int_{R^3} [f_{n_k}(\underline{x},\underline{\xi}',t)\, f_{n_k}(\underline{x},\underline{\xi}'_*,t)\, \Phi(\underline{\xi})\, \psi(\underline{\xi}_*)\, B(\underline{V},\underline{V}.\underline{n})\, d\underline{\xi}_*\, d\underline{n}\, d\underline{\xi}$$

$$\to \int_{R^3} \int_{\mathcal{B}} \int_{R^3} [M(\underline{x},\underline{\xi}',t)\, M(\underline{x},\underline{\xi}'_*,t)\, \Phi(\underline{\xi})\, \psi(\underline{\xi}_*)\, B(\underline{V},\underline{V}.\underline{n})\, d\underline{\xi}_*\, d\underline{n}\, d\underline{\xi}$$

$$(\text{a.e. in } \Omega \times [0,T] \text{ when } k \to \infty) \qquad (3.19)$$

and

$$\int_{R^3}\int_{\mathcal{B}}\int_{R^3} [f_{n_k}(\underline{x},\underline{\xi},t)\, f_{n_k}(\underline{x},\underline{\xi}_*,t)]\, \Phi(\underline{\xi})\, \psi(\underline{\xi}_*)\, B(\underline{V},\underline{V}.\underline{n})\, d\underline{\xi}_*\, d\underline{n}\, d\underline{\xi}$$

$$\rightarrow \int_{R^3}\int_{\mathcal{B}}\int_{R^3} [M(\underline{x},\underline{\xi},t)\, M(\underline{x},\underline{\xi}_*,t)]\, \Phi(\underline{\xi})\, \psi(\underline{\xi}_*)\, B(\underline{V},\underline{V}.\underline{n})\, d\underline{\xi}_*\, d\underline{n}\, d\underline{\xi}$$

$$(\text{a.e. in } \Omega \times [0,T] \text{ when } k \rightarrow \infty) \tag{3.20}$$

It is then possible to extract a subsequence (which we still denote by f_{n_k}) such that :

$$\int_{R^3}\int_{\Omega}\int_{R^3} [f_{n_k}(\underline{x},\underline{\xi}',t)\, f_{n_k}(\underline{x},\underline{\xi}_*',t) - f_{n_k}(\underline{x},\underline{\xi},t)\, f_{n_k}(\underline{x},\underline{\xi}_*,t)]$$

$$\{\log[f_{n_k}(\underline{x},\underline{\xi}',t)\, f_{n_k}(\underline{x},\underline{\xi}_*',t)] - \log[f_{n_k}(\underline{x},\underline{\xi},t)\, f_{n_k}(\underline{x},\underline{\xi}_*,t)]\}$$

$$B(\underline{V},\underline{V}.\underline{n})\, d\underline{\xi}_*\, d\underline{n}\, d\underline{\xi} \rightarrow 0$$

$$(\text{a.e. in } \Omega \times [0,T] \text{ when } k \rightarrow \infty) \tag{3.21}$$

and the same applies to Eq. (3.19) and (3.20) for a dense denumerable set in $C(R^3)$ of nonnegative smooth functions Φ and ψ. But, then, the convexity of the function $C(f,g) = (f - g)(\log f - \log g)$ $(R_+ \times R_+ \rightarrow R_+)$ implies that :

$$[M(\underline{x},\underline{\xi}',t)\, M(\underline{x},\underline{\xi}_*',t) - M(\underline{x},\underline{\xi},t)\, M(\underline{x},\underline{\xi}_*,t)]$$

$$\{\log[M(\underline{x},\underline{\xi}',t)\, M(\underline{x},\underline{\xi}_*',t)] - \log[M(\underline{x},\underline{\xi},t)\, M(\underline{x},\underline{\xi}_*,t)]\}\, B(\underline{V},\underline{V}.\underline{n}) = 0$$

$$(\text{a.e. in } \underline{\xi}_*,\ \underline{n},\ \underline{x},\ \underline{\xi}_*,\ t) \tag{3.22}$$

Then, since $C(f,g)$ is nonnegative, and $B(\underline{V},\underline{V}.\underline{n})$ strictly positive :

$$M(\underline{x},\underline{\xi}',t)\, M(\underline{x},\underline{\xi}_*',t) = M(\underline{x},\underline{\xi},t)\, M(\underline{x},\underline{\xi}_*,t)$$

$$(\text{a.e. in } \underline{\xi}_*,\ \underline{n},\ \underline{x},\ \underline{\xi},\ t) \tag{3.23}$$

Then ([1],[4],[5],[15],[16]) M is a Maxwellian and, moreover, is, thanks to the property of weak stability ([3],[8]), a renormalized solution of the Boltzmann equation, satisfying the boundary condition (2.2). Accordingly $Q(M,M) = 0$, and Eq. (3.16) is satisfied.

The theorem tells us that the solutions of the Boltzmann equation with the boundary conditions (2.5) behave (in the case of a boundary at constant temperature) as Maxwellians satisfying the free transport equation, Eq.

(3.6). These Maxwellians are well known since Boltzmann [13]. In order to recover them quickly, we remark that log M must be a polynomial of the form $a + \underline{b} \cdot \underline{\xi} + c|\underline{\xi}|^2$, and $a, c \in R$ and $\underline{b} \in R^3$ must be such that log M (as well as M) is a function of $\underline{\xi}$ and $\underline{x} - \underline{\xi}t$; this readily implies (see [17]) that :

$$M = \exp[a_0 + \underline{b}_0 \cdot \underline{\xi} + c_0|\underline{\xi}|^2 + d_0|\underline{x} - \underline{\xi}t|^2$$

$$+ \underline{e}_0 \cdot (\underline{x} - \underline{\xi}t) + \underline{f}_0 \cdot (\underline{x} \wedge \underline{\xi}) + g_0\, \underline{\xi} \cdot (\underline{x} - \underline{\xi}t)] \qquad (3.24)$$

where $a_0, c_0, d_0, g_0 \in R$ and $\underline{b}_0, \underline{e}_0, \underline{f}_0 \in R^3$ are constant. Now if we impose the condition that $M(\underline{x},.,t)$ is an L^1 function for any $t > 0$, we see that c_0 must be negative, and $d_0 - g_0$ nonpositive.

We exclude now from our considerations the cases in which the kernel K is a delta function ; in fact, the only significant situations in which a Dirac delta occurs, i.e. the cases of specular and reverse reflections have been treated in detail by Desvillettes [11]. In the other cases, there is only one Maxwellian which is compatible with the boundary conditions, i.e. a Maxwellian with no drift and constant temperature ; this immediately implies that b_0, d_0, $\underline{e}_0$, $\underline{f}_0$, g_0 are zero. Thus M is a uniform Maxwellian, which coincides with M_w, and we have the following result, embodying the results of Desvillettes [11] :

<u>Corollary</u> : The Maxwellian M in the previous theorem is uniform. The only possible exception is provided by the case of a rotationally invariant domain with specular reflection ; in this case, the Maxwellian might describe a solid body rotation.

4. CONCLUDING REMARKS.

This contribution is based on a more detailed paper of the author [17], and should be taken as a step into the direction of the study of the asymptotic trend of the solutions of the Boltzmann equation. We remark, however, that we have left out of the discussion the cases in which the temperature is not constant along the solid boundaries or the latter are moving. In these situations, one should not expect a trend to equilibrium, but at the best (if the boundary conditions are time-independent) a trend toward a steady state. The rigorous discussion of these problems cannot even start at this moment, because we do not possess an existence theorem for solid boundaries in a nonuniform state.

REFERENCES

[1] C. TRUESDELL and R.G. MUNCASTER, *"Fundamentals of Maxwell's kinetic theory of a simple monatomic gas"*, Academic Press, New York, 1980.

[2] C. CERCIGNANI, " H-theorem and trend to equilibrium in the kinetic theory of gases", Arch. Mech., 34, (1982), p. 231.

[3] R. Di PERNA and P. L. LIONS, "On the Cauchy problem for the Boltzmann equation. Global existence and stability", Ann. of Math., 130, (1989), p. 321-366.

[4] C. CERCIGNANI, *"The Boltzmann equation and its applications"*, Springer, New York, 1988.

[5] C. CERCIGNANI, *"Mathematical methods in kinetic theory"*, Plenum Press, New York, 1969 (2nd Edition, 1990).

[6] J. S. DARROZES and J. P. GUIRAUD, "Généralisation formelle du théorème H en présence de parois. Applications", C. R. Acad. Sci., Paris, A 262, (1966), p. 1368-1371.

[7] C. CERCIGNANI, "Scattering kernels for gas-surface interactions", Transport Theory and Stat. Phys., 2, (1972), p. 27-53.

[8] K. HAMDACHE, "Initial boundary value problems for Boltzmann equation. Global existence of weak solutions" (to appear).

[9] L. ARKERYD, "On the Boltzmann equation. Part II : The full initial value problem", Arch. Rational Mech. Anal., 45, (1972), p. 17-34.

[10] L. ARKERYD, "On the long time behaviour of the Boltzmann equation in a periodic box", Technical Report, University of Göteborg, 1988.

[11] L. DESVILLETTES, "Convergence to equilibrium in large time for Boltzmann and BGK equations", Arch. Rational Mech. Anal., 110, (1990), p. 73-91.

[12] G. TOSCANI, " H-theorem and asymptotic trend of the solution for a rarefied gas in the vacuum", Arch. Rational Mech. Anal., 100, (1987), p.1-12.

[13] L. BOLTZMANN, "Uber die Aufstellung und Integration der Gleichungen, welche die Molekularbewegungen in Gasen bestimmen", Sitzungsberichte der Akademie der Wissenschaften, Wien 74, (1876), p. 503-552.

[14] R. DI PERNA and P. L. LIONS, "Global solutions of the Boltzmann equation and the entropy inequality" (to appear).

[15] C. CERCIGNANI, "Are there more than five collision invariants for the nonlinear Boltzmann equation ?", J. Stat. Phys., 58, p. 817-823.

[16] L. ARKERYD and C. CERCIGNANI, "On a functional equation arising in the kinetic theory of gases", Rend. Mat. Acc. Lincei, Ser. 9, 1, (1990), p. 139-149.

[17] C. CERCIGNANI, "Equilibrium states and trend to equilibrium in a gas according to the Boltzmann equation", Rend. Mat., (to appear).

The Dirichlet Boundary Value Problem for B.G.K. Equation

B. Perthame and A. Pham Ngoc Dinh

Département de Mathématiques, Université d'Orléans,
B.P. 6759, F-45067 Orléans Cedex 2, France

Abstract : We prove the global existence of a solution to the B.G.K. model of Boltzmann Equation set in a bounded domain with Dirichlet boundary condition. Our proof combines the averaging lemmas and estimates on the second moment in v of the solution in order to pass to limit (strongly) in the temperature.

I. INTRODUCTION

The B.G.K. model (Bathnagar, Gross and Kroock [1]) is a simplified model of the Boltzmann Equation deviced to have some common properties with the Boltzmann Equation (conservation of mass, momentum, energy; appropriate fluid limit at the inviscid level) while involving a simple collision kernel. This model consists in finding the solution f(x,v,t) of the non-linear kinetic equation

$$\begin{cases} \partial_t f + v.\nabla_x f + f - M[f] = 0 & x \in \Omega,\ v \in \mathbb{R}^N,\ t \geq 0, \\ f(x,v,0) = f_0(x,v) & \\ f(x,v,t) = \varphi(x,v) & \text{for } x \in \partial\Omega,\ v.n(x) \leq 0, \end{cases} \tag{1}$$

where Ω is a smooth domain of $\mathbb{R}^N$, $\partial\Omega$ denotes the boundary of Ω and n(x) is the unit outward normal to $\partial\Omega$ at the point $x \in \partial\Omega$. Here M[f] denotes the so-called "local Maxwellian" associated with f, which is a nonlinear function of f defined as follows

$$M[f] = \frac{\rho(x,t)}{(2\pi\ T(x,t))^{N/2}}\ e^{-|v - u(x,t)|^2/2T(x,t)} \tag{2}$$

where $(\rho,\ u,\ T)(x,t)$ are defined through the moments of f

$$\begin{pmatrix} \rho \\ \rho\, u \\ E \end{pmatrix}(x,t) = \int_{\mathbb{R}^N} \begin{pmatrix} 1 \\ v \\ |v|^2/2 \end{pmatrix} f(x,v,t)\, dv, \tag{3}$$

$$E(x,t) = \left(\frac{1}{2}\rho|u|^2 + \rho\, N\, T\right)(x,t) \tag{4}$$

The temperature $T(x,t)$ may be indifferently written as

$$[\rho\, N\, T](x,t) = \int_{\mathbb{R}^N} (|v-u|^2/2)\, f(x,v,t)\, dv \tag{5}$$

and thus ρ and T remain non-negative as long as f is non-negative i.e. when the initial data f_o and the boundary data φ are non-negative. This shows that when $\rho(x,t) = 0$ for some (x,t) then $f(x,v,t) = 0$ for all v and also we choose $M[f](x,v,t) = 0$ for all v, thus the velocity $u(x,t)$ and temperature $T(x,t)$ need not be defined for such (x,t).

The main difficulty for proving the global existence of a solution to B.G.K. Equation comes from the non-linear term $M[f]$ and we refer to [6] for a previous attempt in solving it. In order to point out this difficulty we may imagine a family $(f_n)_{n\geq 1}$ of solutions; if we want to pass to the limit in (1) we need the strong convergence of ρ_n, u_n, T_n (when the limit $\rho(x,t) \neq 0$ at least) since weak limits are not enough to pass to the limit in $M[f_n]$ as n tends to $+\infty$. This strong compactness can be obtained by the averaging lemmas [4, 5] but the difficulty here is that in (3) the integration is performed in the whole space while averaging lemmas require integration on a bounded set. In order to complete the argument we therefore need to prove first the decay of possible solutions for large v. This was performed when $\Omega = \mathbb{R}^N$ in [8] by providing a third moment in v and this argument is not longer valid in bounded sets. Here we will show that, we can however get an a priori decay on $\int_{|v|\geq R} |v|^2 f\, dv$ as R tends to $+\infty$, which is enough to conclude.

Another B.G.K. model is obtained in replacing the term $f - M[f]$ in (1) by $\rho(x,t)(f - M[f])$; this model has a structure closer to that of the Boltzmann Equation (see [2]) would require to combine the renormalization ideas used to solve the Boltzmann Equation by Di Perna-Lions [3], and estimates on the decay of f for large v. This does not seem possible with the techniques we use, even though global existence for boundary value problems for Boltzmann Equation is proved by Hamdache [7].

The paper is organized as follows. We first state the main a priori estimates for the B.G.K. Equation and our existence result (Section II). In section III, we give the main estimate for large v and complete the existence proof.

II. EXISTENCE RESULT

We give now our precise assumptions and result.

$$\int_{\Omega \times \mathbb{R}^N} f_0(x,v)\left(1+|v|^2+ |\mathrm{Log} f_0|\right) dx\, dv < +\infty, \qquad f_0(x,v) \geq 0, \tag{6}$$

$$\int_{\partial\Omega \times \mathbb{R}^N} \varphi(x,v)\ |v.n|\ (1+|v|^2+ |\mathrm{Log}\varphi|)\ d\sigma(x)\ dv < +\infty, \qquad \varphi(x,v) \geq 0 \tag{7}$$

$$\exists\ \Psi(v) \in L^1(\mathbb{R}^N), \quad |v|^2\varphi(x,v) \leq \Psi(v), \quad \forall\ x \in \partial\Omega, \quad v \in \mathbb{R}^N. \tag{8}$$

Theorem : Under assumptions (6)-(8), there exist a solution to the B.G.K. Equation (1)-(4) which satisfies

$$\int_{\Omega \times \mathbb{R}^N} (1+|v|^2+ |\mathrm{Log} f|)\ f(x,v,t)\ dx\, dv < +\infty, \qquad \forall\ t>0, \quad f(x,v,t) \geq 0 \tag{9}$$

$$\int_0^T \int_{\Sigma^+} v.n(x)\ (1+|v|^2)\ f(x,v,t)\ d\sigma(x)\ dv\ dt < +\infty, \qquad \forall\ T>0 \tag{10}$$

where $\Sigma^+ = \{(x,v) \in \partial\Omega \times \mathbb{R}^N, \quad v.n(x) \geq 0\}$.

In the following we will denote Σ^- the complementary subset of Σ^+ in $\partial\Omega \times \mathbb{R}^N$.

Before going to the proof of this theorem, let us explain why the estimate (9) holds. Considering a solution to (1)-(4), we know that f is non-negative and multiplying (1) by $(1+|v|^2)$ and integrating it, we find

$$\frac{d}{dt}\int_{\Omega \times \mathbb{R}^N} (1+|v|^2)\ f(t)\ dx\, dv + \int_{\Sigma^-} v.n(x)\ \varphi(x,v)\ (1+|v|^2)\ d\sigma(x)\ dv$$
$$\leq - \int_{\Sigma^+} v.n(x)\ f(t)\ (1+|v|^2)\ d\sigma(x)\ dv \quad .$$

This inequality together with (7) just yields (10) and bounds the mass and kinetic energy of f(t). Finally

$$\frac{d}{dt}\int_{\Omega\times\mathbb{R}^N} f(t)\ \mathrm{Log}\ f(t)\ dx\ dv + \int_{\Sigma^-} v.n(x)\ \varphi\ \mathrm{Log}\ \varphi\ d\sigma(x)\ dv \tag{11}$$

$$= -\int_{\Sigma^+} v.n(x)\ f(t)\ \mathrm{Log}\ f(t)\ d\sigma(x)\ dv$$

$$\leqslant \int_{\Sigma^+} v.n(x)\ f(t)\ (\mathrm{Log}\ f(t))_-\ d\sigma(x)\ dv$$

where $(t)_- = \mathrm{Max}(0,-t)$. Now we notice that for $f \leqslant 1$, there is a constant C such that $|f\ \mathrm{Log} f| \leqslant C\sqrt{f}$. Then we bound the r.h.s. (11) by

$$\int_{\Sigma^+\cap\left\{f \geqslant e^{-|v|^2}\right\}} v.n(x)\ |v|^2\ f(x,v,t)\ d\sigma(x)\ dv$$

$$+ C\int_{\Sigma^+\cap\left\{f \leqslant e^{-|v|^2}\right\}} v.n(x)\ \sqrt{f}\ d\sigma(x)\ dv \qquad \in L^1([0,T])$$

using the estimate (10). This proves that $\int_{\Omega\times\mathbb{R}^N} f(t)\ \mathrm{Log} f(t)\ dx\ dv$ is a priori bounded and using the same decomposition as above for $f(x,v,t) \leqslant 1$, we find that $\int_{\Omega\times\mathbb{R}^N} f(t)\ |\mathrm{Log} f(t)|\ dx\ dv$ is also bounded and (9) follows.

III. PROOF OF THE THEOREM

It is enough to prove the following

Proposition :Let φ, f_0^n satisfy (6)-(8), let $f_n(x,v,t)$ be a family of solutions associated to the boundary condition φ, f_0^n (and thus satisfying (9)-(10). If

$$f_0^n \rightharpoonup f_0 \text{ weakly in } L^1(\Omega\times\mathbb{R}^N),$$

$$\int_{\Omega\times\mathbb{R}^N} |v|^2\ f_0^n\ dx\ dv \to \int_{\Omega\times\mathbb{R}^N} |v|^2\ f_0\ dx\ dv \tag{12}$$

as $n \to +\infty$, then there exist a solution f to (1)-(3) such that, extracting a subsequence still denoted f_n,

$$f_n,\ |v|^2 f_n \to f,\ |v|^2 f \text{ in } L^1([0,T]\times\Omega\times\mathbb{R}^N).$$

Indeed, it is possible to build approximate solutions to the model (1)-(3) (see [6] for instance) for which the proof of the proposition applies and which converge to a solution.

Proof of the Proposition : The strong compactness stated in this proposition is obtained through the averaging lemmas (see [4,5]). In order to apply them we notice that

$$\int_{\Omega \times \mathbb{R}^N} M[f_n] \, |\mathrm{Log}\, M[f_n]| \, dx \, dv \leqslant C + \int_{\Omega \times \mathbb{R}^N} f_n \, |\mathrm{Log}\, f_n| \, dx \, dv \tag{13}$$

as in [8]. Thus we obtain from Dunford-Pettis Theorem that f_n, $M[f_n]$ are weakly compact in $L^1([0,T] \times \Omega \times B_R)$ for any ball B_R of $\mathbb{R}^N_v$. And averaging lemmas show that, extracting subsequences we obtain

$$\int_{B_R} \Psi(v) \, f_n(x,v,t) \, dv \xrightarrow[n \to \infty]{} \int_{B_R} \Psi(v) \, f(x,v,t) \, dv \quad \text{in } L^1([0,T] \times \Omega), \tag{14}$$

for $\Psi(v) = 1$, v or $|v|^2$ (in the following we will denote by f_n the subsequences).

To conclude that the integral over converge, it remains to show that

$$\int_0^T \int_\Omega \int_{|v| \geqslant R} |v|^2 \, f_n(x,v,t) \, dx \, dv \, dt \xrightarrow[R \to \infty]{} 0 \quad \text{uniformly in } n \tag{15}$$

If (15) is proved we conclude that

$$\rho_n, \; \rho_n u_n, \; E_n \;\to\; (\rho, \; \rho\, u, \; E) = \int_{\mathbb{R}^N} (1, \; v, \; |v|^2/2) \, f \, dv \tag{16}$$

$$\text{in } L^1([0,T] \times \Omega),$$

and we may pass to the limit following [8] in $M[f_n]$, just getting

$$M[f_n] \;\to\; M[f] \quad \text{in } L^1(dt \otimes (1 + |v|^2) \, dv) \tag{17}$$

and the proposition is proved.

To prove (15), we introduce for $x \in \Omega$,

$$\tau(x,v) = \mathrm{Inf}\,\{t>0; \; x - vt \in \partial\Omega\}$$

and we have, integrating (1),

$$|v|^2 \, f(x,v,t) = |v|^2 \, \varphi(x-v\tau, \; v) \, e^{-\tau} + \int_0^\tau e^{-s} \, |v|^2 \, M[f_n] \, (x-vs,v,t-s) \, ds \tag{18}$$

for $|v|$ large enough so that $\tau < t$. Now, notice that, for same $\tau_0 > 0$,

$$\tau(x,v) = \tau(x,\ v/|v|)/|v| \leqslant \tau_0/|v|$$

and we find

$$\int_\Omega \int_{|v|\geqslant R} |v|^2\ f(x,v,t) \ \leqslant \int_{|v|>R} |v|^2 \sup_{\partial\Omega} \varphi(x,v)\ dv\ |\Omega|$$

$$+\ (\tau_0/|v|) \sup_{[0,T]} \int_{\Omega\times\mathbb{R}^N} |v|^2\ f(x,v,t)\ dx\ dv$$

and thus, using (8) and (9),

$$\int_0^T \int_\Omega \int_{|v|>R} |v|^2\ f_n(x,v,t)\ dx\ dv\ dt \ \leqslant\ T\ |\Omega| \int_{|v|>R} \Psi(v)\ dv + C/R,$$

and (15) is proved, thus concluding the proof of the proposition.

REFERENCES

[1] P.L.BHATNAGAR, E.H.GROSS, M.KROOK, "A model for collision processes in gases", Phys. Rev., 94, (1954), p.511.

[2] C.CERCIGNANI, "*Theorie and applications of the Boltzmann Equation*", Scottish Acad. Press, San Diego, 1975.

[3] R.Di PERNA, P.L.LIONS, "On the Cauchy problem for Boltzmann Equation, global existence and weak stability", Ann. Math., 130, (1989), p.321-366.

[4] F.GOLSE, B.PERTHAME, R.SENTIS, "Un résultat de compacité pour les équations du transport et application au calcul de la valeur propre principale d'un opérateur de transport", C.R. Acad. Sc. Paris, 301, I, (1985), p. 341-344.

[5] F.GOLSE, P.L.LIONS, B.PERTHAME, R.SENTIS, "Regularity of the moments of the solution of a transport equation", J. Funct. Anal., 74, (1988), p. 110-125.

[6] W.GREENBERG, J.POLEWCZAK, "Some remarks about continuity properties of local Maxwellian and an existence theorem for the B.G.K. model of the Boltzmann Equation", J. Statist. Phys., 33, (1983), p. 307-316.

[7] K.HAMDACHE, "Problèmes aux limites pour l'équation de Boltzmann: existence globale de solutions", Comm. in P.D.E., 13, (1988), p. 813-845.

[8] B.PERTHAME, "Global existence to the B.G.K. model of Boltzmann Equation", J. Diff. Eq., 82, (1989), p. 191-205.

Asymptotic Theory of a Steady Flow of a Rarefied Gas Past Bodies for Small Knudsen Numbers

Y. Sone

Department of Aeronautical Engineering, Kyoto University, Kyoto 606, Japan

A survey is made of the asymptotic behavior for small Knudsen numbers of the time-independent solution of the boundary-value problem of the Boltzmann equation over a general domain. Included is the hydrodynamic system (hydrodynamic type equations and their slip boundary conditions) describing the asymptotic behavior, with several new results.

I. INTRODUCTION

The relation between the hydrodynamic system and the Boltzmann system in the boundary-value problem of a steady flow of a slightly rarefied gas is not only of theoretical interest but of practical importance. The Hilbert and the Chapman-Enskog expansions[1] are often mentioned in this connection, but they are not derived in the framework of the boundary-value problem, and the hydrodynamic type equations derived have some awkward properties (see Refs.2 and 3). The formal theory was established about 20 years ago[4-7]; the explicit numerical data such as slip coefficients and Knudsen-layer corrections were also obtained up to the second order of the Knudsen number of the system for the Boltzmann-Krook-Welander (BKW) equation.[4,5] The extension of the theory to the case where there is evaporation or condensation on the boundary surface was also carried out.[8-10] Recently the 1st order slip coefficients and Knudsen-layer corrections were accurately analyzed on the basis of the Boltzmann equation for hard-sphere molecules in Refs. 11 - 13, where an efficient method of computation (numerical kernel method) was proposed. We will give survey of the asymptotic theory of a steady rarefied gas flow past bodies for small Knudsen numbers, with fundamental formulae applicable to practical problems and some new results.

According to the asymptotic theory,[4-7,10] the behavior of a steady flow past bodies with small Knudsen numbers is described as follows. Let f be any physical variable. Then f is expressed in the sum of two terms†

$$f = f_H + f_K, \tag{1}$$

where f_H, whose length scale of variation is much longer than the mean free path, describes the overall behavior of the gas (hydrodynamic part or H-part), and f_K is appreciable only in a thin layer (Knudsen layer), with thickness of the order of the

† The shock layer[7] may appear in the case of Sec. IV.

mean free path, adjacent to the body surface and represents the correction to f_H in the layer (Knudsen-layer correction or K-part).

The f_H and f_K are analyzed by expanding them in terms of the Knudsen number. Each component function of the expansion of the H-part of the velocity distribution function of the gas molecules is expressed in terms of the component functions of the H-parts of the macroscopic variables (the density, velocity, and temperature of the gas) to that order of expansion.† The behavior of the component functions of the H-parts of the macroscopic variables is described by hydrodynamic type equations. The boundary conditions for the hydrodynamic type equations, called slip or jump condition, are obtained by the analysis of the Knudsen layer.

With these general view, we will describe the results of the asymptotic theory according to the physical situation.

We introduce the notations: T_0, p_0, f_0, and ℓ_0 are the temperature, the pressure, the velocity distribution function of gas molecules, and the mean free path of our reference equilibrium state at rest††; R is the specific gas constant (per unit mass); D is the characteristic length of our system; Dx_i is the rectangular space coordinates; $(2RT_0)^{1/2}\zeta_i$ is the molecular velocity; $f_0(1+\phi)$ is the velocity distribution function of gas molecules; $(2RT_0)^{1/2}u_i$ is the gas velocity; $p_0(1+P)$ is the pressure; $T_0(1+\tau)$ is the temperature; $p_0(RT_0)^{-1}(1+\omega)$ is the density; $p_0(\delta_{ij}+P_{ij})$ is the stress tensor[1,14]; $p_0(2RT_0)^{1/2}Q_i$ is the heat flow vector[1,14]. The macroscopic variables u_i, ω, τ etc. are related to the velocity distribution function ϕ as

$$\left.\begin{aligned}
&\omega=\int\phi E d\boldsymbol{\zeta},\ (1+\omega)u_i=\int\zeta_i\phi E d\boldsymbol{\zeta},\ \frac{3}{2}(1+\omega)\tau=\int(\zeta_i^2-\frac{3}{2})\phi E d\boldsymbol{\zeta}-(1+\omega)u_i^2,\\
&P=\omega+\tau+\omega\tau,\qquad P_{ij}=2\int\zeta_i\zeta_j\phi E d\boldsymbol{\zeta}-2(1+\omega)u_iu_j,\\
&Q_i=\int\zeta_i\zeta_j^2\phi E d\boldsymbol{\zeta}-\frac{5}{2}u_i-u_jP_{ij}-\frac{3}{2}Pu_i-(1+\omega)u_iu_j^2,
\end{aligned}\right\} \tag{2}$$

where $E=\pi^{-3/2}exp(-\zeta_i^2)$ and $d\boldsymbol{\zeta}=d\zeta_1 d\zeta_2 d\zeta_3$.

The behavior of the gas ϕ is described by the Boltzmann equation:[14]

$$\zeta_i\frac{\partial\phi}{\partial x_i}=\frac{1}{k}[L(\phi)+J(\phi,\phi)], \tag{3}$$

$$k=\frac{\sqrt{\pi}}{2}\frac{\ell_0}{D}=\frac{\sqrt{\pi}}{2}Kn, \tag{4}$$

where the standard collision integral $J(1+\phi,1+\phi)$ is split into two parts: the linearized

---

† This is the result of analysis but not the assumption in contrast to the Chapman-Enskog expansion.

†† In the system without evaporation and condensation on the boundary, T_0 and p_0 are independent, but in the system of a gas and its condensed phase, p_0 is the saturation gas pressure at temperature T_0. For a hard-sphere molecular gas, $\ell_0=(\sqrt{2}\pi d^2 n_0)^{-1}$, where d is the diameter of a molecule and n_0 is the number density of the gas molecules at the reference state.

operator $L(\phi)$ and the remainder $J(\phi,\phi)$. The complete definition of the collision operators is not given here, but no confusion will take place. Hereafter, k is mainly used instead of the Knudsen number Kn.

On the boundary (body surface), a condition for the reflected molecules is imposed:

$$\phi = \phi_w, \qquad (\zeta_i n_i > 0), \tag{5}$$

where n_i is the unit normal vector to the boundary, pointed to the gas, and ϕ_w is a given function or is related to $\phi(\zeta_i n_i < 0)$.

II. LINEAR THEORY – SMALL REYNOLDS NUMBERS –

In this section we consider the case where the system is nearly in an equilibrium state at rest so that the Boltzmann equation and kinetic boundary condition (e.g., diffuse reflection) can be linearized around the equilibrium state at rest. Thus, when Eq. (2), (3), or (5) is referred to in this section, its linearized form is meant [e.g., Eq. (3) means Eq. (3) without $J(\phi,\phi)$ term].

First we construct the solution ϕ_H of the Boltzmann equation whose length scale of variation is of the order of the characteristic length D of the system $[\partial\phi_H/\partial x_i = O(\phi_H)]$. We expand ϕ_H in a power series of k:

$$\phi_H = \phi_{H0} + \phi_{H1}k + \cdots. \tag{6}$$ †

Substituting Eq. (6) in the Boltzmann equation (3) and arranging the same order terms of k, we obtain a sequence of integral equations for the component functions ϕ_{Hm} of the expansion (6):

$$L(\phi_{H0}) = 0, \qquad L(\phi_{Hm+1}) = \zeta_i \frac{\partial \phi_{Hm}}{\partial x_i}, \tag{7}$$

which can in principle be solved from the lowest order. From the solvability condition†† of Eq. (7):

$$\int (1, \zeta_i, \zeta_i^2)\zeta_j \frac{\partial \phi_{Hm}}{\partial x_j} E d\boldsymbol{\zeta} = 0, \tag{8}$$

we get a sequence of partial differential equations (hydrodynamic type equations) that govern the component functions of the expansions of the macroscopic variables corresponding to Eq. (6):

$$\frac{\partial P_{H0}}{\partial x_i} = 0, \tag{9}$$

$$\frac{\partial u_{iHm}}{\partial x_i} = 0, \qquad \frac{\partial P_{Hm+1}}{\partial x_i} = \gamma_1 \frac{\partial^2 u_{iHm}}{\partial x_j^2}, \qquad \frac{\partial^2 \tau_{Hm}}{\partial x_j^2} = 0, \tag{10}$$

where γ_1 is a constant, related to the collision operator. For hard-sphere molecular gas $\gamma_1 = 1.270042$ (Ref. 15), and for BKW equation $\gamma_1 = 1$ (Ref. 4). The system of equations for u_{iHm}, τ_{Hm}, and P_{Hm+1} is the Stokes system (the solenoidal equation for

---

† Corresponding to the expansion (6), the macroscopic variables ω_H, u_{iH}, etc. defined by Eq. (2) with $\phi = \phi_H$ are also expanded in power series of k in contrast to the Chapman-Enskog expansion.

†† Homogeneous equation $L(\phi) = 0$ has the five independent solutions $1, \zeta_i$, and ζ_i^2.

the velocity, Stokes equation of motion, and Laplace equation for the temperature) irrespective of the order of approximation in the Knudsen number.

The component function ϕ_{Hm}, solution of Eq. (7), is determined by the component functions of the expansions of the macroscopic variables to that order:[4,15]

$$\phi_{H0} = \phi_{eH0}, \tag{11}$$

$$\phi_{H1} = \phi_{eH1} - \zeta_i\zeta_j B(\zeta)\frac{\partial u_{iH0}}{\partial x_j} - \zeta_i A(\zeta)\frac{\partial \tau_{H0}}{\partial x_i}, \tag{12}$$

$$\begin{aligned}\phi_{H2} =& \phi_{eH2} - \zeta_i\zeta_j B(\zeta)\frac{\partial u_{iH1}}{\partial x_j} - \zeta_i A(\zeta)\frac{\partial \tau_{H1}}{\partial x_i} \\ &+ \frac{1}{\gamma_1}\zeta_i D_1(\zeta)\frac{\partial P_{H1}}{\partial x_i} + \zeta_i\zeta_j\zeta_k D_2(\zeta)\frac{\partial^2 u_{iH0}}{\partial x_j \partial x_k} - \zeta_i\zeta_j F(\zeta)\frac{\partial^2 \tau_{H0}}{\partial x_i \partial x_j},\end{aligned} \tag{13}$$

where

$$\phi_{eHm} = \omega_{Hm} + 2\zeta_i u_{iHm} + (\zeta_i^2 - \frac{3}{2})\tau_{Hm}. \tag{14}$$

The functions $A(\zeta)$, $B(\zeta)$, $D_1(\zeta)$, $D_2(\zeta)$, and $F(\zeta)$ of $\zeta(= (\zeta_i^2)^{1/2})$ are obtained numerically for a hard-sphere molecular gas in Refs. 12 and 15.

As is seen from Eqs. (11) - (13), the hydrodynamic part ϕ_H cannot in general be made to satisfy the boundary condition (5)†. Thus, we introduce the Knudsen-layer correction ϕ_K, which is assumed to have the length scale of variation normal to the boundary of the order of ℓ_0 $[kn_i\partial\phi_K/\partial x_i = O(\phi_K)]$ and to be appreciable only near the boundary. That is, we assume

$$\phi = \phi_H + \phi_K, \tag{15}$$

$$\phi_K = \phi_{K0} + \phi_{K1}k + \cdots. \tag{16}$$

Substituting Eq. (15), with Eqs. (6) and (16), in Eq. (3) and arranging the terms with the properties of ϕ_H and ϕ_K in mind, we obtain a sequence of (inhomogeneous) one-dimensional *linearized* Boltzmann equations for ϕ_{Km}:

$$\zeta_i n_i \frac{\partial \phi_{K0}}{\partial \eta} = L(\phi_{K0}), \tag{17}$$

$$\zeta_i n_i \frac{\partial \phi_{K1}}{\partial \eta} = L(\phi_{K1}) - \zeta_i[(\frac{\partial s_1}{\partial x_i})_0\frac{\partial \phi_{K0}}{\partial s_1} + (\frac{\partial s_2}{\partial x_i})_0\frac{\partial \phi_{K0}}{\partial s_2}], \tag{18}$$

$$\cdots\cdots\cdots,$$

$$x_i = n_i k\eta + x_{wi}(s_1, s_2), \tag{19}$$

where x_{wi} is the boundary surface, η is a stretched coordinate normal to the boundary, s_1 and s_2 are (unstretched) coordinates within a parallel surface $\eta = const.$, and $(\ )_0$ denotes that the quantity in ( ) is evaluated at $\eta = 0$. The normal vector n_i is a function

$\dagger$ The lack of freedom in ϕ_{Hm} is due to the fact that the differential operator is multiplied by the small parameter k in Eq. (3).

of s_1 and s_2. The boundary condition for ϕ_{Km} at $\eta = 0$ is

$$\phi_{Km} = \phi_{wm} - \phi_{Hm}, \qquad (\zeta_i n_i > 0), \tag{20}$$

where ϕ_{wm} is defined by

$$\phi_w = \phi_{w0} + \phi_{w1} k + \cdots. \tag{21}$$

The boundary value of ϕ_{Hm}, which is undetermined, is involved in the boundary condition (20). The analysis of ϕ_{Km} under the condition that ϕ_{Km} vanishes rapidly away from the boundary gives conditions among the boundary values of the component functions of the H-parts of macroscopic variables and their derivatives as well as the K-part ϕ_{Km}. These conditions give the boundary conditions for the hydrodynamic type equations. (see Refs. 7, 16, and 17 for the mathematical discussion on the one-dimensional Boltzmann system.)

The boundary condition for these hydrodynamic equations and the Knudsen-layer correction are given as follows. In the following formulae, n_i is the unit normal vector to the body surface pointed to the gas; t_i is a unit tangential vector to the body surface; and u_{wim}, τ_{wm}, and P_{wm} are the component functions of the expansions of u_{wi}, τ_w, and P_w, where $(2RT_0)^{1/2} u_{wi}$ (with $u_{wi} n_i = 0$) is the velocity of the body surface, $T_0(1 + \tau_w)$ is its temperature, and $p_0(1 + P_w)$ is the saturation gas pressure at temperature $T_0(1 + \tau_w)$ [e.g., $\tau_w = \tau_{w0} + \tau_{w1} k + \cdots$].

(i) On the solid boundary where neither evaporation nor condensation takes place:

$$u_{iH0} - u_{wi0} = 0, \qquad \tau_{H0} - \tau_{w0} = 0, \qquad u_{iK0} = \omega_{K0} = \tau_{K0} = 0, \tag{22†}$$

$$\left.\begin{aligned}
&\begin{bmatrix} (u_{iH1} - u_{wi1}) t_i \\ u_{iK1} t_i \end{bmatrix} = S_{ij0} n_i t_j \begin{bmatrix} k_0 \\ Y_0(\eta) \end{bmatrix} + G_{i0} t_i \begin{bmatrix} K_1 \\ \frac{1}{2} Y_1(\eta) \end{bmatrix}, \\
&\begin{bmatrix} u_{iH1} n_i \\ u_{iK1} n_i \end{bmatrix} = 0, \qquad \begin{bmatrix} \tau_{H1} - \tau_{w1} \\ \omega_{K1} \\ \tau_{K1} \end{bmatrix} = -G_{i0} n_i \begin{bmatrix} d_1 \\ \Omega_1(\eta) \\ \Theta_1(\eta) \end{bmatrix},
\end{aligned}\right\} \tag{23}$$

$$\left.\begin{aligned}
(u_{iH2} - u_{wi2}) t_i &= k_0 S_{ij1} n_i t_j + a_1 \frac{\partial S_{ij0}}{\partial x_r} n_j n_r t_i + a_2 \bar{\kappa} S_{ij0} n_i t_j \\
&\quad + a_3 \kappa_{ij} S_{jr0} n_r t_i + a_4 \frac{\partial G_{i0}}{\partial x_j} n_j t_i + a_5 \bar{\kappa} G_{i0} t_i + a_6 \kappa_{ij} G_{j0} t_i - K_1 \frac{\partial \tau_{w1}}{\partial x_i} t_i, \\
u_{iH2} n_i &= b_1 \frac{\partial S_{ij0}}{\partial x_r} n_i n_j n_r + b_2 \Big(\frac{\partial G_{i0}}{\partial x_j} n_i n_j + 2\bar{\kappa} G_{i0} n_i\Big), \\
\tau_{H2} - \tau_{w2} &= -d_1 G_{i1} n_i - d_4 \frac{\partial S_{ij0}}{\partial x_r} n_i n_j n_r - d_5 \bar{\kappa} G_{i0} n_i,
\end{aligned}\right\} \tag{24††}$$

---

† The K-part ϕ_{K0} as well as u_{iK0}, ω_{K0}, etc. vanishes. Then, Eq. (18) is reduced to Eq. (17) with ϕ_{K0} replaced by ϕ_{K1}, and the equation for ϕ_{K2} to Eq. (18) with ϕ_{K0} and ϕ_{K1} replaced, respectively, by ϕ_{K1} and ϕ_{K2}.

†† Owing to limited space, only the slip boundary condition is given. The corresponding Knudsen-layer correction is obtained by replacing the slip coefficients by the Knudsen-layer functions.

$$\bar{\kappa} = \frac{1}{2}(\kappa_1 + \kappa_2), \qquad \kappa_{ij} = \kappa_1 \ell_i \ell_j + \kappa_2 m_i m_j. \tag{25}$$

$$S_{ijm} = -(\frac{\partial u_{iHm}}{\partial x_j} + \frac{\partial u_{jHm}}{\partial x_i}), \qquad G_{im} = -\frac{\partial \tau_{Hm}}{\partial x_i}. \tag{26}$$

where k_0, K_1, a_i, b_i, and d_i are numerical constants; $Y_0(\eta)$, $Y_1(\eta)$, $\Omega_1(\eta)$, and $\Theta_1(\eta)$ are universal functions of η, called Knudsen-layer functions; the κ_1/D and κ_2/D are the principal curvatures of the boundary (κ_1 or κ_2 is taken negative when the corresponding center of curvature lies on the side of the gas); ℓ_i and m_i are the direction cosines of the principal directions corresponding to κ_1 and κ_2 respectively; the quantities with the subscript H, including S_{ijm} and G_{im}, are evaluated on the body surface.

For a hard-sphere molecular gas under the diffuse reflection, the slip (or jump) coefficients k_0, K_1, d_1, and a_4 are:[11,12,15]

$$k_0 = -1.2540, \qquad K_1 = -0.6463, \qquad d_1 = 2.4001, \qquad a_4 = 0.0330, \tag{27}^\dagger$$

and the Knudsen-layer functions $Y_0(\eta)$, $Y_1(\eta)$, $\Omega_1(\eta)$, and $\Theta_1(\eta)$ are tabulated in Refs. 11 and 12 [$(\eta, Y_0(\eta), \frac{1}{2}Y_1(\eta)) = (x_1, -S(x_1), -C(x_1))$ in Ref. 12 and $(\eta, \Omega_1(\eta), \Theta_1(\eta)) = (x_1, \Omega(x_1), \Theta(x_1))$ in Ref. 11]. For the BKW equation with the diffuse reflection, the data of the slip coefficients and Knudsen-layer functions are obtained up to the order of k^2 in Refs. 3 and 4 (see also Ref. 8, where their comprehensive tables are given).

The 0th approximation (Eq. (22)) is the classical gas dynamic condition (nonslip and no Knudsen layer). The 1st order Knudsen layer (Eq. (23)) consists of the Knudsen layers of the three problems over a plane wall: the shear flow[2,12], the thermal creep flow[12,18] (the flow induced over a wall with temperature gradient along it), and the heat flow normal to the wall[11]. The 2nd order condition (Eq. (24)) consists of many terms, among which the terms including the curvature of the body surface and the velocity slip due to thermal stress (a_4 term) should be noted (see Ref. 19). Further, at the 2nd order of the Knudsen number, another boundary layer[20] (S-layer) with thickness of the order of $(the\ mean\ free\ path)^2/(the\ radius\ of\ curvature)$ is present in the bottom of the Knudsen layer. Some changes in the kinetic boundary condition or the molecular model do not alter the fundamental form of the boundary conditions for the hydrodynamic system. In the case with the very small accommodation coefficient, however, the form is subject to considerable change.[21,22]

(ii) On the interface of the gas and its condensed phase:

$$(u_{iH0} - u_{wi0})t_i = 0, \qquad u_{iK0} = 0, \qquad \begin{bmatrix} P_{H0} - P_{w0} \\ \tau_{H0} - \tau_{w0} \\ \omega_{K0} \\ \tau_{K0} \end{bmatrix} = u_{iH0} n_i \begin{bmatrix} C_4^* \\ d_4^* \\ \Omega_4^*(\eta) \\ \Theta_4^*(\eta) \end{bmatrix}, \tag{28}$$

† The result d_1 of Ref. 11 is refined with the finer (M3) lattice system in Ref. 13. For Maxwell type boundary condition, $d_1 = 3.8981, a_4 = 0.0700(\alpha = 0.75); d_1 = 6.8210, a_4 = 0.1547(\alpha = 0.5)$, where α is the accommodation coefficient[15]. The positiveness of a_4 shows that the negative thermophoresis[15,19] occurs for a hard-sphere molecular gas under the Maxwell type boundary condition as well as for the BKW equation.

$$\left.\begin{aligned}&(u_{iH1}-u_{wi1})t_i = k_0 S_{ij0}n_i t_j + K_1 G_{i0}t_i + K_2 t_j \frac{\partial}{\partial x_j}(u_{iH0}n_i),\\&\begin{bmatrix}P_{H1}-P_{w1}\\ \tau_{H1}-\tau_{w1}\end{bmatrix} = u_{iH1}n_i\begin{bmatrix}C_4^*\\ d_4^*\end{bmatrix} - G_{i0}n_i\begin{bmatrix}C_1\\ d_1\end{bmatrix} - S_{ij0}n_i n_j\begin{bmatrix}C_6\\ d_6\end{bmatrix} - 2\bar{\kappa}u_{iH0}n_i\begin{bmatrix}C_7\\ d_7\end{bmatrix},\end{aligned}\right\} \tag{29†}$$

where the quantities with the subscript H, including S_{ijm} and G_{im}, is evaluated on the interface.

For a hard-sphere molecular gas under the conventional kinetic boundary condition†† of evaporation-condensation, the jump coefficients C_4^*, d_4^*, and C_1 are:[11,13]

$$C_4^* = -2.1412, \qquad d_4^* = -0.4557, \qquad C_1 = 1.0947, \tag{30}$$

and the Knudsen-layer functions $\Omega_4^*(\eta)$ and $\Theta_4^*(\eta)$ are tabulated in Ref. 13 $[(\eta, \Omega_4^*(\eta), \Theta_4^*(\eta)) = (x_1, \Omega(x_1), \Theta(x_1))$ in Ref. 13]. The slip coefficients and the Knudsen-layer corrections up to the order of k (those in Eqs. (28) and (29)) are obtained for the BKW equation with the conventional condition[8].

Over the condensed phase where evaporation or condensation is taking place, the Knudsen layer appears in the 0th order of the Knudsen number, and the curvature of the boundary and the S-layer enter in the 1st order of the Knudsen number.[8]

It is shown in Ref. 23 that the force, its moment, the mass transfer, and the energy transfer on a *closed* body in the gas can be obtained by the information of only the hydrodynamic parts of the macroscopic variables. The H-parts of the stress tensor and heat flow vector are given by[15]

$$\left.\begin{aligned}&P_{ijH0} = P_{H0}\delta_{ij}, \qquad P_{ijH1} = P_{H1}\delta_{ij} + \gamma_1 S_{ij0},\\&P_{ijH2} = P_{H2}\delta_{ij} + \gamma_1 S_{ij1} + \gamma_3\frac{\partial^2 \tau_{H0}}{\partial x_i \partial x_j},\\&P_{ijH3} = P_{H3}\delta_{ij} + \gamma_1 S_{ij2} + \gamma_3\frac{\partial^2 \tau_{H1}}{\partial x_i \partial x_j} - \frac{2\gamma_6}{\gamma_1}\frac{\partial^2 P_{H1}}{\partial x_i \partial x_j},\end{aligned}\right\} \tag{31}$$

$$\left.\begin{aligned}&Q_{iH0} = 0, \qquad Q_{iH1} = \frac{5}{4}\gamma_2 G_{i0},\\&Q_{iH2} = \frac{5}{4}\gamma_2 G_{i1} + \frac{\gamma_3}{2\gamma_1}\frac{\partial P_{H1}}{\partial x_i}, \qquad Q_{iH3} = \frac{5}{4}\gamma_2 G_{i2} + \frac{\gamma_3}{2\gamma_1}\frac{\partial P_{H2}}{\partial x_i},\end{aligned}\right\} \tag{32}$$

where γ_i are constants. The γ_3 term in Eq. (31) is called thermal stress. For the BKW equation $\gamma_i = 1$, and for a hard-sphere molecular gas,[15]

$$\gamma_1 = 1.270042, \quad \gamma_2 = 1.922284, \quad \gamma_3 = 1.947906, \quad \gamma_6 = 1.419424. \tag{33}$$

Further, the γ_3 and γ_6 terms in Eq. (31) do not contribute to either the force or its moment on a closed body.[23]

---

† See the footnote to Eq. (24).

†† The gas molecules leaving the condensed phase constitute the corresponding part of the Maxwellian distribution pertaining to the saturated gas at the surface temperature and velocity of the condensed phase.

From the relation between the mean free path and the viscosity, derived from the dimensional form of Eq. (10) or (31), the parameters: Mach number Ma, Reynolds number Re, and Knudsen number Kn are related as

$$Ma \sim Re Kn. \tag{34}$$

Thus the linear theory, where the quantities of the order of Ma^2 and the higher are neglected but the power terms of Kn are retained, can be applied only when $Re \ll 1$. This corresponds to the Stokes system.

III. FINITE REYNOLDS NUMBERS[5,10]

When the Reynolds number of the system is of the order of unity, the Mach number is the same order as the Knudsen number (see Eq. (34)). Thus, the case with a finite Reynolds number cannot be treated by the linearized Boltzmann system. In view of the fact that the Mach number is a measure of deviation from an equilibrium state at rest, we take the case where the deviation of the system from an equilibrium state at rest is of the order of the Knudsen number $[\phi = O(k)]^{\dagger}$ and consider its asymptotic behavior for small Knudsen numbers.

The asymptotic solution of the boundary-value problem is obtained in the form:

$$\phi = \phi_H + \phi_K, \tag{35††}$$

$$\phi_H = \phi_{H1} k + \phi_{H2} k^2 + \cdots, \tag{36}$$

$$\phi_K = \phi_{K1} k + \phi_{K2} k^2 + \cdots, \tag{37}$$

where ϕ_H is the hydrodynamic part and ϕ_K is the Knudsen-layer part. Since we are considering the case where the perturbed distribution ϕ is of the order of k, the series start from the 1st order of k, and ϕ_{Hm} and ϕ_{Km} are of the order of unity.

First we construct ϕ_H as in the linear theory. Its component function ϕ_{Hm} is determined by a similar integral equation to Eq. (7). The difference is in the inhomogeneous term. From the solvability condition of the integral equation, we obtain the system of (slightly compressible) Navier-Stokes type equations. In contrast to the Chapman-Enskog expansion, the orders of the differential equations remain unchanged if we advance the approximation.

$$\frac{\partial P_{H1}}{\partial x_i} = 0, \tag{38}$$

† This requires further restriction besides that on the Mach number. That is, the temperature difference on the boundary should also be of the order of the Knudsen number. The finite boundary temperature difference problem is touched on in Sec. IV. The ordering related to Eq. (34) was first noted and introduced in the asymptotic analysis in Ref. 5.

†† The H-parts of the macroscopic variables, ω_H, u_{iH}, etc., are given by Eq. (2) with $\phi = \phi_H$, but the K-parts ω_K, u_{iK}, etc. are defined as the remainders $\omega - \omega_H$, $u_i - u_{iH}$, etc., which depend on ϕ_H as well as ϕ_K (see Ref. 5) since the problem is nonlinear.

$$\left.\begin{aligned}&\frac{\partial u_{iH1}}{\partial x_i}=0, \qquad u_{jH1}\frac{\partial u_{iH1}}{\partial x_j}=-\frac{1}{2}\frac{\partial P_{H2}}{\partial x_i}+\frac{1}{2}\gamma_1\frac{\partial^2 u_{iH1}}{\partial x_j^2},\\ &u_{jH1}\frac{\partial \tau_{H1}}{\partial x_j}=\frac{1}{2}\gamma_2\frac{\partial^2 \tau_{H1}}{\partial x_j^2},\end{aligned}\right\} \tag{39}$$

$$\frac{\partial u_{jH2}}{\partial x_j}=-u_{jH1}\frac{\partial \omega_{H1}}{\partial x_j}, \tag{40a}$$

$$\begin{aligned}u_{jH1}\frac{\partial u_{iH2}}{\partial x_j}&+(\omega_{H1}u_{jH1}+u_{jH2})\frac{\partial u_{iH1}}{\partial x_j}\\ &=-\frac{1}{2}\frac{\partial}{\partial x_i}[P_{H3}-\frac{1}{6}(\gamma_1\gamma_2-4\gamma_3)\frac{\partial^2\tau_{H1}}{\partial x_j^2}]\\ &\quad+\frac{1}{2}\gamma_1\frac{\partial^2 u_{iH2}}{\partial x_j^2}+\frac{1}{2}\gamma_4\frac{\partial}{\partial x_j}[\tau_{H1}(\frac{\partial u_{iH1}}{\partial x_j}+\frac{\partial u_{jH1}}{\partial x_i})],\end{aligned} \tag{40b}$$

$$\begin{aligned}u_{jH1}\frac{\partial \tau_{H2}}{\partial x_j}&+(\omega_{H1}u_{jH1}+u_{jH2})\frac{\partial \tau_{H1}}{\partial x_j}-\frac{2}{5}u_{jH1}\frac{\partial P_{H2}}{\partial x_j}\\ &=\frac{1}{5}\gamma_1(\frac{\partial u_{iH1}}{\partial x_j}+\frac{\partial u_{jH1}}{\partial x_i})^2+\frac{1}{2}\frac{\partial^2}{\partial x_j^2}(\gamma_2\tau_{H2}+\frac{1}{2}\gamma_5{\tau_{H1}}^2),\end{aligned} \tag{40c}$$

$$\cdots\cdots\cdots\cdots\cdots\cdots,$$

$$P_{H1}=\omega_{H1}+\tau_{H1}, \qquad P_{H2}=\omega_{H2}+\tau_{H2}+\omega_{H1}\tau_{H1}, \ \cdots. \tag{41}$$

where γ_i are numerical constants related to the collision operator. For the BKW equation $\gamma_i = 1$, and for a hard-sphere molecular gas[24],

$$\gamma_4 = 0.635021, \qquad \gamma_5 = 0.961142, \qquad [\gamma_1, \gamma_2 \text{ and } \gamma_3 \text{ are in Eq.(33)}]. \tag{42}$$

As in the linear theory, the Knudsen-layer correction ϕ_K is introduced to make ϕ satisfy the kinetic boundary condition (5) on the body surface. A similar sequence of (inhomogeneous) one-dimensional *linearized* Boltzmann equations to that of Eqs. (17), (18), etc. is obtained for ϕ_{Km}.† From their analysis, we obtain the boundary conditions for the Navier-Stokes type equations (38) - (40c), etc. as well as the Knudsen-layer corrections:

(i) On the solid boundary without evaporation and condensation:[5,10–12]

$$u_{iH1}-u_{wi1}=0, \quad \tau_{H1}-\tau_{w1}=0, \quad u_{iK1}=\omega_{K1}=\tau_{K1}=0, \tag{43}$$ ††

† The difference, coming from the nonlinear term $J(\phi,\phi)$, is in the inhomogeneous term [e.g., $2J((\phi_{H1})_0,\phi_{K1})$ and $J(\phi_{K1},\phi_{K1})$ in the equation for ϕ_{K2} (see Refs. 5 and 10)]. The Knudsen-layer equation for ϕ_{Km} contains ϕ_{Hm-n} in contrast to the linear case of Sec. II.

†† The K-part ϕ_{K1} as well as u_{iK1}, ω_{K1}, etc. vanishes. See footnote to Eq. (22).

$$\left.\begin{aligned}
&\begin{bmatrix}(u_{iH2}-u_{wi2})t_i\\ u_{iK2}t_i\end{bmatrix}=S_{ij1}n_it_j\begin{bmatrix}k_0\\ Y_0(\eta)\end{bmatrix}+G_{i1}t_i\begin{bmatrix}K_1\\ \frac{1}{2}Y_1(\eta)\end{bmatrix},\\
&\begin{bmatrix}u_{iH2}n_i\\ u_{iK2}n_i\end{bmatrix}=0,\qquad \begin{bmatrix}\tau_{H2}-\tau_{w2}\\ \omega_{K2}\\ \tau_{K2}\end{bmatrix}=-G_{i1}n_i\begin{bmatrix}d_1\\ \Omega_1(\eta)\\ \Theta_1(\eta)\end{bmatrix},
\end{aligned}\right\}\tag{44}$$

where the quantities with the subscript H are evaluated on the body surface. Equations (43) and (44) are essentially the same as Eq. (22) and (23) respectively.

(ii) On the interface of the gas and its condensed phase:[9,10,13]

$$\begin{aligned}&(u_{iH1}-u_{wi1})t_i=0,\\ &u_{iK1}=0,\end{aligned}\qquad \begin{bmatrix}P_{H1}-P_{w1}\\ \tau_{H1}-\tau_{w1}\\ \omega_{K1}\\ \tau_{K1}\end{bmatrix}=u_{iH1}n_i\begin{bmatrix}C_4^*\\ d_4^*\\ \Omega_4^*(\eta)\\ \Theta_4^*(\eta)\end{bmatrix},\tag{45}$$

$$\left.\begin{aligned}
&(u_{iH2}-u_{wi2})t_i=k_0S_{ij1}n_it_j+K_1G_{i1}t_i+K_2t_j\frac{\partial}{\partial x_j}(u_{iH1}n_i),\\
&\begin{bmatrix}P_{H2}-P_{w2}\\ \tau_{H2}-\tau_{w2}\end{bmatrix}=u_{iH2}n_i\begin{bmatrix}C_4^*\\ d_4^*\end{bmatrix}-G_{i1}n_i\begin{bmatrix}C_1\\ d_1\end{bmatrix}\\
&\qquad -S_{ij1}n_in_j\begin{bmatrix}C_6\\ d_6\end{bmatrix}-2\bar{\kappa}u_{iH1}n_i\begin{bmatrix}C_7\\ d_7\end{bmatrix}\\
&\qquad +(u_{iH1}n_i)^2\begin{bmatrix}C_8\\ d_8\end{bmatrix}+\tau_{w1}u_{iH1}n_i\begin{bmatrix}C_9\\ d_9\end{bmatrix}+P_{w1}u_{iH1}n_i\begin{bmatrix}C_{10}\\ d_{10}\end{bmatrix},
\end{aligned}\right\}\tag{46}$$

where the quantities with the subscript H are evaluated on the interface. Equation (45) is essentially the same as Eq. (28), but Eq. (46) contains some additional nonlinear terms, compared with Eq. (29).

The hydrodynamic type equations (39) are the Navier-Stokes equations for an incompressible fluid. The next order equations (40a-c) combined with Eqs. (38) and (39) differ a little from the Navier-Stokes equations of a slightly compressible gas. If γ_3 in the numerical coefficient of $\partial^2\tau_{H1}/\partial x_j^2$ in the square brackets of the first term on the right hand side of Eq. (40b) is zero, Eqs. (38) - (40c) are obtained by the Mach number expansion of the Navier-Stokes equations for a compressible gas.† The difference is due to the thermal stress (see Ref. 10). This difference, however, can formally be eliminated by introduction of the new variable P_{H3}^* instead of P_{H3}:

$$P_{H3}^*=P_{H3}+\frac{2}{3}\gamma_3\frac{\partial^2\tau_{H1}}{\partial x_j^2}.\tag{47}$$

Further, the slip boundary conditions (up to the second order of k) do not contain P_{H3} (see Eqs. (43)-(46)). Thus, we conclude:

† Noting that the case $Ma = ck$ with $c = O(1)$ is under consideration, transform the k-expansion to Ma-expansion.

Proposition: *Except for the Knudsen-layer correction, the velocity and the temperature fields of a slightly rarefied gas can be calculated correctly up to the second order of the Knudsen number by the slightly compressible Navier-Stokes equations with the slip boundary conditions. The effect of gas rarefaction comes in through the boundary condition.*
[N. B. In an infinite-domain problem where the pressure is specified at infinity, the pressure modified by Eq.(47) should be used. In most physical problems, however, $\partial^2 \tau_{H1}/\partial x_j^2$ vanishes at infinity and no correction is necessary.]

As in the linear theory, the force, its moment, the mass transfer, and the energy transfer on a closed body in the gas can be obtained by the information of only the hydrodynamic parts of the macroscopic variables, and the proposition in the linear theory that the thermal stress does not contribute to either the force or its moment on a closed body holds with a slight modification (see Ref. 10 for the details).

IV. LARGE REYNOLDS NUMBERS ETC.

The case of finite Mach numbers (or very large Reynolds numbers: $Re \sim k^{-1}$) is considered in Refs. 6 and 7, where the general feature of the flow field is described but the formulae easily applicable to practical problems were not arranged. According to them, the hydrodynamic system is divided into two systems, Euler type system and viscous boundary layer type system. The first one describes the overall behavior of the gas and is derived by the Hilbert expansion (expansion in the power series of k) of the nonlinear Boltzmann equation, where the Euler equation comes first and inhomogeneous Euler type equations follow. The second describes the local behavior in the neighborhood of the body surface of the order of the square root of the mean free path. This, much thicker than the Knudsen layer, corresponds to the viscous boundary layer. The system is expanded in the power series of $k^{1/2}$ and is subject to the Knudsen-layer correction except for the leading term.† The Prandtl boundary layer equation with nonslip boundary condition comes first, and the Prandtl type equations with rarefaction (kinetic) effect terms follow. The boundary condition and Knudsen layer at the 1st approximation ($k^{1/2}$ order) are essentially the same as those (k order) in the linear theory (Eq. (23) in Sec. II). [It should be noted that the hydrodynamic system to be used with this slip condition is not the Navier-Stokes equation but contains a kinetic effect (thermal stress) term.] The single system of hydrodynamic type equations with their boundary conditions is not derived systematically from the system of the Boltzmann equation and its kinetic boundary condition.

When there is no imposed flow (e.g., no uniform flow at infinity, absence of moving walls) but there is a finite temperature difference in the gas caused by nonuniform boundary temperature, the hydrodynamic system is described by a single (k-power series) system[25] of Navier-Stokes type equations, where a thermal stress effect enters in the lowest order approximation and cannot be incorporated in pressure. The flow of the order k is induced by the thermal stress, which was first pointed out by Kogan.[26]

When strong evaporation or condensation is taking place on the body surface, the hydrodynamic system is given by a single system of Euler type, but the Knudsen layer is

† The Euler equation essentially corresponds to the equation for the overall solution up to the 1st approximation ($k^{1/2}$ order) since the Euler type system is k-expansion.

present at the lowest order and is governed by the one-dimensional nonlinear Boltzmann equation.[27]

REFERENCES

1. Grad, H., "Principles of the kinetic theory of gases", Handbuch der Physik, ed. by S. Flügge, Springer-Verlag, Vol. 12, pp. 205-294, 1958.
2. Cercignani, C., "Theory and application of the Boltzmann equation", Chap. V, Sec. 4 and Chap. VI, Sec.4, Scottish Academic, Edinburgh, 1975.
3. Sone, Y., "Analytical studies in rarefied gas dynamics", Rarefied Gas Dynamics, ed. by H. Oguchi, University of Tokyo Press, Tokyo, pp. 71-87, 1984.
4. Sone, Y., "Asymptotic theory of flow of rarefied gas over a smooth boundary I", Rarefied Gas Dynamics, ed. by L. Trilling and H. Y. Wachman, Academic Press, New York, pp. 243-253, 1969.
5. Sone, Y., "Asymptotic theory of flow of rarefied gas over a smooth boundary II", Rarefied Gas Dynamics, ed. by D. Dini, Editrice Tecnico Scientifica, Pisa, pp. 737-749, 1971.
6. Darrozes, J. S., "Approximate solutions of the Boltzmann equation for flows past bodies of moderate curvature", Rarefied Gas Dynamics, ed. by L. Trilling and H. Y. Wachman, Academic Press, New York, pp. 111-120, 1969.
7. Grad, H., "Singular and nonuniform limits of solutions of the Boltzmann equation", Transport Theory, ed. by R. Bellman, et al., American Mathematical Society, Providence, RI, pp. 269-308, 1969.
8. Sone, Y. and Onishi, Y., "Kinetic theory of evaporation and condensation – Hydrodynamic equation and slip boundary condition", J. Phys. Soc. Jpn. **44**, pp. 1981-1994, 1978.
9. Onishi, Y. and Sone, Y., "Kinetic theory of slightly strong evaporation and condensation – Hydrodynamic equation and slip boundary condition for finite Reynolds number", J. Phys. Soc. Jpn. **47**, pp. 1676-1685, 1979.
10. Sone, Y. and Aoki, K., a: "Steady gas flows past bodies at small Knudsen numbers – Boltzmann and hydrodynamic systems", Transp. Theory Stat. Phys. **16**, pp. 189-199, 1987; b: "Asymptotic theory of slightly rarefied gas flow and force on a closed body", Mem. Fac. Eng. Kyoto Univ. **49**, pp. 237-248, 1987.
11. Sone, Y., Ohwada, T., and Aoki, K., "Temperature jump and Knudsen layer in a rarefied gas over a plane wall: Numerical analysis of the linearized Boltzmann equation for hard-sphere molecules", Phys. Fluids A **1**, pp. 363-370, 1989; Erratum: Phys. Fluids A **1**, p.1077, 1989.
12. Ohwada, T., Sone, Y., and Aoki, K., "Numerical analysis of the shear and thermal creep flows of a rarefied gas over a plane wall on the basis of the linearized Boltzmann equation for hard-sphere molecules", Phys. Fluids A **1**, pp. 1588-1599, 1989.
13. Sone, Y., Ohwada, T., and Aoki, K., "Evaporation and condensation on a plane condensed phase: Numerical analysis of the linearized Boltzmann equation for hard-sphere molecules", Phys. Fluids A **1**, pp. 1398-1405, 1989.
14. Sone, Y., "Molecular Gas Dynamics", Handbook of Fluid Dynamics, ed. by Japan Society of Fluid Dynamics, Maruzen, Tokyo, Chap. 14, 1987(in Japanese).

15. Ohwada, T. and Sone, Y., "Higher order hydrodynamic type solutions and thermal stress slip flow ", Proceedings of Fluid Dynamic Conference (Osaka, Japan, 1990) (in Japanese).
16. Bardos, C., Caflisch, R. E., and Nicolaenko, B., "The Milne and Kramers problems for the Boltzmann equation of a hard sphere gas", Comm. Pure Appl. Math. **39**, pp. 323-352, 1986.
17. Cercignani, C., "Half-space problems in the kinetic theory of gases", Trends in Applications of Pure Mathematics to Mechanics, ed. by E. Kröner and K. Kirchgässner, Springer-Verlag, Berlin, pp. 35-50, 1986.
18. Sone, Y., "Thermal creep in rarefied gas", J. Phys. Soc. Jpn. **21**, pp. 1836-1837, 1966.
19. Sone, Y., "Flow induced by thermal stress in rarefied gas", Phys. Fluids **15**, pp. 1418-1423, 1972.
20. Sone, Y., "New kind of boundary layer over a convex solid boundary in a rarefied gas", Phys. Fluids **16**, pp. 1422-1424, 1973.
21. Sone, Y. and Aoki, K., "Slightly rarefied gas flow over a specularly reflecting body", Phys. Fluids **20**, pp. 571-576, 1977.
22. Aoki, K., Inamuro, T., and Onishi, Y., "Slightly rarefied gas flow over a body with small accommodation coefficient", J. Phys. Soc. Jpn. **47**, pp. 663-671, 1979.
23. Sone, Y., "Force, its moment, and energy transfer on a closed body in a slightly rarefied gas", ed. by H. Oguchi, University of Tokyo Press, pp. 117-126, 1984.
24. Ohwada, T., private communication.
25. Sone, Y. and Wakabayashi, M., "Flow induced by nonlinear thermal stress in a rarefied gas", Symposium on "Uchukoko" at Institute of Space Sciences 1988 (in Japanese).
26. Kogan, M. N., Galkin, V. S., and Fridlender, O. G., "Stresses produced in gases by temperature and concentration inhomogeneities", Sov. Phys. Usp. **19**, pp. 420-430, 1976.
27. Aoki, K. and Sone, Y., "Gas flows around the condensed phase with strong evaporation or condensation – Fluid dynamic equation and its boundary condition on the interface and their application –", in this volume.

Computation of Transitional Rarefied Flow

F. Coron[1] and J.P. Pallegoix[2]

[1]ST/S, AEROSPATIALE, B.P. 96, F-78130 Les Mureaux, France
[2]SEGIME, 142, rue Pierre Curie, F-78130 Les Mureaux, France

Abstract : Rarefied flows around reentry vehicles can be evaluated by using a Monte-Carlo numerical simulation. Results for Hermes vehicle are presented and agree fairly well with wind-tunnel data.

As the gas density increases and the flow becomes more transitional, the Monte-Carlo simulation becomes much more expensive both in computational time and memory requirement. When rarefaction effects are effective only in the immediate vicinity of the vehicle, in the Knudsen layer, the flowfield can be numerically determined through the continuum approach by the Navier-Stokes equations with slip boundary conditions at the wall. The variational method developed by F.Golse [1] and S.K.Loyalka [2] gives a practical way to obtain these boundary conditions in the case of gas with internal energy.

I MONTE-CARLO SIMULATION

1) Description of the method

Rarefied flows are described by a density distribution f which satisfies the Boltzmann equation under the assumption of molecular chaos and for a dilute gas (mean spacing between molecules much larger than the mean particle "diameter" or distance of interaction) (see Chapman and Cowling [3]). The Monte -Carlo methods are the most widely used numerical methods in order to compute rarefied flows (see Bird [4]). In these methods, a large number of particles is used (far less however than the real physical number).

In the presented algorithm, twenty particles are introduced in every cell at the beginning of the simulation. The velocity of each particle is randomly chosen according to the Maxwellian velocity distribution at infinity. The evolutions of the position, the velocity and the internal energy of these particles are numerically performed during successive time steps according to the following transport and collision process which are activated alternatively :

a) **Transport process**. The Maxwellian state at infinity gives the number of particles which are to be injected at the boundary of the computational domain. The velocity and the internal energy of these particles are randomly chosen with the same method used for the initialisation.

During this transport step, particles do not interact with each other and move according to their own velocity. The particles which leave the computational domain are removed. A particle which hits the vehicle surface, can be specularly reflected (with a certain probability) or otherwise accomodated by the surface. Its velocity and internal energy then change to different values randomly chosen according to the equilibrium state of the wall. The impact of the particles on the surface occuring during the simulation is used to get the aerodynamical forces on the vehicle.

A grid is introduced for the collision step. Voronoi's polyhedrons are used. Every particule represents a given number of particles in the physical space, this representativity is constant in a cell and is related to the volume of the cell in which the particle is located. Therefore, when a particle goes from one cell to another, it should be either duplicated or removed with a probability related to the representativity ratio of the two cells (see Lengrand [5]).

b) **Collision process**. During the collision process, only particles belonging to the same cell interact. The Variable Hard Sphere model is considered for the collision cross section and the time counter introduced by Bird [4] is used. The kinetic and internal energy exchanges are simulated with the Larsen Borgnakke model which sets the probability for a collision to be inelastic.

2) Description of the computational case

The flowfield around Hermes for a Mach number M=20.2 and a Knudsen number at infinity $Kn_\infty \cong 0.0173$ is computed for three different angles of attack 30, 40 and 50 degrees. Complete accomodation at the wall is assumed. On figure 1, the body mesh (34×31 cells on the body) is represented. There are 16 layers of cells from the body to the external surface of the computational domain.

After the transient phase there are more than 500 000 particules in the computational domain and therefore an average of 30 particles per cell. The time step (1.5 microsecond) is smaller than the mean time between collisions; 250 time steps are used and time averaging is performed when stationary state is obtained. This computation requires 6 CPU Hours on IBM 3090 (now a vectorised version of this code has been developed and is used on CRAY XMP 116 at Aérospatiale Les Mureaux). More details concerning the numerical method are given in [6].

On Figure 2, the time evolution of aerodynamic coefficients is plotted for an angle of attack of 30 degrees. This shows that stationary values are reached after some iterations. On Figures 3, 4, 5 comparisons between numerical results and wind tunnel data obtained at SR3 (SESSIA Chalais Meudon, see [7]) are represented. The Euler description of the flow is not suitable for these rarefied conditions (Figures 3, 4, 5). The Monte-Carlo results agree quite well with wind tunnel data except for high angles of attack. However for the higher angle of attack the discrepancy can arise from :

- the experimental data : the non uniformity of the flow field in the wind tunnel cross section may have an importance as well as perhaps the interaction of the walls of the tunnel;
- the computation : the base of Hermes (which was not taken into account in this simulation) is to be considered and numerical results obtained by Gropengiesser, Neunzert and Struckmeier prove that some aerodynamic coefficients are very sensitive to the wall/gas surface interaction (see [8]).

II SLIP BOUNDARY CONDITIONS

Slip boundary conditions were studied by a lot of different authors specially Grad [9], Cercignani [10], Sone [11]... These conditions are related to linear half space problems (see [12] for example) and therefore can be deduced from the knowledge of the collision operator and the wall/gas surface interaction at the microscopic level. However, the precise description of this microscopic behavior is in practical cases not well known and for this reason empirical models are often introduced (for example the Larsen Borgnakke model for internal energy and Maxwell accomodation for the wall interaction). In fact, a variational method was developed by F.Golse [1] and S.K.Loyalka [2] to get an approximation of the asymptotic states of linear half space problems (and therefore the coefficients in the slip boundary conditions). As a first approximation, the results obtained by this method depend only on the collision cross section via macroscopic properties (viscosity, thermal conductivity) which are better known. We show here briefly how this method could be used in a very practical way in a given case.

There are no universal model and appropriate scaling to take into account any chemical or physical situation at the kinetic level. In any practical case there are assumptions (equilibrium or nonequilibrium between kinetic, vibrational energy; continuous or discrete internal energy description; dissociation; frozen or equilibrium flow...) which lead to a given model. The formulation of the Euler or Navier-Stokes equations is related to the assumptions made on the choice of the considered unknown. For example, the Euler system is related to the equilibrium state at the kinetic level. The collision operator gives the relaxation law towards this local in space equilibrium at a short characteristic time scale $1/\theta$.

If we differentiate this equilibrium state with respect to the "macroscopic" parameters (density, mean velocity, temperatures,...) and multiply it by the velocity, we get the Sonine polynomials. Assuming, as a first approximation, that these polynomials are eigenfunctions of the linearised Boltzmann operator, the Chapman-Enskog expansion could be obtained very simply. The eigenvalues are related to the viscosity and thermal conductivity present in the Navier-Stokes equations.

In the Maxwell method, the boundary conditions for Navier-Stokes are obtained from the Chapman-Enskog expansion and the kinetic boundary conditions by assuming that half fluxes are conserved in the knudsen layer. This hypothesis is made in order to avoid the computation of the kinetic boundary layer (Knudsen layer).

Remark : If the kinetic boundary conditions imply some relations on the total flux of the distribution function, these relations are automaticaly transferred at the Navier-Stokes level. For example, without ablation, the particle flux being zero at the wall, we get rigorously that the mean velocity has no component normal to the wall. In the same way, if the wall is adiabatic, we get a condition on the temperature gradient at the wall.

At the kinetic level, the gas-surface interaction is usely described by accomodation coefficients. More generally, the kinetic boundary condition prescribes the distribution of the particles leaving the surface in term of incoming distribution (see Cercignani [10]).

The Maxwell method and the variational method use only the way the linearised Boltzmann operator L acts on the Sonine polynomials.

Example : Gas with internal energy "frozen case"

We use the BGKM model described by Brun and Larini [13] using a continuous description of the internal energy $\varepsilon > 0$. The Knudsen layer is described as a first approximation by the 1 dimensional problem

$$v_x \frac{\partial}{\partial x} f(x,v,\varepsilon) = \theta(f^0 - f)$$

$$f^0 = n \left(\frac{m}{2\pi kT}\right)^{3/2} \exp\left(-\frac{mv^2}{2kT}\right) \frac{\exp(-\varepsilon/kT')}{kT'}$$

x is the direction normal to the wall, n, T and T' are respectively the number density, the translational and the internal temperature of f. We do not consider a mean velocity and we assume $\theta >> 1$. We want to compute the internal temperature jump at the wall; the accomodation coefficient being β. Considering only the term with depending on temperature, the Enskog Morse expansion writes

$$f = f^{0} \left\{1 - \frac{1}{\theta}\left[v_x \left(\frac{mv^2}{2kT} - \frac{5}{2}\right)\frac{1}{T}\frac{dT}{dx} + v_x \frac{\varepsilon - kT'}{kT'}\frac{1}{T'}\frac{dT'}{dx}\right]\right\}$$

Linearising around the Maxwellian at the wall and considering only the term related to internal energy, we get:

$$f = f_w \left\{1 + \frac{T' - T'_w}{T'_w}\left(\frac{\varepsilon - kT'_w}{kT'_w}\right) - \frac{1}{\theta} v_x \frac{\varepsilon - kT'_w}{kT'_w}\frac{1}{T'_w}\frac{dT'}{dx}\right\} + \ldots$$

(index w referring to wall quantities).

This expansion does not satisfy the kinetic boundary condition. Therefore a kinetic layer term χ satisfying the kinetic equation (linearised around the Maxwellian at the wall) is to be added to this expansion in order to satisfy the boundary condition at x=0. Therefore

$$\chi(0,v.n>0) = (1-\beta)\,\chi(0,Rv) - f(0,v) + (1-\beta)\, f(0,Rv) + \alpha\,\beta$$

with $Rv=v-2(v.n)n$ (**n** is the normal at the wall pointing towards the fluid and β is the accomodation coefficient). The coefficient α is related to the mass flux of f for particles going towards the wall. We introduce the asymptotic limit of χ

$$\lim_{x \to +\infty} \chi(x,v) = \ell_\infty \left(\frac{\varepsilon - kT'_w}{kT'_w}\right) f_w$$

The internal temperature jump is obtained by writing that ℓ_∞ must be 0. To obtain this limit, we make the following two steps :

First step : we assume that

$$\chi(0,v.n<0) = \ell \left(\frac{\varepsilon - kT'_w}{kT'_w}\right) f_w$$

and we compute ℓ so that the following quantity which does not depend on x and cancels at infinity

$$\int_{\varepsilon=0}^{+\infty}\int_{v\in\mathbb{R}^3} v_x \frac{\varepsilon - kT'_w}{kT'_w}\,\chi\,dv\,d\varepsilon$$

is 0 at x=0. We thus get

$$\ell = -\frac{T' - T'_w}{T'_w} + \frac{1}{\theta}\,\frac{2-\beta}{\beta}\sqrt{\frac{2kT_w}{m}}\,\frac{1}{T'_w}\,\frac{\sqrt{\pi}}{2}\,\frac{dT'}{dx}$$

The Maxwell slip boundary condition is obtained by prescribing l=0 (see Larini and Brun [13]).

Second step : from the knowledge of χ at x=0, we compute ℓ_∞ using the fact that the quantity

$$\int_{\varepsilon=0}^{+\infty}\int_{v\in\mathbb{R}^3} v_x \chi\, L^{-1}\left(v_x \frac{\varepsilon - kT'_w}{kT'_w} f_w\right) dv\,d\epsilon$$

is conserved (L is the linearised BGKM operator).
Because

$$L^{-1}\left(v_x \frac{\varepsilon - kT'_w}{kT'_w} f_w\right) = v_x \frac{\varepsilon - kT'_w}{kT'_w} f_w$$

we obtain

$$\ell_\infty = 0 \iff \Delta T' = \frac{1}{\theta}\sqrt{\frac{2kT}{m}}\,\frac{\sqrt{\pi}}{2}\,\frac{2 - \left(2 - \frac{4}{\pi}\right)\beta + \left(\frac{1}{2} - \frac{2}{\pi}\right)\beta^2}{\beta}\,\frac{dT'}{dx}$$

Larini and Brun [13] have found

$$\Delta T' = \frac{1}{\theta}\sqrt{\frac{2kT}{m}}\,\frac{2 - 0.852\,\beta}{\beta}\,\frac{\sqrt{\pi}}{2}\,\frac{dT'}{dx}$$

The relative differences of the slip coefficients obtained by the Maxwell approximation and the variational method compared to the results obtained by Larini and Brun are plotted on figure 6. For β=1, the Maxwell approximation leads to an error of 13% whereas the error of the variational method is always smaller than 2%.

REFERENCES

[1] f. GOLSE : "Knudsen layer from a computational view point", Preprint in Habilitation (Univ. Paris VII), June 1988.

[2] S. K. LOYALKA : "Approximate method in kinetic theory", The Physics of Fluids, Vol 14, N° 11 (1971).

[3] S. CHAPMAN, T. COWLING : *"The Mathematical Theory of Nonuniform Gases"*, Cambridge University Press, (1952).

[4] G. A. BIRD : *"Molecular Gas Dynamics"*, Oxford University Press, (1976).

[5] J. C. LENGRAND : "Mise en oeuvre de la méthode de Monte-Carlo pour la simulation numérique directe d'un écoulement de gaz raréfiés", Rapport du laboratoire d'Aérothermique, Meudon, avril 1988.

[6] J. F. PALLEGOIX : "Calcul des coefficients aérodynamiques en écoulement de transition, comparaison calcul mesure sur Hermès 0.0", Rapport Aérospatiale TK/AT n° 46684, July 1990.

[7] J. ALLEGRE, X. HERIARD DUBREUILH, M. RAFFIN : "Etude de la stabilité longitudinale en écoulement raréfié de l'avion spatial Hermès", Rapport SESSIA n° 516/90.905, April 1990.

[8] F. GROPENGIESSER, H. NEUNZERT, J. STRUCKMEIER, B. WIESEN : "Rarefied gas flow around a disc with different angles of attack", Proc. 17th RGD Symposium, Aachen, 1990.

[9] H. GRAD : *"Principles of Kinetic Theory of Gases"*, Handbuch der Physik, Vol. 12 (1958).

[10] C. CERCIGNANI : *"Theory and Applications of the Boltzmann Equation"*, Scottish Academic Press, (1975).

[11] Y. SONE, T. OHWADA, K. AOKI : "Temperature jump and Knudsen layer in a rarefied gas over a plane wall", Phys. of Fluids A, Vol. 1, N°2, February 1989.

[12] F. CORON : "Derivation of slip boundary conditions for the Navier-Stokes system from the Boltzmann Equation", J. of Stat. Physics, Vol. 56, N°3/4, (1989).

[13] M. LARINI, R. BRUN : "Discontinuités de température pariétales dans un gaz polyatomique hors d'équilibre", Int. J. Heat Mass Transfer, Vol.16, (1973).

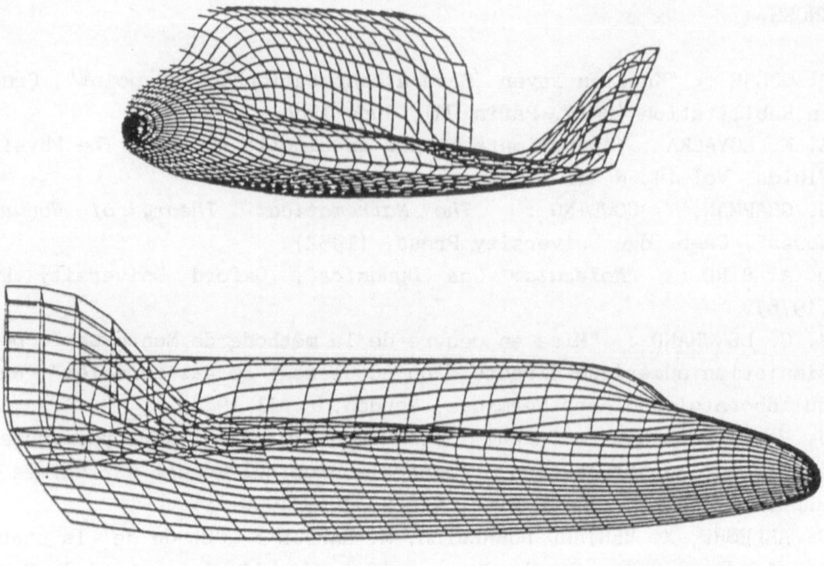

Figure 1 : Body mesh

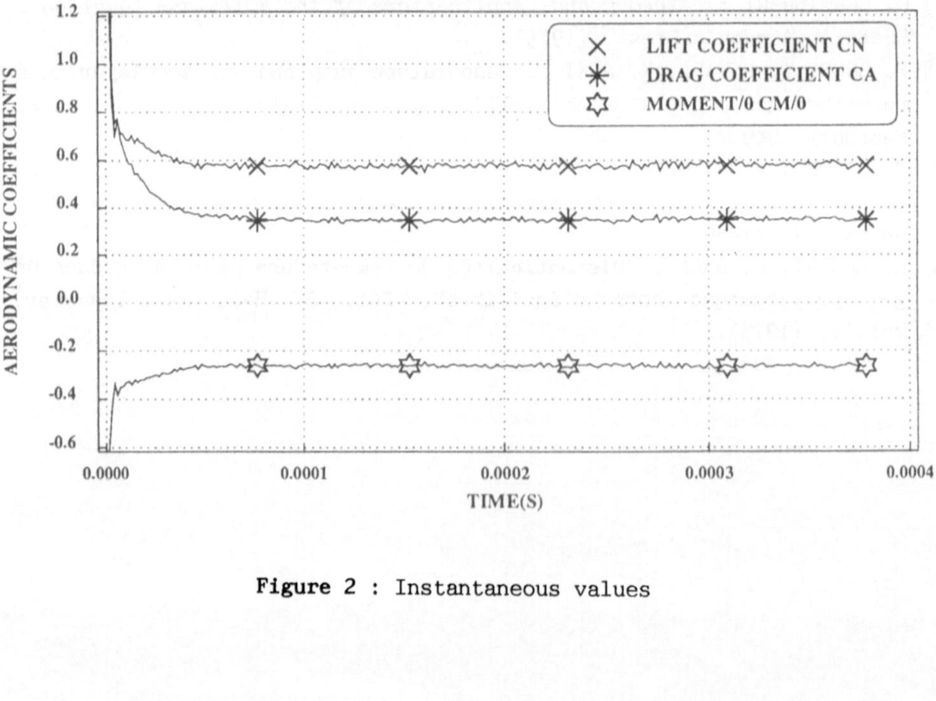

Figure 2 : Instantaneous values

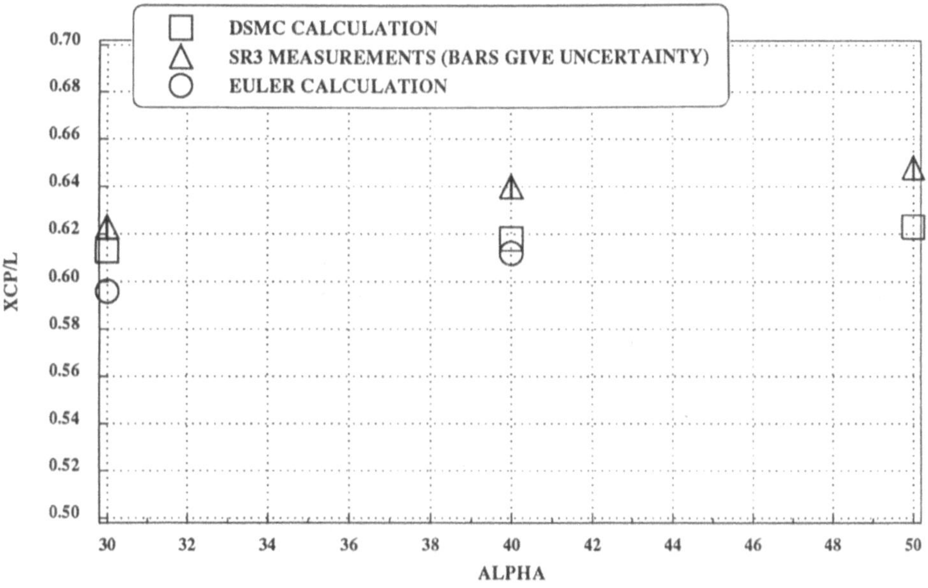

Figure 3 : XCP/L = F(ALPHA)

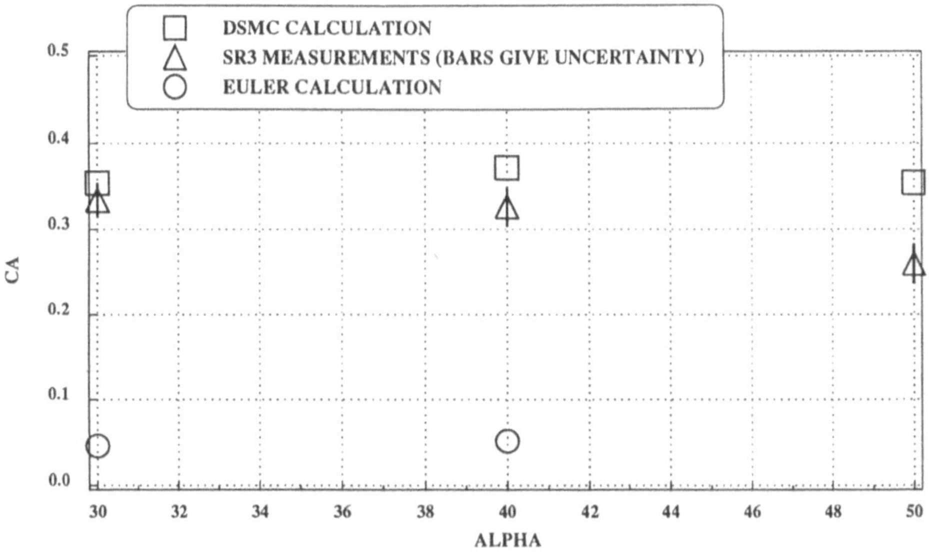

Figure 4 : CA = F(ALPHA)

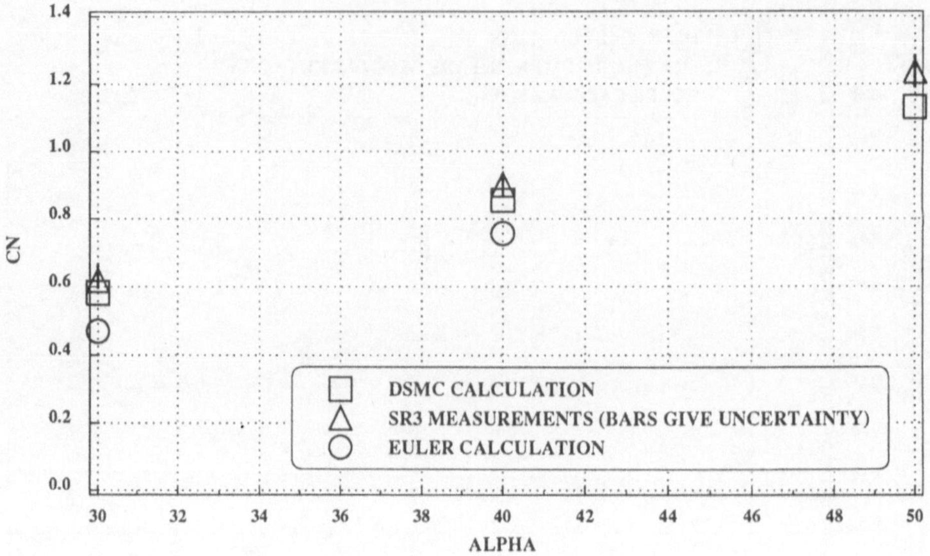

Figure 5 : CN = F(ALPHA)

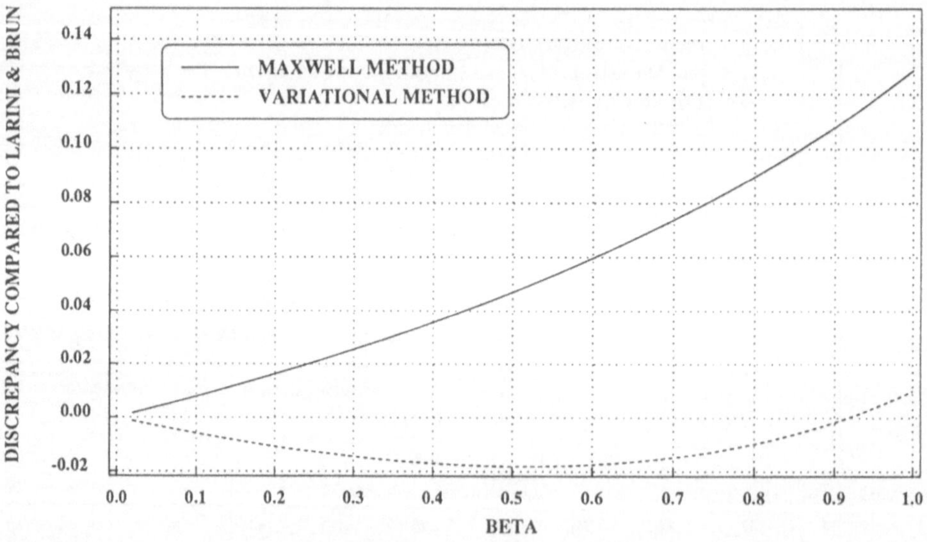

Figure 6 : Slip coefficient

Gas Flows Around the Condensed Phase with Strong Evaporation or Condensation – Fluid Dynamic Equation and Its Boundary Condition on the Interface and Their Application –

K. Aoki and Y. Sone

Department of Aeronautical Engineering, Kyoto University, Kyoto 606, Japan

A gas in contact with its condensed phase is considered, and a steady gas flow around the condensed phase, on the surface of which strong evaporation or condensation is taking place, is investigated on the basis of the kinetic theory. The system of fluid dynamic equation and its boundary condition on the interface of the gas and its condensed phase that describes the gas flow in the continuum flow limit is derived systematically with the aid of the recent results of numerical analysis of the half-space problem of evaporation and condensation. The effect of condensation factor in the kinetic boundary condition on the fluid dynamic boundary condition is also discussed. As an application of the system, the gas flow between two parallel condensed phases with different temperatures is investigated.

I. INTRODUCTION

We consider a system composed of a gas and its condensed phase of a smooth shape, on the surface of which evaporation or condensation is taking place. When the Knudsen number of the system (the mean free path of the gas molecules divided by the characteristic length of the system) is small, a general theory[1–4] describing the steady behavior of the gas, established by a systematic analysis of the Boltzmann equation, is available for weak evaporation and condensation (i.e., for small Mach numbers) (see also Refs. 5 and 6). According to the theory, the flow field is described by the solution of the fluid-dynamic type equations (the Stokes equations for small Reynolds numbers and the Navier-Stokes type equations for finite Reynolds numbers) with the appropriate jump boundary conditions and by the Knudsen-layer corrections to the solution near the condensed phase. However, for strong evaporation and condensation (i.e., the case of finite Mach numbers), little is known about the fluid dynamic equation and its boundary condition to be used even in the continuum flow limit (i.e., at the level of classical fluid dynamics).

The aim of the present study is to establish the fluid dynamic system describing the steady gas flow around the condensed phase with strong evaporation or condensation in the continuum flow limit. That is, we will derive the correct system of fluid dynamic equation and its boundary condition on the interface of the gas and its condensed phase by a systematic analysis of the Boltzmann equation with the aid of the recent results of numerical analysis of the half-space problem of evaporation and condensation[7–12].

At first we analyze the problem under the conventional kinetic boundary condition for evaporation and condensation, where the gas molecules leaving the condensed phase are assumed to constitute the corresponding part of the Maxwellian distribution pertaining to the saturated gas at the surface temperature and velocity of the condensed phase. In Sec. IV a generalization of this condition is discussed. A simple application of the fluid dynamic equation and its boundary condition is given in Sec. V.

II. BASIC EQUATION

The Boltzmann equation in a steady state is written in the following nondimensional form.

$$\zeta_i \frac{\partial f}{\partial x_i} = \frac{1}{k} J(f,f), \tag{1}$$

$$k = (\sqrt{\pi}/2)Kn = (\sqrt{\pi}/2)(\ell_0/L), \tag{2}$$

where $(2RT_0)^{1/2}\zeta_i$ is the molecular velocity, Lx_i is the space rectangular coordinates, $\rho_0(2RT_0)^{-3/2}f$ is the velocity distribution function, ρ_0 is the reference density, T_0 is the reference temperature, R is the gas constant per unit mass, L is the characteristic length of the system, ℓ_0 is the mean free path of the reference equilibrium state at rest at density ρ_0 and temperature T_0,† Kn is the corresponding Knudsen number, and $J(f,f)$ is the standard collision term. The complete definition of $J(f,f)$ is not given here, but no confusion will be expected (cf. Ref. 13).

The gas density $\rho_0\omega$, the gas flow velocity $(2RT_0)^{1/2}u_i$, the gas temperature $T_0\tau$, and the gas pressure p_0P, where $p_0(= R\rho_0T_0)$ is the reference pressure, are given by the moments of f as

$$\left.\begin{aligned} &\omega = \int f d\boldsymbol{\zeta}, \qquad u_i = \frac{1}{\omega}\int \zeta_i f d\boldsymbol{\zeta}, \\ &\tau = \frac{2}{3}\frac{1}{\omega}\int (\zeta_i - u_i)^2 f d\boldsymbol{\zeta}, \qquad P = \omega\tau, \qquad d\boldsymbol{\zeta} = d\zeta_1 d\zeta_2 d\zeta_3. \end{aligned}\right\} \tag{3}$$

The conventional boundary condition on the condensed phase is

$$f = f_w, \qquad f_w = \frac{1}{\pi^{3/2}}\frac{P_w}{\tau_w^{5/2}} exp(-\frac{(\zeta_i - u_{wi})^2}{\tau_w}), \qquad \text{for} \quad \zeta_i n_i > 0, \tag{4}$$

where n_i is the unit normal vector (pointed to the gas) to the surface of the condensed phase, $T_0\tau_w$ is the temperature of the condensed phase, p_0P_w is the saturation gas pressure at temperature $T_0\tau_w$,†† and $(2RT_0)^{1/2}u_{wi}$ is the velocity of the condensed phase ($u_{wi}n_i = 0$).

---

† $\ell_0 = (2/\sqrt{\pi})(2RT_0)^{1/2}\nu_0^{-1}$, where ν_0 is the mean collision frequency of the gas molecules at the reference equilibrium state. For hard-sphere molecular gas, $\ell_0 = (\sqrt{2}\pi d^2\rho_0 m^{-1})^{-1}$, where d is the diameter of a molecule and m is the mass of a molecule.

†† The saturated gas pressure p_0P_w (or p_w, which enters in Sec. III B) is related to $T_0\tau_w$ by the Clausius-Clapeyron relation. In the present analysis, however, this relation is never used. Thus, P_w and τ_w can be chosen freely in the results and discussion (e.g., the last paragraph of Sec. V), and the relation may be introduced when necessary.

We will analyze Eq. (1) with boundary condition (4) for small k to derive the fluid dynamic system in the continuum flow limit.

III. ANALYSIS

A. *Fluid Dynamic Equation*

First we look for a moderately varying solution f_H (Hilbert solution) whose length scale of variation is of the order of the characteristic length L [i.e., $\partial f_H/\partial x_i = O(f_H)$] by the standard Hilbert expansion[13]

$$f_H = f_{H0} + f_{H1}k + \cdots. \tag{5}$$

Correspondingly, the macroscopic variables ω_H, u_{iH}, τ_H, and P_H defined by Eq. (3) with $f = f_H$ are expanded as

$$\Phi_H = \Phi_{H0} + \Phi_{H1}k + \cdots, \quad (\Phi_H = \omega_H, u_{iH}, \tau_H, \text{ or } P_H), \tag{6}$$

where, for example,

$$\left.\begin{aligned} \omega_{H0} &= \int f_{H0}d\boldsymbol{\zeta}, \qquad u_{iH0} = \frac{1}{\omega_{H0}}\int \zeta_i f_{H0}d\boldsymbol{\zeta}, \\ \tau_{H0} &= \frac{2}{3}\frac{1}{\omega_{H0}}\int (\zeta_i - u_{iH0})^2 f_{H0}d\boldsymbol{\zeta}, \qquad P_{H0} = \omega_{H0}\tau_{H0}. \end{aligned}\right\} \tag{7}$$

Substituting Eq. (5) into Eq. (1) and equating the coefficients of the same power of k, we obtain

$$J(f_{H0}, f_{H0}) = 0 \tag{8}$$

$$2J(f_{H0}, f_{H1}) = \zeta_i \frac{\partial f_{H0}}{\partial x_i}, \tag{9}$$

$$2J(f_{H0}, f_{Hn}) = \zeta_i \frac{\partial f_{Hn-1}}{\partial x_i} - \sum_{m=1}^{n-1} J(f_{Hm}, f_{Hn-m}), \qquad (n = 2, 3, \cdots), \tag{10}$$

where $J(f, g)$ is the conventional bilinear form that reduces to the collision term $J(f, f)$ when $g = f$. The definition of $J(f, g)$ is also omitted here (cf. Ref. 13). Equation (8) shows that f_{H0} is a local Maxwellian, i.e.,

$$f_{H0} = \frac{1}{\pi^{3/2}}\frac{P_{H0}}{\tau_{H0}^{5/2}} exp(-\frac{(\zeta_i - u_{iH0})^2}{\tau_{H0}}). \tag{11}$$

The functions f_{Hn} ($n = 1, 2, \cdots$) are, in principle, determined successively from f_{H0} by the sequence of integral equations (9) and (10). These integral equations, however, have solutions only when the respective inhomogeneous terms satisfy the solvability condition[13]:

$$\int (1, \zeta_i, \zeta_i^2)[\text{Inhomogeneous term}]d\boldsymbol{\zeta} = 0. \tag{12}$$

The solvability condition of the integral equation for f_{Hn} gives the partial differential equations governing the macroscopic variables Φ_{Hn-1}. The differential equations for the leading term Φ_{H0}, obtained from the solvability condition of Eq. (9), are the Euler equations for a compressible gas, i.e.,

$$\left.\begin{aligned} &\frac{\partial(\rho v_j)}{\partial X_j}=0, \qquad \rho v_j\frac{\partial v_i}{\partial X_j}+\frac{\partial p}{\partial X_i}=0,\\ &v_j\frac{\partial}{\partial X_j}\left(\frac{3}{2}RT+\frac{1}{2}v_i^2+\frac{p}{\rho}\right)=0, \qquad p=R\rho T, \end{aligned}\right\} \tag{13}$$

where

$$\left.\begin{aligned} &\rho=\rho_0\omega_{H0}, \qquad v_i=(2RT_0)^{1/2}u_{iH0}, \qquad T=T_0\tau_{H0},\\ &p=p_0P_{H0}, \qquad X_i=Lx_i. \end{aligned}\right\} \tag{14}$$

For convenience of practical application, we gave Eq. (13) in the dimensional form (The ρ, v_i, T, and p are, respectively, the density, flow velocity, temperature, and pressure of the gas corresponding to the k^0-order Hilbert solution).

The Hilbert solution, in general, cannot be made to satisfy the kinetic boundary condition (4) because it is a solution in simple series expansion of k of the singular equation (1), where the differential operator is multiplied by the small parameter k. In fact, if we adjust f_{H0} to Eq. (4), we have, on the condensed phase,

$$P_{H0}=P_w, \qquad u_{iH0}=u_{wi}, \qquad \tau_{H0}=\tau_w. \tag{15}$$

Equation (13) with Eq. (15) forms an overdetermined system [Consider the gas between two parallel condensed phases with different temperatures. In this problem P_{H0}, u_{iH0}, τ_{H0} are constant (cf. Sec. V) and cannot satisfy Eq. (15) on both of the condensed phases].

B. *Jump Boundary Condition*

We now try to find the solution satisfying the boundary condition in the form

$$f=f_H+f_K, \tag{16}$$

where f_K is a rapidly varying correction term, which has a length scale of variation of the order of the mean free path ℓ_0 normal to the surface of the condensed phase [i.e., $kn_i(\partial f_K/\partial x_i)=O(f_K)$] and is appreciable only in a thin layer with thickness of the order of ℓ_0 adjacent to the condensed phase (Knudsen layer). The Knudsen-layer variable (a stretched normal coordinate) η is introduced by

$$x_i=k\eta n_i+x_{Bi}(s_1,s_2), \tag{17}$$

where x_{Bi} is the surface of the condensed phase, and s_1 and s_2 are (unstretched) coordinates within a parallel surface $\eta=const$. The Knudsen layer correction f_K is expanded as

$$f_K=f_{K0}+f_{K1}k+\cdots. \tag{18}$$

Substituting Eq. (16) with Eqs. (5) and (18) into Eq. (1) and arranging the terms in the order of k, we obtain a sequence of equations for f_{Km}. In the ordering we take into account the length scale of variation of f_K and the properties of f_H, especially the fact that f_H can be expressed as

$$f_H = (f_{H0})_B + [(f_{H1})_B + (\frac{\partial f_{H0}}{\partial x_i})_B n_i \eta]k + \cdots, \tag{19}$$

in the Knudsen layer[14,15], where $(\ \ )_B$ indicates the value on the condensed phase. The equation for the leading term f_{K0} is given by the spatially one-dimensional Boltzmann equation

$$\zeta_i n_i \frac{\partial f^*}{\partial \eta} = J(f^*, f^*), \tag{20}$$

if we introduce f^* by

$$f^* = (f_{H0})_B + f_{K0}. \tag{21}$$

The corresponding boundary condition for f^* on the condensed phase is, from Eq. (4),

$$f^* = f_w, \qquad (\text{for} \quad \zeta_i n_i > 0, \quad \text{at} \quad \eta = 0). \tag{22}$$

The condition as $\eta \to \infty$ is given by

$$f^* \to (f_{H0})_B, \qquad (\text{as} \quad \eta \to \infty), \tag{23}$$

because f_K vanishes as $\eta \to \infty$.

The boundary value problem (20), (22), and (23), which is equivalent to a steady gas flow problem over its plane condensed phase, contains the parameters τ_w, P_w, and u_{wi} (with $u_{wi} n_i = 0$) in f_w and $(\tau_{H0})_B$, $(P_{H0})_B$, and $(u_{iH0})_B$ in $(f_{H0})_B$. By proper arrangement of the variables in the problem, we can reduce these parameters to $(\tau_{H0})_B/\tau_w$, $(P_{H0})_B/P_w$, $(u_{iH0}n_i)_B/\tau_w^{1/2}$, and $(u_{iH0} - u_{jH0}n_j n_i)_B/\tau_w^{1/2} - u_{wi}/\tau_w^{1/2}$ (or $(u_{iH0}t_i)_B/\tau_w^{1/2} - u_{wi}t_i/\tau_w^{1/2}$, where t_i is an arbitrary unit tangential vector to the surface of the condensed phase).† Various numerical examinations of the boundary value problem on the basis of the Boltzmann-Krook-Welander (BKW) equation show that the solution exists only when the parameters satisfy some conditions. The conditions give some relations between the surface condition of the condensed phase and the values of u_{iH0}, τ_{H0}, and P_{H0} on the interface, that is, the boundary condition of the Euler equation (13) on the interface of the gas and its condensed phase.

The boundary value problem (20), (22), and (23) is studied numerically in detail for the special case $(u_{iH0}t_i)_B - u_{wi}t_i = 0$ on the basis of the BKW equation in Refs. 7 - 12. Fortunately our recent numerical study, done in collaboration with H. Sugimoto and K. Nishino, shows that the solution for the general case exists only when $(u_{iH0}t_i)_B - u_{wi}t_i = 0$, if evaporation is taking place on the interface ($(u_{iH0}n_i)_B \geq 0$ or $(P_{H0})_B/P_w \leq 1$). Thus we can derive the result for the general case of evaporation from that of Refs. 8 - 11. When condensation is taking place, the result (solution and the relations among

† Except for a hard-sphere molecular gas and the BKW model, there is another parameter in the system: the ratio of the characteristic scale of the intermolecular potential and mRT_0. (Y. Sone, Lecture note on molecular gas dynamics, 1987)

the parameters) depends on $(u_{iH0}t_i)_B - u_{wi}t_i$, but only the result for $(u_{iH0}t_i)_B = u_{wi}t_i$ is available now (Refs. 7, 9, and 12).†

Let $T_w = T_0\tau_w$ (the temperature of the condensed phase), $p_w = p_0P_w$ (the saturation gas pressure at temperature T_w), $v_{wi} = (2RT_0)^{1/2}u_{wi}$ (the velocity of the condensed phase ($v_{wi}n_i = 0$)), and $M_n = (6/5)^{1/2} \mid u_{iH0}n_i \mid \tau_{H0}^{-1/2} = \mid v_in_i \mid (\frac{5}{3}RT)^{-1/2}$ (the local Mach number corresponding to the normal speed $\mid v_in_i \mid$). Then the following condition holds on the condensed phase.

(a) The boundary condition on the interface where evaporation is taking place ($v_in_i \geq 0$ or $p/p_w \leq 1$):

$$M_n \leq 1, \qquad \frac{p}{p_w} = h_1(M_n), \qquad \frac{T}{T_w} = h_2(M_n), \qquad (v_i - v_{wi})t_i = 0. \tag{24}$$

(b) The boundary condition (applicable to the limited case: $v_it_i = v_{wi}t_i$) on the interface where condensation is taking place ($v_in_i < 0$, or $p/p_w > 1$):

$$M_n < 1, \qquad \frac{p}{p_w} = F_s(M_n, \frac{T}{T_w}), \tag{25}$$

or p/p_w, T/T_w, and M_n take arbitrary values satisfying

$$M_n \geq 1, \qquad \frac{p}{p_w} \geq F_b(M_n, \frac{T}{T_w}). \tag{26}$$

The numerical data of h_1 and h_2 are given in Fig. 1 and Table I, and those of F_s and F_b are given in Figs. 2, 3 and Tables II, III. The data in Tables II and III are those arranged for neat values of M_n and T/T_w by interpolation from the actual numerical data in Ref. 12 for convenience of practical use. Some of the data in Table I are also the result of interpolation from the data in Refs. 8, 9, and 11. No steady evaporation takes place for $M_n > 1$.[8,10,11] Both of the functions $F_s(M_n, T/T_w)$ and $F_b(M_n, T/T_w)$ are almost independent of T/T_w except for M_n close to 1. They coincide at $M_n = 1$, i.e., $\lim_{M_n \to 1} F_s(M_n, T/T_w) = F_b(1, T/T_w)$. The $F_b(1, T/T_w)$ is estimated as[9,12]: $17.1 < F_b(1, 0.5) < 17.5$, $13.4 < F_b(1,1) < 13.6$, $11.8 < F_b(1,2) < 12.0$, and $11.6 < F_b(1,4) < 11.8$.

C. *Summary*

We have analyzed the Boltzmann system in the continuum limit and derived the fluid dynamic equation (13) and its boundary condition (24) - (26) on the interface of the gas and its condensed phase where strong evaporation or condensation is taking place. The data for h_1, h_2, F_s, and F_b in Tables I - III and Figs. 1 -3 are based on the BKW equation, and the condition (25) and (26) on the surface with condensation is applicable only to the case with $v_it_i = v_{wi}t_i$. The generalization of the latter will be published near future. In a high-speed gas flow (with finite Mach numbers) past a solid boundary without either evaporation or condensation, the Knudsen layer, which is the effect of gas rarefaction, does not appear in the continuum flow limit, but there is a thin

† We have some partial data for the general case, which show their weak dependence on $(u_{iH0}t_i)_B - u_{wi}t_i$. The detailed data will be presented near future in a joint paper with Sugimoto and Nishino.

layer in which viscosity and thermal conductivity are important between the outer inviscid region and the boundary (Prandtl boundary layer). In the present case with strong evaporation or condensation, the Knudsen layer appears even in the continuum flow limit, and the viscous layer is blown away from the evaporating surface by strong convection and is merged with the Knudsen layer on the surface with condensation. Therefore, the Knudsen layer is matched directly to the outer inviscid region. A shock layer may exist in the inviscid region.

IV. EFFECT OF CONDENSATION FACTOR

The fluid dynamic boundary condition (24) - (26) was derived on the basis of the conventional kinetic condition for the molecules leaving the condensed phase [Eq. (4)], in which the velocity distribution of the outgoing molecules is determined only by the condition of the condensed phase and is independent of the velocity distribution of the incoming molecules. Recently, Wortberg et al.[16] presented experimental data showing the deviation from this condition on a certain solid surface. The generalized boundary condition suggested by the experiment is described as follows. The velocity distribution function for the outgoing molecules is expressed by the sum of two terms: α_c times of the distribution of the conventional boundary condition and $(1-\alpha_c)$ times of the distribution of the diffuse reflection, where α_c $(0 < \alpha_c \leq 1)$ is a constant called condensation factor. That is, the (nondimensional) velocity distribution function for the outgoing molecules is given by replacing P_w in the conventional condition (4) by the following $\hat{P}_w$:

$$\hat{P}_w = \alpha_c P_w - (1-\alpha_c) 2\sqrt{\pi}\tau_w^{1/2} \int_{\zeta_i n_i < 0} \zeta_i n_i (f)_B d\boldsymbol{\zeta}. \tag{27}$$

This condition has been applied to half-space problem of strong evaporation and condensation in Refs. 10, 11, 17, and 18. In Ref. 17 was given the idea that a solution under the generalized condition is derived from a solution under the conventional condition by a simple conversion. The explicit form of the conversion formula was given in Refs. 10 and 11 for the evaporation case (See Eqs. (8) and (9) in Ref. 11). The same conversion formula holds also in the case of condensation except for the additional positiveness condition on $\hat{P}_w$.† With these results, we can express the fluid dynamic boundary condition under the generalized kinetic condition in terms of the functions h_1, h_2, etc. introduced in Eqs. (24) - (26), i.e.,

(a') When evaporation is taking place, the following relation holds.

$$M_n \leq 1, \quad \frac{p}{p_w} = \frac{\alpha_c}{\alpha_c - (1-\alpha_c)K_e(M_n)} h_1(M_n), \quad \frac{T}{T_w} = h_2(M_n), \quad (v_i - v_{wi})t_i = 0, \tag{28}$$

where

$$K_e(M_n) = -\left(\frac{10\pi}{3}\right)^{1/2} \frac{h_1(M_n)}{[h_2(M_n)]^{1/2}} M_n. \tag{29}$$

The $K_e(M_n)$ and p/p_w [Eq. (28)] are, respectively, shown in Figs. 11 and 12 of Ref. 11. The present $M_n, p/p_w$, and $K_e(M_n)$ correspond, respectively, to $M_\infty, p_\infty/p_w^G$, and $-(K(M_\infty)+1)$ in these figures.

† The velocity distribution function cannot be nonnegative when $\hat{P}_w < 0$.

(b') When condensation is taking place,[†] p/p_w, T/T_w, and M_n satisfy

$$M_n < 1, \qquad \frac{p}{p_w} = \frac{\alpha_c}{\alpha_c - (1-\alpha_c)K_s(M_n, T/T_w)} F_s(M_n, \frac{T}{T_w}), \tag{30}$$

with the restriction

$$K_s(M_n, \frac{T}{T_w})[1 + K_s(M_n, \frac{T}{T_w})]^{-1} < \alpha_c, \tag{31}$$

where

$$K_s(M_n, \frac{T}{T_w}) = (\frac{10\pi}{3})^{1/2} F_s(M_n, \frac{T}{T_w})(\frac{T}{T_w})^{-1/2} M_n, \tag{32}$$

or take arbitrary values satisfying

$$M_n \geq 1, \qquad \frac{p}{p_w} > \frac{\alpha_c}{\alpha_c - (1-\alpha_c)K_b(M_n, T/T_w)} F_b(M_n, \frac{T}{T_w}), \tag{33}$$

with the restriction

$$K_b(M_n, \frac{T}{T_w})[1 + K_b(M_n, \frac{T}{T_w})]^{-1} < \alpha_c, \tag{34}$$

where

$$K_b(M_n, \frac{T}{T_w}) = (\frac{10\pi}{3})^{1/2} F_b(M_n, \frac{T}{T_w})(\frac{T}{T_w})^{-1/2} M_n. \tag{35}$$

The restrictions (31) and (34) are the consequence of the positiveness of $\hat{P}_w$ in Eq. (27).

V. APPLICATION TO TWO-SURFACE PROBLEM

As an application of the fluid dynamic equation and its boundary condition, let us consider a gas between its two parallel condensed phases at rest and with different temperatures T_{w1} and T_{w2} ($T_{w1} > T_{w2}$). We denote the saturation gas pressure at temperature T_{w1} by p_{w1} and that at T_{w2} by p_{w2} ($p_{w1} > p_{w2}$). Then, evaporation (condensation) takes place on the hotter (colder) condensed phase, and the gas flows from the hotter condensed phase to the colder.

In this one-dimensional case, p, T, $M = (\frac{5}{3}RT)^{-1/2} v_1$ are all constant from Eq. (13) (X_1 axis is taken from the hotter to the colder condensed phase, normal to them). On the hotter condensed phase, the condition for evaporation (24) is applied, i.e.,

$$p/p_{w1} = h_1(M), \qquad T/T_{w1} = h_2(M), \qquad (M \leq 1). \tag{36}$$

On the colder condensed phase, the condition for condensation (25) or (26) holds, i.e.,

$$p/p_{w2} = F_s(M, T/T_{w2}), \qquad (M < 1), \tag{37}$$

or

$$p/p_{w2} \geq F_b(M, T/T_{w2}), \qquad (M \geq 1). \tag{38}$$

From Eq. (36) a flow with $M > 1$ never occurs. When $M < 1$, eliminating p and

---

[†] As the boundary condition (b) in Sec. III B, the condition to be given is applicable to the limited case: $v_i t_i = v_{wi} t_i$.

T from Eqs. (36) and (37), we have

$$(p_{w1}/p_{w2})h_1(M) = F_s(M, (T_{w1}/T_{w2})h_2(M)). \tag{39}$$

The M is determined by this equation, and then p and T are obtained from Eq. (36). When $M = 1$, Eqs. (36) and (38) lead to

$$(p_{w1}/p_{w2})h_1(1) \geq F_b(1, (T_{w1}/T_{w2})h_2(1)). \tag{40}$$

From our data (Figs. 1 - 3 and Tables I - III), $F_s(M, (T_{w1}/T_{w2})h_2(M))/h_1(M)$ increases monotonically from 1 to $F_b(1, (T_{w1}/T_{w2})h_2(1))/h_1(1)$ as M increases from 0 to 1. Thus, from Eq. (39), we find that M increases monotonically from 0 to 1 as p_{w1}/p_{w2} increases from 1 to $F_b(1, (T_{w1}/T_{w2})h_2(1))/h_1(1)$. For larger p_{w1}/p_{w2}, $M = 1$ (the flow remains sonic). In Table IV, we show some results obtained by solving Eq. (39) by interpolation on the basis of the data of Tables I and II. Since $F_s(M_n, T/T_w)$ in Eq. (25) does not depend much on T/T_w, the solution M of Eq. (39) is insensitive to T_{w1}/T_{w2} and is mostly determined by p_{w1}/p_{w2}.

REFERENCES

1. Y. Sone and Y. Onishi, J. Phys. Soc. Jpn. **44**, 1981, (1978).
2. Y. Onishi and Y. Sone, J. Phys. Soc. Jpn. **47**, 1676, (1979).
3. Y. Sone and K. Aoki, Transp. Theory Stat. Phys. **16**, 189, (1987); Mem. Fac. Eng. Kyoto Univ. **49**, 237, (1987).
4. Y. Sone, T. Ohwada, and K. Aoki, Phys. Fluids A **1**, 1398, (1989).
5. Y. Sone, in Rarefied Gas Dynamics, edited by A. E. Beylich (VCH Verlags-gesellschaft) (to be published).
6. Y. Sone, in this volume.
7. Y. Sone, K. Aoki, and I. Yamashita, in Rarefied Gas Dynamics, edited by V. C. Boffi and C. Cercignani (Teubner, Stuttgart, 1986), Vol. II, p. 323.
8. Y. Sone and H. Sugimoto, J. Vacuum Soc. Jpn. **31**, 420 (1988).
9. Y. Sone, K. Aoki, H. Sugimoto, and T. Yamada, Theor. Appl. Mech. Bulgarian Acad. Sci. **19**, No. 3, 89 (1988).
10. H. Sugimoto and Y. Sone, J. Vacuum Soc. Jpn. **32**, 214 (1989).
11. Y. Sone and H. Sugimoto, in Adiabatic Waves in Liquid-Vapor Systems, edited by G. E. A. Meier and P. A. Thompson (Springer, Berlin, 1990), p. 294.
12. K. Aoki, Y. Sone, and T. Yamada, Phys. Fluids A **2**, No. 10 (1990).
13. H. Grad, in Handbuch der Physik, edited by S. Flügge (Springer, Berlin, 1958), Vol. 12, p. 205.
14. Y. Sone, in Rarefied Gas Dynamics, edited by D. Dini (Editrice Tecnico Scientifica, Pisa, 1971), p. 737.
15. Y. Sone and K. Yamamoto, J. Phys. Soc. Jpn. **29**, 495 (1970); Y. Sone and Y. Onishi, J. Phys. Soc. Jpn. **47**, 672 (1979).
16. R. Mager, G. Adomeit, and G. Wortberg, in Rarefied Gas Dynamics: Physical Phenomena, edited by E. P. Muntz, D. P. Weaver, and D. H. Campbell (AIAA, Washington, D. C., 1989), p. 460.
17. M. N. Kogan and N. K. Makashev, Fluid Dynamics **6**, 913 (1971).
18. A. P. Kryukov, Fluid Dynamics **23**, 320 (1988).

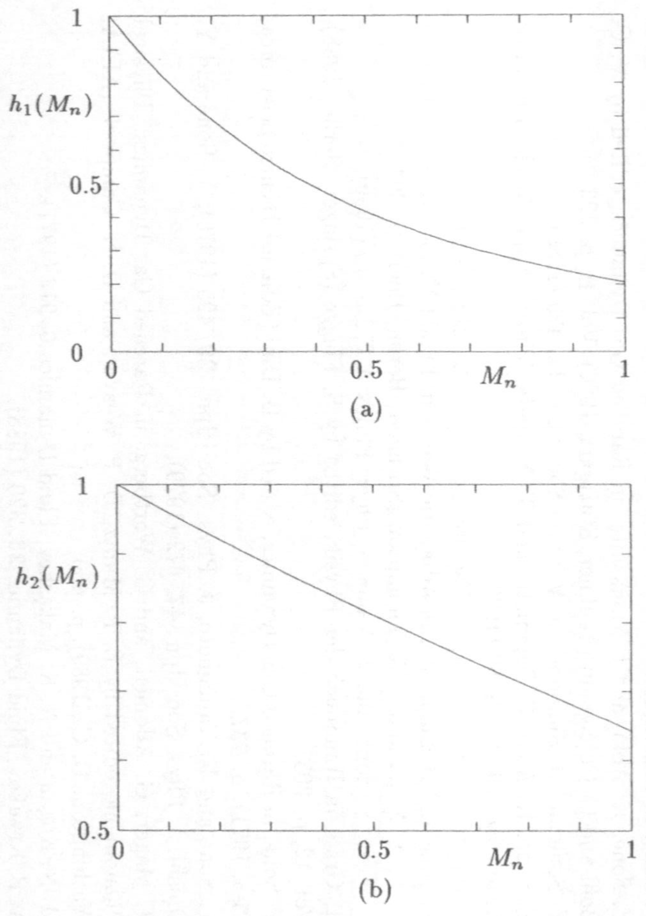

Fig. 1. $h_1(M_n)$ and $h_2(M_n)$ (Ref. 8).

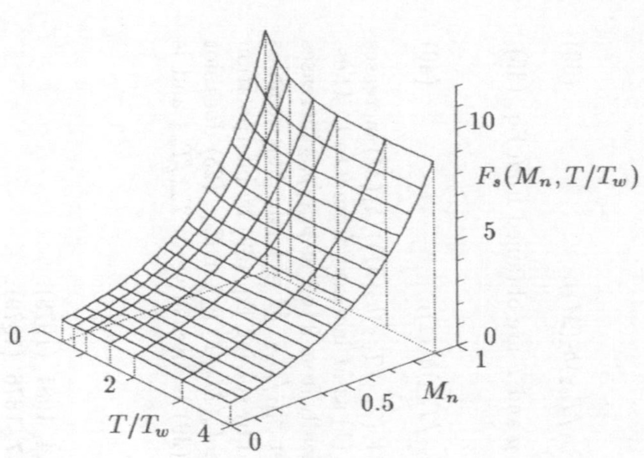

Fig. 2. $F_s(M_n, T/T_w)$ (Ref. 12).

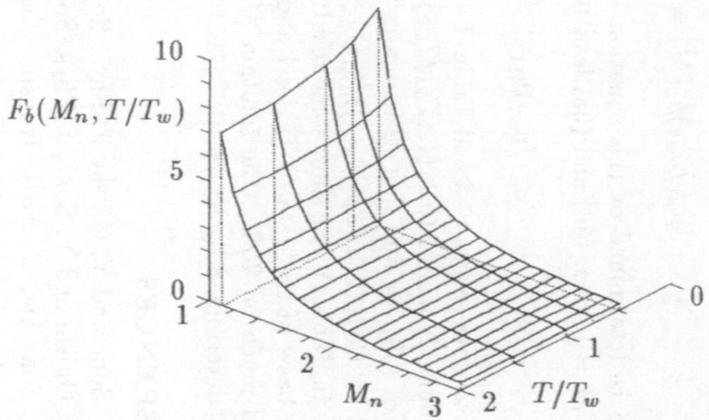

Fig. 3. $F_b(M_n, T/T_w)$ (Ref. 12).

Table I. $h_1(M_n)$ and $h_2(M_n)$ (Refs. 8, 9, and 11).

M_n	$h_1(M_n)$	$h_2(M_n)$	M_n	$h_1(M_n)$	$h_2(M_n)$
0.00	1.0000	1.0000	0.50	0.4178	0.8113
0.05	0.9083	0.9798	0.55	0.3867	0.7938
0.10	0.8267	0.9599	0.60	0.3585	0.7765
0.15	0.7539	0.9404	0.65	0.3331	0.7594
0.20	0.6891	0.9212	0.70	0.3099	0.7424
0.25	0.6309	0.9022	0.75	0.2888	0.7256
0.30	0.5789	0.8836	0.80	0.2695	0.7088
0.35	0.5321	0.8652	0.85	0.2519	0.6923
0.40	0.4900	0.8470	0.90	0.2357	0.6758
0.45	0.4520	0.8290	0.95	0.2210	0.6595
			1.00	0.2075	0.6434

Table II. $F_s(M_n,\ T/T_w)$ (Ref. 12). The symbol * indicates the limiting value as $M_n \to 0$ ($F_s(0,\ T/T_w)$ is not defined except for $T/T_w = 1$).

	$F_s(M_n,\ \frac{T}{T_w})$						
M_n	$\frac{T}{T_w} = 0.5$	$\frac{T}{T_w} = 0.75$	$\frac{T}{T_w} = 1.0$	$\frac{T}{T_w} = 1.5$	$\frac{T}{T_w} = 2.0$	$\frac{T}{T_w} = 3.0$	$\frac{T}{T_w} = 4.0$
0.00	1.000*	1.000*	1.000	1.000*	1.000*	1.000*	1.000*
0.05	1.114	1.104	1.104	1.106	1.112	1.119	1.130
0.10	1.232	1.221	1.220	1.225	1.233	1.250	1.267
0.15	1.367	1.354	1.352	1.359	1.370	1.396	1.421
0.20	1.525	1.506	1.502	1.511	1.526	1.559	1.592
0.25	1.707	1.679	1.673	1.683	1.701	1.742	1.785
0.30	1.914	1.878	1.869	1.879	1.900	1.951	2.002
0.35	2.146	2.106	2.092	2.103	2.130	2.186	2.254
0.40	2.427	2.369	2.350	2.359	2.385	2.454	2.521
0.45	2.757	2.675	2.649	2.654	2.685	2.761	2.838
0.50	3.122	3.031	2.998	2.995	3.026	3.111	3.212
0.55	3.583	3.449	3.396	3.389	3.423	3.513	3.612
0.60	4.092	3.942	3.869	3.849	3.870	3.975	4.107
0.65	4.734	4.525	4.424	4.385	4.411	4.509	4.639
0.70	5.527	5.225	5.077	5.014	5.029	5.124	5.284
0.75	6.411	6.074	5.873	5.758	5.747	5.838	5.993
0.80	7.626	7.105	6.826	6.640	6.597	6.666	6.829
0.85	9.092	8.385	8.040	7.695	7.602	7.630	7.758
0.90	11.11	9.993	9.443	8.968	8.790	8.754	8.902

Table III. $F_b(M_n, T/T_w)$ (Ref. 12).

M_n	$F_b(M_n, \frac{T}{T_w})$ $\frac{T}{T_w}=0.5$	$\frac{T}{T_w}=0.75$	$\frac{T}{T_w}=1.0$	$\frac{T}{T_w}=1.5$	$\frac{T}{T_w}=2.0$
1.1	9.009	8.130	7.703	7.331	7.210
1.2	5.586	5.185	5.002	4.864	4.850
1.3	3.825	3.614	3.526	3.477	3.498
1.4	2.793	2.673	2.629	2.619	2.650
1.5	2.137	2.064	2.042	2.051	2.085
1.6	1.692	1.647	1.638	1.654	1.686
1.7	1.376	1.348	1.346	1.367	1.396
1.8	1.143	1.126	1.129	1.151	1.180
1.9	0.9666	0.9573	0.9623	0.9852	1.011
2.0	0.8296	0.8252	0.8318	0.8545	0.8797
2.1	0.7209	0.7199	0.7274	0.7496	0.7733
2.2	0.6331	0.6346	0.6428	0.6640	0.6861
2.3	0.5611	0.5644	0.5730	0.5936	0.6141
2.4	0.5014	0.5059	0.5145	0.5342	0.5534
2.5	0.4513	0.4566	0.4653	0.4841	0.5023
2.6	0.4086	0.4146	0.4231	0.4411	0.4583
2.7	0.3722	0.3799	0.3872	0.4044	0.4208
2.8	0.3407	0.3474	0.3557	0.3723	0.3879
2.9	0.3133	0.3202	0.3283	0.3443	0.3591
3.0	0.2893	0.2963	0.3043	0.3196	0.3338

Table IV. The gas flow owing to evaporation and condensation between two parallel plane condensed phases with different temperatures.

$\frac{T_{w1}}{T_{w2}}$	$\frac{p_{w1}}{p_{w2}}$	M	$\frac{T}{T_{w2}}$	$\frac{p}{p_{w2}}$	$\frac{v_1}{(2RT_{w2})^{1/2}}$
1.2	2	0.178	1.12	1.43	0.171
1.2	4	0.354	1.04	2.11	0.329
1.2	6	0.456	0.992	2.69	0.415
1.2	10	0.581	0.940	3.69	0.514
2	2	0.175	1.86	1.44	0.218
2	6	0.455	1.66	2.69	0.534
2	10	0.582	1.57	3.68	0.665
2	20	0.751	1.45	5.77	0.825
4	6	0.443	3.33	2.74	0.738
4	10	0.573	3.14	3.73	0.928
4	20	0.748	2.91	5.79	1.16
4	30	0.848	2.77	7.58	1.29

Part II

Discrete Kinetic Theory

Fluid Dynamic Limits of Discrete Velocity Kinetic Equations

C. Bardos[1], *F. Golse*[1], *and D. Levermore*[2]

[1]Département de Mathématiques, Université Paris VII, F-75251 Paris Cedex 05, France

[2]Department of Mathematics, University of Arizona, Tucson, AZ 85721, USA

Abstract

The connection between discrete velocity kinetic theory and fluid dynamics is systematically described. Conditions that formally lead to generalized compressible Euler equations or to generalized incompressible Navier-Stokes equations are given. These conditions are related to an H-theorem. It is proven that a large class of polynomial collision operators in semidetailed balance satisfies this H-theorem. Finally, results are given concerning the global validity in time of the convergence for the case where the formal scaling of the kinetic equation leads to the linearized incompressible Navier-Stokes limit.

I. Introduction

The analytical theory of discrete velocity kinetic models has undergone substantial developments (cf. the articles of Platkowski and Illner [PI] and Cabannes [C2]). Among others, two reasons for this developments should be mentioned:

i) the success of the lattice gas automata, both in theory and in simulation (cf. for instance [FHP] and [FD'HHLPR]);

ii) the major contributions of H. Cabannes, R. Gatignol and their group in "le Laboratoire de Modélisation en Mécanique de l'Université Pierre et Marie Curie".

Therefore, as a testimony of our friendship and in consideration of his a this long term program, we dedicate the present paper to Henry Cabannes. It presents the extension of some of our work on the fluid dynamic limits of kinetic equations to these discrete velocities models.

In [BGL1] we considered the relations between the continuous kinetic equations (Boltzmann equation, BGK model etc...) and the classical macroscopic equations of fluid dynamic that are obtained when the Knudsen number goes to zero. The following facts were observed.

i) To obtain a compressible Euler equation the only hypotheses needed are the existence of conserved quantities and the fact that the collision operator satisfies an H-theorem.

* Research partly supported by the DRET Contract 88.34.129

ii) To obtain a fluid equation with a finite Reynolds number it is compulsory to consider a fluid with a vanishing Mach number, in which case the macroscopic limit will describe an incompressible gas. The structure of this macroscopic equation depends not only on the conserved quantities and the H-theorem but also on properties of the Fréchet derivative of the collision operator.

As is shown below, these observations apply to the discrete velocity models too.

Consider a gas of particles confined to a domain $X \subset R^D$ whose velocities are restricted to a discrete set $V \subset R^D$ of finite cardinality $|V|$ (although some of the results will generalize to the countable case). At the kinetic level this gas is described by a nonnegative function $F(t,x,v)$ that represents the density of particles with position x and velocity v in the single particle phase space $X \times V$ at time t. It is common in the literature to view F as a vector-valued function of (t,x) indexed by v (i.e. with values in $R^{|V|}$). However, we feel that the former notation is both more computationally convenient and more clearly draws the similarities between these and classical kinetic theories; these similarities are essential to understanding the universality in derivations of fluid dynamic limits.

The evolution of $F(t,x,v)$ is assumed to be governed by a kinetic equation of the form

$$\partial_t F + v\cdot\nabla F = C(F) . \tag{1}$$

Here the interaction of particles through collisions is modelled by the operator C that acts only on the variable v and is generally nonlinear. Indeed, it is a nonlinear operator acting on the space R_+^V of functions defined in V with values in R_+. It will be kept abstract in sections II and III, being defined only by properties given below. In section IV a class of generalized Boltzmann collision operators will be considered that models a wide variety of collisional processes involving arbitrary numbers of particles.

In section II we give sufficients conditions on the discrete velocity collision operator to obtain a generalized compressible Euler system in the limit when the Knudsen number goes to zero. In Section III we give sufficient conditions for obtaining the incompressible Naviers-Stokes limit. Section IV is devoted to the analysis of a class of collision operators satisfying the assumptions of sections II and III. As in [G] or in [CG1,2], these collision operators are polynomials discribing binary or multiple collisions. For these operators, the property of being in semidetailed balance implies an H-theorem and that they satisfy the hypotheses of sections II and III. This is a generalisation of some of the results of [G] and also applies to the 14 velocity models introduced by Cabannes [C1]. In the derivations described in sections II and III some crucial assumptions about the convergence of moments are made. The purpose of the section V is to comment on these assumptions and sketch some convergence results.

II. The Compressible Euler Limit

In this section the sum of any scalar or vector valued function $f(v)$ over the variable v will be denoted by $\langle f\rangle$;

$$\langle f\rangle = \sum_{v\in V} f(v) . \tag{2}$$

Concepts of conservation are central to the existence of fluid dynamic limits. Two of them will play a central role in this paper, the first is introduced below while the second appears in section IV.

Definition 1. A mapping $e : V \mapsto R$ (alternatively a vector $e \in R^V$) is said to be a conserved quantity for the collision operator C if

$$\langle e\, C(F) \rangle = 0 ,$$

for every $F \in R^V_+$.

Remark 1. In physically motivated examples these usually include quantities associated with classical conservation laws like those of mass, momentum and energy. However we will not be limited by insisting that the conserved quantities be the physical ones. Classical equations of fluid dynamics can be derived from kinetic equations with collision operators that posses nonstandard conservations laws [BGL3].

The set of all conserved quantities of C is a linear subspace of R^V which is assumed to be nontrivial. Let N be the dimension of this space and $\{e_i(v) \,:\, 1 \le i \le N\}$ a basis. The vector-valued map from V to R^N whose components are these basis vectors is denoted $\vec{e}(v)$. Vectors in R^N will be denoted with arrows and the Euclidean inner product of two such vectors, $\vec{\alpha}$ and $\vec{\beta}$, is denoted by $\vec{\alpha} \odot \vec{\beta}$. Associated with the conserved quantities $\vec{e}$ are the N independent local conservation laws

$$\partial_t \langle \vec{e}\, F \rangle + \nabla \cdot \langle v\, \vec{e}\, F \rangle = 0 , \tag{3}$$

The functions $\langle \vec{e}\, F \rangle$ and $\langle v\, \vec{e}\, F \rangle$ are called respectively the vectors of conserved densities and fluxes associated with $\vec{e}$.

Additionally, the collision operator C is assumed to have the dissipation property

$$\langle C(F) \log F \rangle \le 0 , \tag{4}$$

for every $F \in R^V_+$; here the quantity appearing on the left is called the entropy dissipation rate. This property implies the local entropy inequality

$$\partial_t \langle F(\log F - 1) \rangle + \nabla \cdot \langle v\, F(\log F - 1) \rangle = \langle C(F) \log F \rangle \le 0 . \tag{5}$$

The functions $\langle F(\log F - 1) \rangle$ and $\langle v\, F(\log F - 1) \rangle$ are called respectively the entropy density and entropy flux.

Finally, the positive equilibria of C are assumed to be characterized by the vanishing of the entropy dissipation rate and given by the class of "Maxwellian" distributions with the form

$$F = M(\vec{\beta}) \equiv \exp\left(\vec{\beta} \odot \vec{e}\right) , \tag{6}$$

for some $\vec{\beta} \in R^N$. More precisely, for every $F \in R^V_+$ the following properties are assumed to be equivalent:

$$\begin{aligned} &(i) \quad C(F)=0\,,\\ &(ii) \quad \langle C(F)\log F\rangle = 0\,, \\ &(iii) \quad F \text{ is a "Maxwellian" with the form (6).} \end{aligned} \tag{7}$$

This assumption about C, like the previous one, merely abstracts some of the consequences of Boltzmann's celebrated H-theorem [Ce].

It is important to understand the relation between the "Maxwellians", parametrized by $\vec{\beta}$, and the conserved densities. Note that for all finite velocity models $M(\vec{\beta})$ is a well-defined velocity distribution for every $\vec{\beta} \in R^N$. The map from R^N into R^N defined by $\vec{\beta} \mapsto \vec{\rho} = \langle \vec{e}\, M(\vec{\beta})\rangle$ can be written as

$$\vec{\rho} = \langle \vec{e}\, M(\vec{\beta})\rangle = \partial_{\vec{\beta}}\sigma^*(\vec{\beta})\,, \tag{8}$$

where $\sigma^* : R^N \to R_+$ is defined by $\sigma^*(\vec{\beta}) \equiv \langle M(\vec{\beta})\rangle$. The second derivative of the function σ^* is the positive definite matrix $\langle \vec{e} \otimes \vec{e} M(\vec{\beta})\rangle$; thus σ^* is strictly convex and the map (8) from $\vec{\beta}$ to $\vec{\rho}$ is open and one to one. Let $U \subset R^N$ be the image of the map $\partial_{\vec{\beta}}\sigma^*$. The inverse map is then given by

$$\vec{\beta} = \partial_{\vec{\rho}}\sigma(\vec{\rho})\,, \tag{9}$$

where $\sigma : U \to R$ is the Legendre transform of σ^* which is defined for every $\vec{\rho} \in U$ by the relation

$$\sigma^*(\vec{\beta}) + \sigma(\vec{\rho}) = \vec{\beta} \odot \vec{\rho}, \tag{10}$$

where $\vec{\beta}$ is determined by (8). Since

$$\sigma(\vec{\rho}) = \langle (\vec{\beta} \odot \vec{e} - 1) M(\vec{\beta})\rangle = \langle M(\vec{\beta})(\log M(\vec{\beta}) - 1)\rangle\,,$$

it follows that $\sigma(\vec{\rho})$ is the entropy density of the "Maxwellian" associated with the conserved densities $\vec{\rho} \in U$.

The conserved fluxes of the "Maxwellian" corresponding to the $\vec{\beta}$ determined by (8) for any $\vec{\rho} \in U$ defines a map $\vec{v} : U \to R^{D\times N}$ that can be written as

$$\vec{v}(\vec{\rho}) = \langle v\vec{e}\, M(\vec{\beta})\rangle = \partial_{\vec{\beta}}\tau^*(\vec{\beta})\,, \tag{11}$$

where $\tau^* : R^N \to R^D$ is defined by $\tau^*(\vec{\beta}) \equiv \langle vM(\vec{\beta})\rangle$. Now define a function $\tau : U \to R^D$ for every $\vec{\rho} \in U$ by the relation

$$\tau^*(\vec{\beta}) + \tau(\vec{\rho}) = \vec{\beta} \odot \vec{v}(\vec{\rho})\,, \tag{12}$$

where $\vec{\beta}$ is determined by (8). Since

$$\tau(\vec{\rho}) = \langle v(\vec{\beta} \odot \vec{e} - 1) M(\vec{\beta})\rangle = \langle vM(\vec{\beta})(\log M(\vec{\beta}) - 1)\rangle\,,$$

it follows that $\tau(\vec{\rho})$ is the entropy flux of the "Maxwellian" associated with the conserved densities $\vec{\rho} \in U$.

In order to obtain the formal "Euler" limit of the kinetic equation (1), a small dimensionless parameter ϵ called the Knudsen number is introduced which represents the ratio of the microscopic spatial-temporal collisional scales to those of macroscopic interest. The "Euler" limit is then given by the following theorem.

Theorem I. Given a collision operator C that satisfies the above hypotheses, let F_ϵ be a sequence of nonnegative solutions of the equation

$$\partial_t F_\epsilon + v \cdot \nabla F_\epsilon = \frac{1}{\epsilon} C(F_\epsilon), \tag{13}$$

such that the F_ϵ converge in the sense of distributions and almost everywhere to a positive function F as ϵ goes to zero. Furthermore, assume that as ϵ tends to zero the conserved densities and fluxes converge in the sense of distributions according to

$$\langle \vec{e} F_\epsilon \rangle \to \langle \vec{e} F \rangle , \qquad \langle v\, \vec{e} F_\epsilon \rangle \to \langle v\, \vec{e} F \rangle ;$$

the entropy densities and fluxes converge in the sense of distributions according to

$$\langle F_\epsilon (\log F_\epsilon - 1) \rangle \to \langle F(\log F - 1) \rangle , \quad \langle v F_\epsilon (\log F_\epsilon - 1) \rangle \to \langle v F(\log F - 1) \rangle ;$$

and, as distributions, the entropy dissipation rates satisfy

$$\limsup_{\epsilon \to 0} \langle C(F_\epsilon) \log F_\epsilon \rangle \leq \langle C(F) \log F \rangle .$$

Then the limit F is a local "Maxwellian" distribution,

$$F(t,x,v) = M\big(\partial_{\vec{\rho}} \sigma(\vec{\rho}(t,x))\big) = \exp\Big(\partial_{\vec{\rho}} \sigma(\vec{\rho}(t,x)) \odot \vec{e}(v)\Big) , \tag{14}$$

where the functions $\vec{\rho}(t,x) \in U$ solve the compressible "Euler" system

$$\partial_t \vec{\rho} + \nabla \cdot \vec{v}(\vec{\rho}) = 0 , \tag{15}$$

and satisfy the entropy inequality

$$\partial_t \sigma(\vec{\rho}) + \nabla \cdot \tau(\vec{\rho}) \leq 0 , \tag{16}$$

in the sense of distributions.

Proof. Multiplying (13) by $\epsilon \log F_\epsilon$ and integrating over v gives the entropy relation

$$\epsilon \Big(\partial_t \langle F_\epsilon (\log F_\epsilon - 1) \rangle + \nabla \cdot \langle v F_\epsilon (\log F_\epsilon - 1) \rangle \Big) = \langle C(F_\epsilon) \log F_\epsilon \rangle .$$

Letting ϵ go to zero above and using the convergence assumptions for the theorem shows that the limiting distribution F must satisfy

$$0 = \limsup_{\epsilon \to 0} \langle C(F_\epsilon) \log F_\epsilon \rangle \leq \langle C(F) \log F \rangle .$$

But the entropy dissipation rate of $C(F)$ is nonpositive by assumption (4), so that the above inequality implies $\langle C(F) \log F \rangle = 0$. The characterization of equilibria (7) then gives, that for almost every (t,x), the distribution F is a solution of the equation $C(F)=0$ and is a "Maxwellian" distribution with the form (14).

The system of local conservation laws

$$\partial_t\langle\vec{e}\,F_\epsilon\rangle+\nabla\cdot\langle v\vec{e}\,F_\epsilon\rangle=0\,,$$

is not closed since the fluxes can not be written as a function of the densities. However, if the convergence assumptions are used, one can pass to the limit of ϵ going to zero and replace F_ϵ by F, as given by (14), in these equations. A system of N equations for the N unknowns $\vec{\rho}$ is obtained which is just the compressible "Euler" system (15).

Finally, the entropy dissipation property (4),

$$\langle C(F_\epsilon)\log F_\epsilon\rangle\leq 0\,,$$

leads to the inequality

$$\partial_t\langle F_\epsilon(\log F_\epsilon-1)\rangle+\nabla\cdot\langle vF_\epsilon(\log F_\epsilon-1)\rangle\leq 0\,.$$

Once again using the convergence hypothesis along with the form of F given by (14), this inequality yields the entropy inequality (16).

Remark 2. The entropy density $\sigma(\vec{\rho})$ is a strictly convex function of $\vec{\rho}$ that satisfies the condition to be a Lax entropy; the hyperbolicity of the "Euler" system (15) is thus ensured.

III. The Incompressible "Navier-Stokes" Limit

In this section the connection between the Boltzmann equation and the incompressible Navier-Stokes equations which was derived in our previous paper [BGL1] is adapted to discrete velocity kinetic equations. As in the continous case, this connection relies on a choice of a scaling. In order to realize an incompressible limit it is natural to consider distributions that are perturbations about a given absolute "Maxwellian" (independent of space and time) and to consider long time scales.

This scaling is quantifed in terms of the (small) Knudsen number ϵ; the timescale considered is of order ϵ^{-1} while the distance to the absolute "Maxwellian" is of order ϵ^r for some $r\geq 1$. Denote this absolute "Maxwellian" by M. Solutions F_ϵ to the equation

$$\epsilon\,\partial_t F_\epsilon+v\cdot\nabla F_\epsilon=\frac{1}{\epsilon}C(F_\epsilon)\,, \tag{17}$$

are then sought in the form

$$F_\epsilon=M(1+\epsilon^r g_\epsilon)\,. \tag{18}$$

The basic case $r=1$ is the unique scaling compatible with the usual incompressible "Navier-Stokes" equations while $r>1$ leads to the "Stokes" equations.

Associated with the absolute "Maxwellian" M are a weighted average and inner product that are given respectively by

$$\langle g\rangle_M \equiv \frac{\langle M g\rangle}{\langle M\rangle}, \qquad (g|h)_M \equiv \langle g\, h\rangle_M \,, \tag{19}$$

for any $g, h \in R^V$. Henceforth the components of $\vec{e}$ are assumed to be chosen orthonormal with respect to this inner product and the subspace of R^V that they span will be denoted E. Thus $g \in E \subset R^V$ if and only if $g = \vec{\beta} \odot \vec{e}$ for some $\vec{\beta} \in R^N$, in which case $\vec{\beta} = \langle \vec{e}\, g\rangle_M$.

Denote by L and Q the first two derivatives of the operator $G \mapsto M^{-1}C(MG)$ at $G = 1$:

$$L(g) = \frac{1}{M} DC(M)\cdot(Mg)\,, \qquad Q(g,g) = \frac{1}{M} D^2C(M){:}(Mg \vee Mg)\,, \tag{20}$$

where $\vee$ is the usual symmetric tensor product over R^V. Taylor's formula then gives

$$\frac{1}{M} C(M(1+\epsilon g)) = \epsilon L(g) + \epsilon^2 \tfrac{1}{2} Q(g,g) + O(\epsilon^3)\,. \tag{21}$$

Some properties of L and Q can be readily inferred from those of C.

Since $\vec{e}$ is a vector of conserved quantities of C, it follows from (21) that

$$\langle \vec{e} L(g)\rangle_M = 0\,, \qquad \langle \vec{e} Q(g,g)\rangle_M = 0\,, \tag{22}$$

for every $g \in R^V$. The first of these identities implies the inclusion $E \subset Ker(L^T)$, where L^T denotes the adjoint of L with respect to the inner product defined in (19).

Examining the second variation of the entropy dissipation rate in the light of the dissipation property (4) yields that

$$\langle g\, L(g)\rangle_M \leq 0\,, \tag{23}$$

for every $g \in R^V$. This shows that $L + L^T$, which is twice the symmetric part of L, must be Hermitian nonpositive and implies the inclusion $Ker(L) = Ker(L^T) \subset Ker(L + L^T)$.

Finally, by the equilibrium characterization (7), any $g \in E$ will satisfy the identity $C(M \exp(\epsilon g)) = 0$; expanding this through order ϵ^2 gives the formulas

$$L(g) = 0\,, \qquad L(g^2) + Q(g,g) = 0\,, \tag{24}$$

for every $g \in E$. The first of these identities implies the inclusion $E \subset Ker(L)$, the second plays an important role in the computation of the "Navier-Stokes" equations.

In order to obtain the incompressible "Navier-Stokes" limit described below an additional hypothesis is needed about the linear operator L. Namely, that the preceding inferred inclusions are equalities:

$$E = Ker(L) = Ker(L^T) = Ker(L + L^T)\,. \tag{25}$$

The orthogonal projection onto this subspace will be denoted P; by the alternative theorem $P^{\perp}$ is the orthogonal projection onto the range of these operators, $E^{\perp} = Ran(L) = Ran(L^T) = Ran(L + L^T)$. The inverse of L^T restricted to $E^{\perp}$ is denoted by L^{-T}.

Theorem II. Let F_ϵ be a sequence of nonnegative solutions of the scaled kinetic equation (16) such that, when it is written in the form $F_\epsilon = M(1 + \epsilon^r g_\epsilon)$, the sequence g_ϵ converges in the sense of distributions and almost everywhere to a function g as ϵ goes to zero. Furthermore, assume that as ϵ goes to zero the moments

$$\langle \vec{e} g_\epsilon \rangle_M , \quad \langle v\, \vec{e}\, g_\epsilon \rangle_M , \quad \langle L^{-T}(P^{\perp}(v\,\vec{e}))v\, g_\epsilon \rangle_M , \quad \langle L^{-T}(P^{\perp}(v\,\vec{e}))Q(g_\epsilon, g_\epsilon) \rangle_M ,$$

converge in the sense of distributions to the corresponding moments

$$\langle \vec{e} g \rangle_M , \quad \langle v\, \vec{e}\, g \rangle_M , \quad \langle L^{-T}(P^{\perp}(v\,\vec{e}))v\, g \rangle_M , \quad \langle L^{-T}(P^{\perp}(v\,\vec{e}))Q(g, g) \rangle_M ,$$

and that all formally small terms vanish. Then the limit g has the form

$$g(t,x,v) = \vec{e}(v) \odot \vec{\rho}(t,x) , \tag{26}$$

where $\vec{\rho} = \langle \vec{e} g \rangle_M$ satifies

$$\langle v \cdot \nabla \vec{e} \otimes \vec{e} \rangle_M \odot \vec{\rho} = 0 . \tag{27}$$

Moreover, $\vec{\rho}$ is a weak solution of the equations

$$\partial_t \vec{\rho} + \nabla^2 : [\langle L^{-T}(P^{\perp}(v\,\vec{e})) \otimes (v\,\vec{e}) \rangle_M \odot \vec{\rho}] + \nabla \cdot \langle P^{\perp}(v\,\vec{e})(\vec{e} \odot \vec{\rho})^2 \rangle_M = \langle v \cdot \nabla \vec{e} \otimes \vec{e} \rangle_M \odot \vec{\pi} , \tag{28}$$

when $r = 1$, and of the equations

$$\partial_t \vec{\rho} + \nabla^2 : [\langle L^{-T}(P^{\perp}(v\,\vec{e})) \otimes (v\,\vec{e}) \rangle_M \odot \vec{\rho}] = \langle v \cdot \nabla \vec{e} \otimes \vec{e} \rangle_M \odot \vec{\pi} , \tag{29}$$

when $r > 1$.

Remark 3. The systems (27,28) and (27,29) are generalizations of the Navier-Stokes and Stokes systems respectively. The condition (27) generalizes the $\nabla u = 0$, $\nabla(\rho + \theta) = 0$ conditions that arise in the classical incompressible Navier-Stokes limit [BGL1], $\vec{\pi}$ is the Lagrange multiplier corresponding to the constraint (27) and is the generalization of the classical pressure term, the quadratic term in (28) corresponds to the classical convection terms, and the linear operator

$$\vec{\rho} \mapsto -\langle L^{-T}(P^{\perp}(v \cdot \nabla \vec{e})) \otimes P^{\perp}(v \cdot \nabla \vec{e}) \rangle_M \odot \vec{\rho} ,$$

that appears in (28) and (29) is dissipative. Following the strategy of Leray [L] suggested by these structural similarities, one can easily prove [BGL3] the existence of global weak solutions of the system (27,28) in any space dimension D for any initial data in

$$H = \{ \vec{\eta} \in L^2(X, R^N) : \langle v \cdot \nabla \vec{e} \otimes \vec{e} \rangle_M \odot \vec{\eta} = 0 \} . \tag{30}$$

For $D \leq 2$ this solution turns out to be smooth. The system (27,29) which generalizes the Stokes system always has a smooth solution.

Proof. Setting the form of the solution (18) into the scaled kinetic equation (17) and Taylor expanding the collision operator gives

$$\epsilon\,\partial_t g_\epsilon + v\cdot\nabla g_\epsilon = \frac{1}{\epsilon}L(g_\epsilon) + \epsilon^{r-1}\tfrac{1}{2}Q(g_\epsilon, g_\epsilon) + O(\epsilon^{2r-1})\,. \tag{31}$$

Multiplying this by ϵ, letting ϵ go to zero, and using the moment convergence assumptions yields the relation

$$L(g) = 0\,.$$

This implies that $g \in Ker(L)$ and thus can be written according to the formula (26).

The scaled local conservation laws associated with the expanded kinetic equation (31) are

$$\epsilon\,\partial_t\langle \vec{e}\,g_\epsilon\rangle_M + \nabla\cdot\langle v\,\vec{e}\,g_\epsilon\rangle_M = 0\,. \tag{32}$$

Letting ϵ tend to zero, using the moment convergence assumptions, and applying formula (26) yields condition (27).

As was done in our paper devoted to the classical continuous Boltzmann equation [BGL1], equation (32) is divided by ϵ and the flux is decomposed in the form

$$\partial_t\langle \vec{e}\,g_\epsilon\rangle_M + \frac{1}{\epsilon}\nabla\cdot\langle P^\perp(v\,\vec{e})g_\epsilon\rangle_M + \frac{1}{\epsilon}\nabla\cdot\langle P(v\,\vec{e})g_\epsilon\rangle_M = 0\,. \tag{33}$$

The moment convergence assumptions imply that the time derivative term is well behaved as ϵ tends to zero. The flux terms are handled separately.

The second flux term of (33) has the form

$$\frac{1}{\epsilon}\nabla\cdot\langle P(v\,\vec{e})g_\epsilon\rangle_M = \langle v\cdot\nabla\vec{e}\otimes\vec{e}\rangle_M \odot \langle \vec{e}\,g_\epsilon\rangle_M\,,$$

and so will be eliminated upon integrating (33) against test functions in $C^\infty \cap H$, all of which satisfy condition (27). This is analogous to the classical (since the contribution of Leray [L]) way of handling the pressure term in most treatements of the incompressible Navier-Stokes equations.

The limit of first flux term in (33) is computed using the fact that $P^\perp(v\,\vec{e}) \in Ran(L^T)$ along with the expanded kinetic equation (31) to obtain

$$\begin{aligned}\frac{1}{\epsilon}\langle P^\perp(v\,\vec{e})g_\epsilon\rangle_M &= \langle L^{-T}(P^\perp(v\,\vec{e}))\frac{1}{\epsilon}L(g_\epsilon)\rangle_M \\ &= \langle L^{-T}(P^\perp(v\,\vec{e}))v\cdot\nabla g_\epsilon\rangle_M - \epsilon^{r-1}\langle L^{-T}(P^\perp(v\,\vec{e}))Q(g_\epsilon, g_\epsilon)\rangle_M + O(\epsilon)\,.\end{aligned}$$

By the moment convergence assumptions, this term will tend to the expression

$$\langle L^{-T}(P^\perp(v\,\vec{e}))v\cdot\nabla g\rangle_M - \langle L^{-T}(P^\perp(v\,\vec{e}))Q(g,g)\rangle_M\,, \tag{34}$$

when $r = 1$, and to the expression

$$\langle L^{-T}(P^\perp(v\,\vec{e}))v\cdot\nabla g\rangle_M\,, \tag{35}$$

when $r > 1$. Substituting formula (26) for g into the linear term of (34) and (35) yields the dissipative term that appears in (28) and (29).

The limiting quadratic term of (34) must be evaluated further in order to bring it into the form that appears in (28). Since $g \in Ker(L)$, the second identity of (24) can be employed to show

$$\begin{aligned} -\langle L^{-T}(P^{\perp}(v\,\vec{e}))Q(g,g)\rangle_M &= \langle L^{-T}(P^{\perp}(v\,\vec{e}))L(g^2)\rangle_M \\ &= \langle P^{\perp}(v\,\vec{e})g^2\rangle_M \,. \end{aligned}$$

After applying formula (26) the proof that $\vec{\rho}$ satisfies equation (28) or (29) is complete.

IV. Generalized Boltzmann Collision Operators

The purpose of this section is to show that the hypotheses used in the sections II and III are satisfied by a large class of operators. Indeed, collision operators $C(F)$ that are multivariable polynomials with respect to the values of the functions $F \in R_+^V$ will be considered in this section. It is shown below (Theorem III) that the property of semidetailed balance (which generalizes that of detailed balance) is sufficient to recover the properties related to the H-theorem.

The notation we use is not common in the literature so it is described here in detail. The set of all possible collisions is doubly multiindexed by $s, s' \in N^V$; the values of the multiindex s indicate a state of $s(v)$ particles with velocity v before the collision while the values of the multiindex s' similarly indicate the postcollisional state. The rate at which the collision indexed by s and s' takes place is assumed to be proportional to F^s, which is to be understood as in the classical notation of multiindices,

$$F^s \equiv \prod_{v\in V} (F(v))^{s(v)} \,.$$

The class of collision operators considered here is written according to the formula

$$C(F) = \sum_{s,s'\in N^V} (s' - s)\, F^s \alpha(s,s') \,. \tag{36}$$

In (36) the proportionality coefficients $\alpha(s,s')$ of the collision rates are nonnegative; it will be assumed that only a finite number of them are positive. Because of their structural similarity to Boltzmann's classical operator, we refer to them as generalized Boltzmann collision operators.

In this notation the notions of detailed and semidetailed balance [G] are rendered as

Definition 2. The operator C is said to be in detailed balance if the relation

$$\alpha(s,s') = \alpha(s',s) \tag{37}$$

is true for all $s, s' \in N^V$. It is said to be in semidetailed balance if the relation

$$\sum_{s'\in N^V} \alpha(s,s') = \sum_{s'\in N^V} \alpha(s',s) \,, \tag{38}$$

is true for all $s \in N^V$.

The notion of semidetailed balance is clearly a weakening of the classical detailed balance condition. Its use requires the following technical property.

Lemma. If the collision operator is in semidetailed balance then the formula

$$\sum_{s,s' \in N^V} \eta(s)\, \alpha(s,s') = \sum_{s,s' \in N^V} \eta(s')\, \alpha(s,s') \tag{39}$$

is true for any $\eta : N^V \to R$.

Proof. Multiply (38) by $\eta(s)$, sum over s, and exchange the role s and s' on the right hand side.

According to Definition 1, $e \in R^V$ is a conserved quantity for the collision operator C provided $\langle e\, C(F)\rangle = 0$ for every $F \in R_+^V$. For collision operators given by formula (36) this becomes

$$0 = \sum_{s,s' \in N^V} \langle (s'-s)e\rangle\, F^s \alpha(s,s'),$$

for every $F \in R_+^V$. But the family of polynomials parametrized by s and given by $F \mapsto F^s$ is linearly independent; thus the above equality holds if and only if the coefficient of each F^s vanishes:

$$\sum_{s' \in N^V} \langle (s'-s)e\rangle\, \alpha(s,s') = 0, \tag{40}$$

for all $s \in N^V$.

Another notion of conservation is one that holds for individual collisions.

Definition 3. A vector $e \in R^V$ is said to be a microscopicly conserved quantity for the collision operator C given by formula (36) if

$$\langle (s'-s)e\rangle\, \alpha(s,s') = 0, \tag{41}$$

for all $s, s' \in N^V$.

While it is clear from comparing (40) and (41) that any microscopicly conserved quantity is also a conserved quantity, the converse is generally not true. However it turns out to be true for the large class of operators which are in semidetailed balance.

Proposition. For any collision operator C given by the formula (36) that is in semidetailed balance, any conserved quantity is also a microscopic conserved quantity.

Proof. Let $e \in R^V$ be a conserved quantity, multiplying the equation (41) by $\langle se\rangle$ and summing over s gives

$$0 = \sum_{s,s' \in N^V} \langle (s'-s)e\rangle \langle se\rangle\, \alpha(s,s') = -\sum_{s,s' \in N^V} \left(\langle se\rangle^2 - \langle s'e\rangle\langle se\rangle\right) \alpha(s,s'). \tag{42}$$

Applying formula (39) of the Lemma to half of the first term inside of the last sum of (42) gives

$$\begin{aligned} 0 &= \sum_{s,s'\in N^V} \left(\tfrac{1}{2}\langle se\rangle^2 + \tfrac{1}{2}\langle s'e\rangle^2 - \langle s'e\rangle\langle se\rangle\right)\alpha(s,s') \\ &= \sum_{s,s'\in N^V} \tfrac{1}{2}\langle (s-s')e\rangle^2 \alpha(s,s') . \end{aligned} \tag{43}$$

Since each term of this last sum is nonnegative then all of them must be equal to zero. But that implies (41) is satisfied and shows that e is also microscopicly conserved.

Generalized Boltzmann collision operators in semidetailed balance satisfy the following H-theorem which yields the properties required in section II.

Theorem III. If the collision operator C given by the formula (36) is in semidetailed balance then it has the dissipation property (4),

$$\langle C(F) \log F\rangle \leq 0 ,$$

for every $F \in R_+^V$ and satisfies the charactization of equilibria (7) which states that for every $F \in R_+^V$ the following are equivalent:

$$\begin{aligned} &(i) \quad C(F) = 0 , \\ &(ii) \quad \langle C(F) \log F\rangle = 0 , \\ &(iii) \quad F = M(\vec{\beta}) \quad \text{for some } \vec{\beta} \in R^N . \end{aligned} \tag{44}$$

Here

$$M(\vec{\beta}) \equiv \exp\left(\vec{\beta} \odot \vec{e}\right) ,$$

where $\vec{e}$ is a basis of the N dimensional space of conserved quantities.

Proof. Using the fundamental property of logarithms,

$$\langle (s-s') \log F\rangle = -\log\left(\frac{F^{s'}}{F^s}\right) ,$$

along with formula (39) of the Lemma with $\eta(s) = F^s$,

$$0 = \sum_{s,s'\in N^V} F^{s'} \alpha(s,s') - \sum_{s,s'\in N^V} F^s \alpha(s,s') ,$$

one obtains

$$\begin{aligned} -\langle C(F) \log F\rangle &= \sum_{s,s'\in N^V} \langle (s-s') \log F\rangle\, F^s \alpha(s,s') \\ &= \sum_{s,s'\in N^V} \left(\frac{F^{s'}}{F^s} - 1 - \log\left(\frac{F^{s'}}{F^s}\right)\right) F^s \alpha(s,s') . \end{aligned} \tag{45}$$

Every term in this last sum is nonnegative since $y - 1 - \log y \geq 0$ for all $y \in R_+$, so the collision operator C satisfies the dissipation property.

The characterization of equilibria (44) is argued as follows: (i) implies (ii) implies (iii) implies (i). The first implication is obvious. Assuming (ii), the last sum in (45) is zero and each of its nonnegative terms must vanish. This gives the formula

$$\left(\frac{F^{s'}}{F^s} - 1 - \log\left(\frac{F^{s'}}{F^s}\right)\right) F^s \alpha(s,s') = 0\,,$$

for all $s, s' \in N^V$; which then implies

$$\langle (s-s') \log F \rangle \, \alpha(s,s') = 0\,,$$

for all $s, s' \in N^V$. Thus $\log F$ satisfies (41) and is therefore a microscopicly conserved quantity; (iii) then follows. Finally, using the fact that all conserved quantites are microscopicly conserved and employing formula (39) of the Lemma, it is easy to show that $C(M(\vec{\beta})) = 0$ for every $\vec{\beta} \in R^N$. This completes the proof of the Theorem III.

All that remains is to show the additional hypothesis (25) required in section III is satisfied by these same generalized Boltzmann collision operators.

Corollary. If the collision operator C given by formula (36) is in semidetailed balance and L is its linearization defined by (20) for some "Maxwellian" M then

$$E = Ker(L) = Ker(L^T) = Ker(L + L^T)\,.$$

Proof. By Theorem III all that needs to be proved is that $Ker(L + L^T) \subset E$. A direct calculation following definition (20) yields

$$L(g) = \frac{1}{M} \sum_{s,s' \in N^V} (s'-s) \langle s\, g \rangle M^s \alpha(s,s')\,.$$

If $g \in Ker(L + L^T)$ then applying the semidetailed balance property as in the proof of the Propositon shows

$$\begin{aligned} 0 = -\langle g\, L(g) \rangle_M &= - \sum_{s,s' \in N^V} \langle (s'-s) g \rangle \langle s\, g \rangle M^s \alpha(s,s') \\ &= \sum_{s,s' \in N^V} \tfrac{1}{2} \langle (s'-s) g \rangle^2 M^s \alpha(s,s')\,. \end{aligned}$$

Since each term of this last sum is nonnegative then all of them must be equal to zero. But that implies g satisfies (41) and shows that g must be a conserved quantity ($g \in E$).

V. Remarks on the Convergence of the Fluid Dynamic Limit

As was observed in [BGL1], any proof concerning the fluid dynamical limit for (discrete or continous) a kinetic model will, as a by-product, give an existence proof for the correponding macroscopic equation. Some uniform regularity estimates would likely be needed for obtaining the limit of any nonlinear terms. These estimates, if they exist, must be sharp because it is known (and is proven by Sideris [S] for a very general situation) that the solutions of the compressible nonlinear Euler equations (and more generally of any strictly hyperbolic system) become singular after a finite time. Existence and convergence results have to be in agreement with these observations.

It is easy to prove that any finite velocity model has a unique smooth solution during a finite time that depends on the size of the initial data and on the collision operator. To the best of our knowledge it has not been proven that the time of existence of this smooth solution is independent of ϵ for the solutions of (13) and (17) in sections II and III. However, such a result should be easy to obtain following the ideas of Nishida [N]. It could be used to prove the assumed convergence of the moments.

Existence of global solutions for discrete kinetic models have been obtained by several authors (cf. for instance Cabannes [C2], Bony [B], and the review article of Platowski and Illner [PI]). These results cannot be used here because they are based on the dispersive properies of the linearized equations and concern small (with respect to the collision operator) pertubations of the vacuum.

In [BGL2] we prove with compactness assumptions the global in time convergence of DiPerna-Lions renormalized solutions of the Boltzmann equation to the Leray solution of the incompressible Navier-Stokes equations. The proof relies on two basic ingredients: the entropy inequality and the averaging lemma (cf. [GLPS] or [DiPL]). The averaging lemma is valid for continuous solutions and has no counterpart for discrete velocity models (except in one space dimension (cf. Tartar [T]). However, in the proof of Theorem III pointwise convergence is only needed to evaluate the limit of the nonlinear moments

$$\langle L^{-T}(P^{\perp}(v\,\vec{e}))\,Q(g_\epsilon, g_\epsilon)\rangle_M\ ;$$

this term disappears in the case $r > 1$. On the other hand from section IV one deduces that an entropy inequality similar to the one used in [BGL2] remains valid for a large class of collision operators. With this observation one could show a result of the following type. Assume that g_ϵ is a family of weak global solutions of (17) with independant of ϵ initial data. Assume furthermore that the collision operator satisfies the hypotheses of Theorem III and that $r > 1$ then this family converges in the sense of Theorem III to the solution of the corresponding Stokes equation.

REFERENCES

[BGL1] C. Bardos, F. Golse and D. Levermore, *Fluid Dynamic Limits of Kinetic Equations I:Formal Derivations*, (submitted to J. of Stat. Phys.).

[BGL2] C. Bardos, F. Golse and D. Levermore, *Fluid Dynamic Limits of Kinetic Equations II: Convergence Proofs for the Boltzmann Equation*, (submitted to Annals of Math.).

[B] J. M. Bony, *Exitence Globale à données de Cauchy Petites Pour les Modèles Discrets de l'Equation de Boltzmann*, (Preprint Centre de Mathématiques Ecole Polytechnique 91128 Palaiseau).

[C1] H. Cabannes, *Etude de la propagation des ondes dans un gaz à quatorze vitesses*, J. de Méc., **14**, (1975), 705-744.

[C2] H. Cabannes, *The discrete Boltzmann Equation (theory and applications)*, Univ. of California, Berkeley, CA, 1980.

[CG] F. Coulouvrat and R. Gatignol, *Description hydrodynaique d'un gaz en théorie cinétique discrète:les modèles réguliers* , C.R. Acad. Sci. Paris, **306**, (1988), 393-398.

[DiPL] R.J. DiPerna and P.-L. Lions, *On the Cauchy Problem for the Boltzmann Equation: Global Existence and Weak Stability Results*, Annals of Math., **130**, (1989), 321-366.

[FHP] U. Frisch, B.Hasslacher and Y. Pommeau, *Lattice Gas Automata for the Navier Stokes Equation*, Phys. Rev. Lett., **56** (1988), 1505.

[FD'HHLPR] U. Frisch, B.Hasslacher, D. D'Humières, P. Lallemand, Y. Pommeau and J. P. Rivet, *Lattice Gas Hydrodynamic s in Two and three dimensions*, J. Stat. Phys ???.

[G] R.Gatignol, *Théorie Cinétique des Gaz à Répartition Discrète de Vitesses*, Lecture Notes in Physics, **36**, (1975), Springer Verlag.

[GC] R. Gatignol and F. Coulouvrat *Description hydrodynamique d'un gaz en théorie cinétique discrète:le modèle général* , C.R. Acad. Sci. Paris, **306**, (1988), 169-174.

[GLPS] F. Golse, P.-L. Lions, B. Perthame and R. Sentis, *Regularity of the Moments of the Solution of a Transport Equation*, J. of Funct. Anal., **76** (1988), 110-125.

[L] J. Leray, *Sur le Mouvement d'un Liquide Visqueux Emplissant l'Espace*, Acta Mathematica, **63**, (1934), 193-248.

[N] T. Nishida, *Fluid Dynamical Limit of the Nonlinear Boltzmann Equation to the Level of the Incompressible Euler Equation*, Comm. Math. Phys., **61** (1978), 119-148.

[PI] T.Platkowski and R. Ilner, *Discrete Velocity Models of the Boltzmann Equation: a Survey of the Mathematicals Aspects of ther Theory*, SIAM Review., **30**, (1988), 213-255.

[S] T. Sideris, *Formation of Singularities in Three Dimensionnal Compressible Fluids*, Comm. Math. Phys., **101** (1985), 475-485.

[T] L.Tartar, *Existense globale pour un système semi linéaire de la théorie cinétique des gas* ,Séminaire Goulaouic Schwartz, Ecole Polytechnique Palaiseau, Fance (1975), 475-485.

On the Euler Equation in Discrete Kinetic Theory

S. Kawashima[1] *and N. Bellomo*[2]

[1]Faculty of Engineering 36, Kyushu University, Fukuoka 812, Japan
[2]Dipartimento di Matematica, Politecnico di Torino,
Corso degli Abruzzi 24, I-10129 Torino, Italy

1. Introduction

The aim of this note is to investigate some mathematical structure of the Euler equation derived from the discrete Boltzmann equation as the first order approximation of the Chapman-Enskog expansion. The analysis developed in this article would be a basis for the study of nonlinear waves in discrete kinetic theory (cf. [2,7,3] for shock waves, [10] for rarefaction waves, and [6] for diffusion waves).

We shall consider the discrete Boltzmann equation in one-space dimension:

$$c_i \left(\frac{\partial F_i}{\partial t} + v_i \frac{\partial F_i}{\partial x} \right) = Q_i(F), \quad i = 1, \dots, m, \tag{1.1}$$

where c_i is a positive constant, F_i represents the mass density of gas particles with the velocity v_i in x-direction (v_i is assumed to be real constant), and $Q_i(F)$ the term related to binary and multiple collisions of gas particles; multiple collision case is not excluded in our study. For concrete expressions of $Q_i(F)$, we refer the reader to [4,5,1].

Let us denote by M_0 the totality of vectors $\phi = (\phi_i)_{i=1}^m$ satisfying

$$\sum_i \phi_i Q_i(F) = 0 \quad \text{for any} \quad F = (F_i)_{i=1}^m, \tag{1.2}$$

where (and in what follows) $\sum_i$ means the summtation taken over all $i = 1, \dots, m$. M_0 is a subspace of $\mathbf{R}^m$. We make the following assumption on (1.1).

Basic Assumption: (i) The velocity set $\{v_i\}_{i=1}^m$ contains at least 3 different values. (ii) The space M_0 defined above is of 2-dimension and is spanned by $(1)_{i=1}^m$ and $(v_i)_{i=1}^m$, where $(1)_{i=1}^m$ means the vector whose components are all equal to 1.

Under this assumption, we can show that the Euler equation derived from (1.1) is strictly hyperbolic and genuinely nonlinear (in the sense of [9]). Moreover, based on this fact, we can analyze elementary shock waves for our Euler equation in detail. Our study generalizes some part of the work [2] for the Broadwell model to a large class of models satisfying the above Basic Assumption.

2. Derivation of the Euler equation

Basic Assumption (ii) implies that (1.1) admits two independent conservation laws of the form

$$\frac{\partial}{\partial t}\sum_i c_i F_i + \frac{\partial}{\partial x}\sum_i c_i v_i F_i = 0,$$
$$\frac{\partial}{\partial t}\sum_i c_i v_i F_i + \frac{\partial}{\partial x}\sum_i c_i v_i^2 F_i = 0. \tag{2.1}$$

These equations represent the conservation of mass and momentum (in x-direction), respectively. We introduce

$$\rho = \sum_i c_i F_i, \quad \rho u = \sum_i c_i v_i F_i, \quad \rho\sigma = \sum_i c_i v_i^2 F_i, \tag{2.2}$$

and rewrite (2.1) simply as

$$\rho_t + (\rho u)_x = 0, \quad (\rho u)_t + (\rho\sigma)_x = 0. \tag{2.3}$$

Here ρ and u are the mass density and the velocity, respectively, in macroscopic sense. The quantity σ is not a function of (ρ, u) in general, and therefore (2.3) is a non-closed system.

In order to reduce (2.3) to a closed system, we apply the Chapman-Enskog expansion to (1.1). It is known ([4,8,5]) that, at the first order approximation of the expansion, $F = (F_i)_{i=1}^m$ must be a Maxwellian, namely, F needs to satisfy $Q_i(F) = 0$ for all i. This means ([4,5,1]) that $(\log F_i)_{i=1}^m$ is in the subspace M_0. Therefore, by Basic Assumption (ii), we have $\log F_i = \alpha + \beta v_i$ for all i, where $\alpha, \beta \in \mathbf{R}$. Consequently, we arrive at the expression

$$F_i = e^{\alpha + \beta v_i}, \quad i = 1, \ldots, m. \tag{2.4}$$

We substitute (2.4) into (2.2) and obtain $\rho = e^\alpha G(\beta)$ together with

$$u = G'(\beta)/G(\beta) \equiv U(\beta), \tag{2.5}$$

$$\sigma = G''(\beta)/G(\beta), \tag{2.6}$$

where (and in what follows) the prime denotes the differentiation with respect to a single variable, and where we put

$$G(\beta) = \sum_i c_i e^{\beta v_i}. \tag{2.7}$$

We shall show that the function $\beta \to u$ defined by (2.5) has the inverse. To this end, we compute the derivative $U'(\beta)$:

$$U'(\beta) = (G''(\beta)G(\beta) - G'(\beta)^2)/G(\beta)^2 \equiv H(\beta)/G(\beta)^2. \tag{2.8}$$

By a straightforward computation, using (2.7), we find that

$$H(\beta) = \frac{1}{2}\sum_{ij} c_i c_j (v_i - v_j)^2 e^{\beta(v_i + v_j)}, \tag{2.9}$$

where the summation is taken over all $i, j = 1, \ldots, m$. It then follows from Basic Assumption (i) that $H(\beta)$ is strictly positive for all $\beta \in \mathbf{R}$; this is indeed true if the set $\{v_i\}_{i=1}^m$ contains at least 2 different values. Therefore, $u = U(\beta)$ is a strictly increasing function of $\beta \in \mathbf{R}$ onto (u_0, u_1), where

$$u_0 = \lim_{\beta \to -\infty} U(\beta), \quad u_1 = \lim_{\beta \to +\infty} U(\beta). \tag{2.10}$$

Consequently, the inverse function, denoted by $\beta = \beta(u)$, exists on (u_0, u_1) and satisfies $\beta'(u) = 1/U'(\beta(u))$.

Now, substituting $\beta = \beta(u)$ into (2.6), we obtain

$$\sigma = G''(\beta(u))/G(\beta(u)) \equiv \sigma(u). \tag{2.11}$$

Thus, at the first order approximation of the chapman-Enskog expansion, σ becomes a function of $u \in (u_0, u_1)$ through the relation (2.11). Consequently, (2.3) is reduced to

$$\rho_t + (\rho u)_x = 0, \quad (\rho u)_t + (\rho\sigma(u))_x = 0. \tag{2.12}$$

This closed system is called the Euler equation associated with (1.1). Note that the domain of definition of (2.12) is $\Omega = \{\rho > 0, u \in (u_0, u_1)\}$.

3. Structure of the Euler equation

We rewrite the Euler equation (2.12) as

$$\begin{pmatrix} \rho \\ u \end{pmatrix}_t + \begin{pmatrix} u & \rho \\ (\sigma(u) - u^2)/\rho & \sigma'(u) - u \end{pmatrix} \begin{pmatrix} \rho \\ u \end{pmatrix}_x = 0 \tag{3.1}$$

We denote the coefficient matrix in (3.1) by $A(\rho, u)$. The characteristic equation for $A(\rho, u)$ is

$$\det(\lambda - A(\rho, u)) = \lambda^2 - \sigma'(u)\lambda - (\sigma(u) - u\sigma'(u)) = 0. \tag{3.2}$$

The roots of (3.2), i.e., the eigenvalues of $A(\rho, u)$ depend only on u, so that we denote them by $\lambda_1(u)$ and $\lambda_2(u)$. By a simple computation, we have

$$\lambda_1(u) = J_-(u, \sigma'(u)), \quad \lambda_2(u) = J_+(u, \sigma'(u)), \tag{3.3}$$

where

$$\begin{aligned} J_\pm(u, \eta) &= \frac{1}{2}\{\eta \pm \sqrt{\eta^2 + 4(\sigma(u) - u\eta)}\} \\ &= u + \frac{1}{2}\{(\eta - 2u) \pm \sqrt{(\eta - 2u)^2 + 4(\sigma(u) - u^2)}\}, \quad \eta \in \mathbf{R}. \end{aligned} \tag{3.4}$$

We shall show that

$$\lambda_1(u) < u < \lambda_2(u), \quad 2\lambda_1(u) < \sigma'(u) < 2\lambda_2(u), \tag{3.5}$$

for all $u \in (u_0, u_1)$; this in particular implies that the Euler equation (2.12) is strictly hyperbolic on Ω. To see this, we combine (2.11), (2.5) and (2.8), obtaining $\sigma(u) - u^2 = H(\beta)/G(\beta)^2$ with $\beta = \beta(u)$. This shows that

$$\sigma(u) > u^2, \tag{3.6}$$

for all $u \in (u_0, u_1)$, because $H(\beta)$ is strictly positive. Substituting (3.6) into (3.4) yields

$$J_-(u,\eta) < u < J_+(u,\eta), \quad 2J_-(u,\eta) < \eta < 2J_+(u,\eta), \tag{3.7}$$

for all $u \in (u_0, u_1)$ and $\eta \in \mathbf{R}$. Therefore, the desired inequalities in (3.5) follow from (3.3) and (3.7) with $\eta = \sigma'(u)$.

Next we shall show that both characteristic fields are genuinely nonlinear in the sense of Lax [9]. We denote the right eigenvector of $A(\rho, u)$ for $\lambda_j(u)$ by $r_j(\rho, u)$, $j = 1, 2$. What we need to check is:

$$\nabla\lambda_j(u) \cdot r_j(\rho, u) \neq 0, \quad j = 1, 2, \tag{3.8}$$

for all $(\rho, u) \in \Omega$, where ∇ denotes the gradient with respect to (ρ, u). We know that

$$r_j(\rho, u) = \begin{pmatrix} \rho \\ \lambda_j(u) - u \end{pmatrix}, \quad j = 1, 2. \tag{3.9}$$

Therefore,

$$\nabla\lambda_j(u) \cdot r_j(\rho, u) = \lambda_j'(u)(\lambda_j(u) - u), \quad j = 1, 2. \tag{3.10}$$

On the other hand, noting that $\lambda = \lambda_j(u)$ satisfies the characteristic equation (3.2), we differentiate (3.2) (with $\lambda = \lambda_j(u)$ being substituted) with respect to u. This yields

$$\lambda_j'(u) = \sigma''(u)(\lambda_j(u) - u)/(2\lambda_j(u) - \sigma'(u)), \quad j = 1, 2. \tag{3.11}$$

Therefore, since we have (3.5), for the proof of (3.8), it is enough to show the following:

$$\sigma''(u) > 0, \quad \text{or equivalently,} \quad \lambda_j'(u) > 0, \quad j = 1, 2, \tag{3.12}$$

for all $u \in (u_0, u_1)$.

To prove (3.12), we differetiate (2.11) with respect to u and use $\beta'(u) = 1/U'(\beta)$ and (2.8), obtaining

$$\sigma'(u) = (G'''(\beta)G(\beta) - G''(\beta)G'(\beta))/H(\beta) = H'(\beta)/H(\beta), \tag{3.13}$$

with $\beta = \beta(u)$, where the last equality follows from a straightforward computation using (2.7) and (2.9). We further differentiate (3.13) with respect to u. This yields

$$\sigma''(u) = (H''(\beta)H(\beta) - H'(\beta)^2)G(\beta)^2/H(\beta)^3, \tag{3.14}$$

with $\beta = \beta(u)$. By a straightforward computation, using (2.9), we find that

$$\begin{aligned} &H''(\beta)H(\beta) - H'(\beta)^2 \\ &= \frac{1}{8}\sum_{ijkl} c_i c_j c_k c_l (v_i - v_j)^2 (v_k - v_l)^2 (v_i + v_j - v_k - v_l)^2 e^{\beta(v_i+v_j+v_k+v_l)}, \end{aligned} \tag{3.15}$$

where the summation is taken over all $i, j, k, l = 1, \ldots, m$. By virtue of Basic Assumption (i), the right hand side of (3.15) is strictly positive for all $\beta \in \mathbf{R}$. Therefore, recalling the strict positivity of $H(\beta)$, we conclude that $\sigma''(u) > 0$ for all $u \in (u_0, u_1)$. Thus (3.12) (and hence (3.8)) has been verified.

These observations are summarized as follows.

Proposition 3.1. *The Euler equation (2.12) is strictly hyperbolic on the whole domain of definition* $\Omega = \{\rho > 0, u \in (u_0, u_1)\}$. *Moreover, both characteristic fields are genuinely nonlinear on* Ω *in the sense of Lax [9].*

4. Shock waves for the Euler equation

In this section we shall study elementary shock waves for the Euler equation (2.12). Let $\rho_\pm, u_\pm, s$ be constants such that $(\rho_\pm, u_\pm) \in \Omega$ and $(\rho_+, u_+) \neq (\rho_-, u_-)$. Put

$$(\rho, u) = \begin{cases} (\rho_-, u_-), & x/t < s, \\ (\rho_+, u_+), & x/t > s. \end{cases} \tag{4.1}$$

It is well known ([9]) that the (ρ, u) defined by (4.1) becomes a weak solution of (2.12) if and only if the following Rankine-Hugoniot condition holds:

$$\begin{aligned} -s(\rho_+ - \rho_-) \ &+ \ (\rho_+ u_+ - \rho_- u_-) = 0, \\ -s(\rho_+ u_+ - \rho_- u_-) \ &+ \ (\rho_+ \sigma(u_+) - \rho_- \sigma(u_-)) = 0. \end{aligned} \tag{4.2}$$

In addition, such a solution becomes physically reasonable one if the following Lax shock condition is satisfied:

$$\begin{aligned} &\lambda_1(u_+) < s < \lambda_1(u_-) \quad \text{and} \quad s < \lambda_2(u_+), \quad \text{or} \\ &\lambda_2(u_+) < s < \lambda_2(u_-) \quad \text{and} \quad \lambda_1(u_-) < s, \end{aligned} \tag{4.3}$$

where $\lambda_j(u)$ are the eigenvalues given in (3.3). The discontinuity (at $x/t = s$) in (4.1) is called a shock wave when (4.2) and (4.3) hold true.

In what follows, we shall show that for any given $(\rho_-, u_-) \in \Omega$, there exist $(\rho_+, u_+) \in \Omega$ (with $u_+ < u_-$) and s satisfying (4.2) and (4.3). Firstly, we rewrite (4.2) as

$$\begin{aligned} \rho_+(s - u_+) \ &= \ \rho_-(s - u_-), \\ \rho_+(s u_+ - \sigma(u_+)) \ &= \ \rho_-(s u_- - \sigma(u_-)), \end{aligned} \tag{4.4}$$

and combine these two equtions to eliminate ρ_+ and ρ_-. After some arrangement, this yields

$$s^2 - \mu s - (\sigma(u) - u\mu) = 0, \quad \text{with} \quad u = u_\pm, \tag{4.5}$$

where

$$\mu = (\sigma(u_+) - \sigma(u_-))/(u_+ - u_-). \tag{4.6}$$

(Notice that $\sigma(u_+) - u_+\mu = \sigma(u_-) - u_-\mu$ by the definition (4.6) of μ.) The equation (4.5) is solved as

$$s = s_1 \equiv J_-(u_\pm, \mu), \quad s = s_2 \equiv J_+(u_\pm, \mu), \tag{4.7}$$

where $J_\pm(u, \eta)$ are given in (3.4). (Notice that $J_\pm(u_+, \mu) = J_\pm(u_-, \mu)$.) Therefore we see from (3.7) that

$$s_1 < u_\pm < s_2, \quad 2s_1 < \mu < 2s_2. \tag{4.8}$$

We substitute (4.7) into the first equation of (4.4) to obtain

$$\rho_+/\rho_- = (s_j - u_-)/(s_j - u_+), \quad j = 1, 2. \tag{4.9}$$

Thus we know that for any given $(\rho_-, u_-) \in \Omega$, the Rankine-Hugoniot relation (4.2) is solved with respect to s and ρ_+ explicitly as in (4.7) and (4.9) by regarding u_+ ($\neq u_-$) as a parameter.

Secondly, we characterize the Lax shock condition (4.3).

Lemma 4.1. *Suppose that $(\rho_\pm, u_\pm) \in \Omega$ and s satisfy the Rankine-Hugoniot condition (4.2), where $(\rho_+, u_+) \neq (\rho_-, u_-)$. Then the Lax shock condition (4.3) is equivalent to*

$$u_+ < u_-. \tag{4.10}$$

Proof. That (4.10) follows from (4.3) is obvious because both $\lambda_j(u)$ are strictly increasing functions of $u \in (u_0, u_1)$ by (3.12). We shall prove the converse assertion. Let $u_+ < u_-$. Then, taking into account of the fact that $\sigma(u)$ is a strictly convex function of $u \in (u_0, u_1)$ by (3.12), we see from (4.6) that

$$\sigma'(u_+) < \mu < \sigma'(u_-). \tag{4.11}$$

Next, recalling the expressions (4.7) and (3.3) for s_j and λ_j, respectively, we investigate the dependence of $J_\pm(u, \eta)$ upon $\eta \in \mathbf{R}$. Since both $s = J_\pm(u, \eta)$ solve the equation $s^2 - \eta s - (\sigma(u) - u\eta) = 0$, we have

$$J_\pm(u, \eta)^2 - \eta J_\pm(u, \eta) - (\sigma(u) - u\eta) = 0, \tag{4.12}$$

for all $u \in (u_0, u_1)$ and $\eta \in \mathbf{R}$. By differentiating with respect to η, we obtain

$$\partial J_\pm(u, \eta)/\partial\eta = (J_\pm(u, \eta) - u)/(2J_\pm(u, \eta) - \eta). \tag{4.13}$$

This combined with (3.7) shows that $\partial J_\pm(u, \eta)/\partial\eta > 0$. Therefore, for each fixed $u \in (u_0, u_1)$, both $J_\pm(u, \eta)$ are strictly increasing functions of $\eta \in \mathbf{R}$. Hence we have from (4.11),

$$J_\pm(u, \sigma'(u_+)) < J_\pm(u, \eta) < J_\pm(u, \sigma'(u_-)), \tag{4.14}$$

for all $u \in (u_0, u_1)$. We put $u = u_\pm$ in (4.14) and then use (3.3), (4.7) to obtain

$$\lambda_j(u_+) < s_j < \lambda_j(u_-), \quad j = 1, 2, \tag{4.15}$$

which proves a part of (4.3).

It remains to check the inequalities $s_1 < \lambda_2(u_+)$ and $\lambda_1(u_-) < s_2$. To this end, we note that $s = s_j$ and $\lambda = \lambda_k(u_\pm)$ satisfy the equations (4.5) and (3.2) (both with $u = u_\pm$), respectively, where $j, k = 1, 2$. We subtrust these two equations, obtaining

$$(s_j - \lambda_k(u))(s_j + \lambda_k(u) - \sigma'(u)) = (\mu - \sigma'(u))(s_j - u), \quad \text{with} \quad u = u_\pm, \tag{4.16}$$

where $j, k = 1, 2$. Since $\lambda_1(u) + \lambda_2(u) = \sigma'(u)$, we see that $s_j + \lambda_k(u) - \sigma'(u) = s_j - \lambda_j(u)$ where $k \neq j$. Therefore, applying (4.8), (4.11) and (4.15) to (4.16), we obtain the desired inequalities $s_1 < \lambda_2(u_+)$ and $\lambda_1(u_-) < s_2$. This completes the proof of Lemma 4.1.

All these observations are summarized as follows.

Proposition 4.2. *For any given $(\rho_-, u_-) \in \Omega$, there exist $(\rho_+, u_+) \in \Omega$ (with $u_+ < u_-$) and s satisfying the Rankine-Hugoniot condition (4.2) and the Lax shock condition (4.3). Furthermore, this solution is given explicitly as in (4.7), (4.9) by regarding $u_+ (< u_-)$ as a parameter.*

Finally, we note that the Lax shock condition (4.3), which is equivalent to (4.10), is also equivalent to

$$(s - u_\pm)(\rho_+ - \rho_-) < 0, \tag{4.17}$$

since we have $(s - u_\pm)(\rho_+ - \rho_-) = \rho_\mp(u_+ - u_-)$ from the first equation of (4.2).

References

[1] N. Bellomo and S. Kawashima, The discrete Boltzmann equation with multiple collisions: Global existence and stability for the initial value problem, J. Math. Phys., **31**(1990), 245-253.

[2] R. E. Caflisch, Navier-Stokes and Boltzmann shock profiles for a model of gas dynamics, Comm. Pure Appl. Math., **32**(1979), 521-554.

[3] R. E. Caflisch and T.-P. Liu, Stability of shock waves for the Broadwell equations, Commun. Math. Phys., **114**(1988), 103-130.

[4] R. Gatignol, Théorie cinétique de gaz à répartition discrète de vitesses, Lecture Notes in Physics **36**, Springer-Verlag, New York, 1975.

[5] R. Gatignol and F. Coulouvrat, Constitutive laws for discrete velocity models of gas, in Discrete Kinetic Theory, Lattice Gas Dynamics and Foundations of Hydrodynamics (edited by R. Monaco), 121-145, World Scientific, Singapore, 1989.

[6] S. Kawashima, Large-time behavior of solutions of the discrete Boltzmann equation, Commun. Math. Phys., **109**(1987), 563-589.

[7] S. Kawashima and A. Matsumura, Asymptotic stability of traveling wave solutions of systems for one-dimension gas motion, Commun. Math. Phys., **101**(1985), 97-127.

[8] S. Kawashima and Y. Shizuta, The Navier-Stokes equation in the discrete kinetic theory, J. Méc. Théor. Appl., **7**(1988), 597-621.

[9] P.D. Lax, Hyperbolic systems of conservation laws, II, Comm. Pure Appl. Math., **10**(1957), 537-566.

[10] A. Matsumura, Asymptotics toward rarefaction wave of solutions of the Broadwell model of a discrete velocity gas, Japan J. Appl. Math., **4**(1987), 489-502.

Existence globale et diffusion en théorie cinétique discrète

J.-M. Bony

Centre de Mathématiques, Ecole Polytechnique,
F-91128 Palaiseau Cedex, France

1 Introduction

L'objet de ce travail est de présenter, en renvoyant éventuellement à [3] [4] [5] pour les démonstrations, les résultats d'existence de solutions globales bornées du problème de Cauchy que nous avons obtenus pour les modèles généraux de la théorie cinétique discrète des gaz dans les deux cas suivants : données petites en dimension quelconque d'espace et données quelconques en dimension 1.

Dans ces modèles (voir [9] [6]), on suppose que les molécules ne peuvent prendre qu'un nombre fini de vecteurs vitesse $C_i \in \mathbf{R}^n$, où i parcourt un ensemble fini I. Il est naturel de supposer (sauf en dimension 1, voir ci-dessous), que l'indexation est bijective, c'est à dire que

$$i \neq j \Rightarrow C_i \neq C_j \tag{1}$$

On note $u_i(t,x)$ la densité des molécules, à l'instant t et au point $x \in \mathbf{R}^n$, qui sont animées de la vitesse C_i. L'analogue de l'équation de Boltzmann est le système suivant

$$\frac{\partial u_i}{\partial t} + C_i \cdot \nabla u_i = Q_i(u) , \tag{2}$$

en notant ∇ le gradient par rapport aux variables d'espace. Le terme d'interaction non-linéaire, en supposant que les seules interactions sont binaires et à distance nulle, est de la forme suivante

$$Q_i(u) = \sum_{j,k,l \in I} \left(A_{ij}^{kl} u_k u_l - A_{kl}^{ij} u_i u_j \right) , \tag{3}$$

où les coefficients A_{ij}^{kl} (probabilités de transition), qui décrivent la proportion de couples de molécules de vitesses (C_k, C_l) transformées en molécules de vitesses (C_i, C_j) par unité de temps, vérifient les conditions

$$A_{ij}^{kl} \geq 0 \quad , \quad A_{ij}^{kl} = A_{ji}^{kl} = A_{ij}^{lk} \tag{4}$$

et (les particules de même vitesse n'interagissent pas entre elles)

$$A_{ij}^{kl} \neq 0 \Rightarrow k \neq l . \tag{5}$$

Nous obtiendrons nos résultats sous ces seules hypothèses, c'est-à-dire en ne faisant appel ni à la microréversibilité (et à l'inégalité d'entropie qui en résulte) ni à la conservation de l'énergie ni même, sauf dans la seconde partie, à celle de la quantité de mouvement.

Dans une première partie, nous démontrons l'existence de solutions globales, uniformément bornées dans l'espace-temps, du problème de Cauchy défini par (2) (3) et

$$u_i(0,x) = f_i(x) \ , \tag{6}$$

où les f_i sont des fonctions données définies dans $\mathbf{R}^n$ et suffisamment petites. Cette condition de 'petitesse' qui est mesurée par les normes introduites dans la section 2, exige non seulement que les f_i soient petites en norme L^1 et L^∞ (ce qui, même pour le modèle de Broadwell, n'entraîne pas que les u_i soient bornées, voir [12]) mais encore que leurs intégrales sur certaines sous-variétés linéaires soient également petites. Elle est automatiquement vérifiée lorsque les f_i appartiennent à l'espace de Sobolev H^s, $s > n/2$, et que leur norme y est suffisamment petite. Elle est également vérifiée si les quantités $\|f_i(x)(1+|x|)^\sigma\|_{L^\infty}$, $\sigma > n$, sont assez petites.

Ces résultats étaient connus pour le modèle de Broadwell [13] [11]. Par contre, en ce qui concerne les modèles généraux, nous avons montré dans [4] que les hypothèses des théorèmes fréquemment cités de [11] et de [10] n'étaient essentiellement jamais réalisées pour des données de Cauchy non nulles.

Ce théorème d'existence globale nous fournit le groupe à un paramètre des opérateurs G_t qui font passer de l'état du système à l'instant 0 à son état à l'instant t. Ces opérateurs sont des bijections bilipschitziennes définies dans un voisinage de l'origine de l'espace fonctionnel traduisant les conditions de petitesse évoquées ci-dessus.

Nous étudions ensuite dans la section 3 le comportement asymptotique des solutions, en le comparant aux évolutions libres définies par $v_i(t,x) = g_i(x - C_i t)$. Nous démontrons que, étant donnée une telle évolution libre, il existe une solution de (2) qui lui est asymptotique pour $t \to -\infty$ (ce qui, dans d'autres contextes physiques, s'appellerait existence des opérateurs d'ondes) et que cette dernière est asymptotique, pour $t \to +\infty$, à une autre évolution libre $h_i(x - C_i t)$ (complétude asymptotique). Cela nous permet de définir l'opérateur de diffusion (ou scattering), qui à g fait correspondre h.

Tous ces résultats sont valables pour des termes d'interaction notablement plus généraux que (3). En particulier, ils s'appliquent à des données non nécessairement positives, et persistent si on renverse le sens du temps.

Dans une seconde partie, nous nous intéressons à des données de Cauchy non nécessairement petites en dimension 1 d'espace. Il ne s'agit pas d'un cas particulier du cas précédent en raison de la présence éventuelle de vitesses multiples. Ces modèles unidimensionnels sont en effet obtenus en considérant des modèles tridimensionnels pour des données de Cauchy ne dépendant que d'une variable. L'équation ne fait apparaître que les projections des C_i selon cette variable, projections qui peuvent fort bien être égales pour deux indices différents.

De très nombreux résultats (voir [1] [2] [6] [8] [13] [14] [15] [16]... et leurs bibliographies) ont été consacrés à l'existence globale en dimension 1 pour des modèles plus ou moins généraux. Les premiers résultats d'existence de solutions globalement bornées dans l'espace-temps sont dûs à J. T. Beale [1] [2], d'abord pour le modèle de Broadwell, puis pour des modèles généraux du type (3) où les seules restrictions portent sur les vitesses multiples.

Nous décrivons dans la section 4 notre résultat [3] valable pour tous les modèles unidimensionnels conservant la quantité de mouvement : si les données de Cauchy sont positives sommables et bornées, la solution du problème de Cauchy existe dans tout l'espace-temps et y est globalement bornée. Un argument simple permet d'en déduire l'existence globale de solutions non nécessairement bornées pour des données de Cauchy positives et localement bornées. Enfin, nous obtenons également des résultats, moins précis que pour les données petites, sur le comportement asymptotique des solutions.

2 Existence globale à données petites

Nous considérerons le problème de Cauchy associé à (2) (6), mais avec des termes d'interaction plus généraux que ceux définis par (3)

$$Q_i(u) = \sum_{j,k\in I} B_i^{jk} u_j u_k \tag{7}$$

où les B_i^{jk} sont des nombres réels quelconques assujettis à la seule condition

$$B_i^{jk} \neq 0 \Rightarrow j \neq k\ . \tag{8}$$

Nous aurons besoin des concepts géométriques suivants, liés aux caractéristiques de l'équation (2). Dans toute la suite, l'expression *p-plan* signifiera *sous-variété affine de dimension p*.

Définition 2.1 *Pour $i \in I$, on notera D_i le vecteur de $\mathbf{R}^{n+1}$ dont les composantes sont* $(1, C_i)$.
(a) *Nous dirons qu'un p-plan Π de $\mathbf{R}^{n+1}$, pour $p = 1, \dots, n+1$, est* caractéristique *s'il existe p vecteurs linéairement indépendants parmi les D_i qui soient parallèles à Π.*
(b) *Nous dirons qu'un p-plan π de $\mathbf{R}^n$, pour $p = 0, \dots, n$ est* de type trace *s'il est l'intersection de $\{t = 0\}$ avec un $(p+1)$-plan caractéristique. Il est équivalent de dire qu'il existe $p+1$ indices $i_0, \dots, i_p$ tels que les vecteurs $C_{i_1} - C_{i_0}, \dots, C_{i_p} - C_{i_0}$ soient linéairement indépendants et parallèles à π.*
(c) *Pour chaque (direction de) p-plan de type trace π, nous noterons $J(\pi)$ le sous-ensemble de I constitué des i tels qu'il existe un $(p+1)$-plan caractéristique Π s'appuyant sur π avec D_i parallèle à Π.*

Les 1-plans caractéristiques sont bien évidemment les droites caractéristiques au sens habituel. Nous avons représenté sur la figure un 2-plan caractéristique Π et un 1-plan de type trace π. Dans ce cas, l'ensemble $J(\pi)$ contient au moins les indices i et j, mais il peut en contenir d'autres.

On voit facilement qu'un point quelconque $m \in \mathbf{R}^n$ est un 0-plan de type trace et que l'on a $J(m) = I$. D'autre part, sous l'hypothèse suivante

$$\textit{Les vecteurs } D_i \textit{ engendrent } \mathbf{R}^{n+1}\ , \tag{9}$$

le n-plan $\mathbf{R}^n$ est de type trace, et on a $J(\mathbf{R}^n) = I$. Le lecteur pourra faire sienne cette hypothèse — vérifiée par tous les modèles physiquement raisonnables — dès à présent, mais nous ne l'utiliserons que dans la section 3.

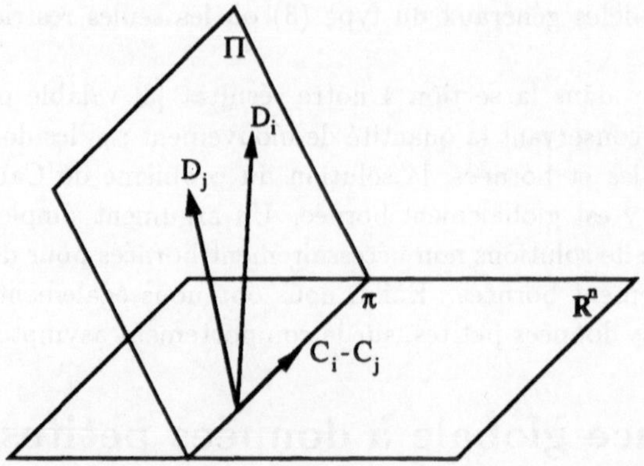

Nous pouvons maintenant définir les normes qui exprimeront la petitesse des données de Cauchy et l'espace fonctionnel qui leur est associé.

Définition 2.2 *Soit* $f = \big(f_i(x)\big)_{i\in I}$ *une (classe de) fonction mesurable dans* $\mathbf{R}^n$. *Pour* $p = 0, \ldots, n$, *on pose*

$$\mathcal{N}_p(f) = \sup_{i\in J(\pi)} \text{sup ess} \left\{ \int_\pi |f_i(x)|\ d^p x \ \Big|\ \pi \text{ p-plan de type trace} \right\} \tag{10}$$

On notera $\mathcal{E}$ *l'espace constitué des* f *pour lesquels on a* $\mathcal{N}_p(f) < \infty$, $p = 0, 1, \ldots, n$. *C'est un espace de Banach pour la norme* $\sum_0^n \mathcal{N}_p(\cdot)$. *On notera* $\mathcal{E}_0$ *l'adhérence dans* $\mathcal{E}$ *de l'ensemble des* f *dont les composantes sont continues à support compact.*

Nous avons noté $d^p x$ la mesure de Lebesgue p-dimensionnelle sur le p-plan π. Les bornes supérieures essentielles figurant au membre de droite de (10) sont définies précisément comme suit. Il n'existe qu'un nombre fini de directions de p-plans de type trace, et chacune de celles-ci peut s'identifier à $\mathbf{R}^p$ pour un choix convenable de coordonnées $\mathbf{R}^n = \mathbf{R}^{n-p}_{x'} \times \mathbf{R}^p_{x''}$. Il s'agit alors des normes, dans $L^\infty(\mathbf{R}^{n-p}_{x'})$ des fonction $x' \to \|f_i(x',\cdot)\|_{L^1(\mathbf{R}^p)}$.

On voit facilement que la norme $\mathcal{N}_0(f)$ est équivalente à la norme $\sum \|f_i\|_{L^\infty}$. D'autre part, sous l'hypothèse (9), la norme $\mathcal{N}_n(f)$ est équivalente à $\sum \|f_i\|_{L^1}$.

Quant aux normes intermédiaires, pour $1 \le p \le n-1$, ce sont donc des sommes de normes $L^\infty(\mathbf{R}^{n-p}_{x'}; L^1(\mathbf{R}^p_{x''}))$ de certains des f_i, pour des coordonnées identifiant un p-plan de type trace à $\mathbf{R}^p$. Pour le modèle de Broadwell, elles apparaissent explicitement dans [13] [11], et Illner [12] a montré que la petitesse des normes L^1 et L^∞ n'est pas suffisante, même pour ce modèle très simple, pour obtenir l'existence de solutions globalement bornées.

En notant $C_0(\mathbf{R}^n)$ l'espace des fonctions continues tendant vers 0 à l'infini, il est facile de décrire l'espace $\mathcal{E}_0$. Il s'agit des $f = (f_i)_{i\in I}$ qui appartiennent à $C_0(\mathbf{R}^n)$ (et à $L^1(\mathbf{R}^n)$ sous l'hypothèse (9)), et qui pour chaque direction π de p-plan de type trace (identifié à $\mathbf{R}^p$ comme ci-dessus) et pour chaque $i \in J(\pi)$ vérifient $f_i \in C_0(\mathbf{R}^{n-p}_{x'}; L^1(\mathbf{R}^p_{x''}))$.

Nous pouvons maintenant énoncer le résultat fondamental de cette section : existence de solutions globalement bornées pour des données initiales dont les normes dans $\mathcal{E}$ sont assez petites.

Théorème 2.3 *Sous l'hypothèse (8), le problème de Cauchy (2) (7) (6) possède une unique solution définie et bornée dans* $\mathbf{R}^{n+1}$ *entier, appartenant pour chaque t à* $\mathcal{E}$ *(resp.* $\mathcal{E}_0$*) pourvu que les données f appartiennent à* $\mathcal{E}$ *(resp.* $\mathcal{E}_0$*) et y aient une norme assez petite.*

Plus précisément, il existe des constantes δ_0 *et* C*, ne dépendant que des* C_i *et des* B_i^{jk} *telle que, si les* $f_i \in L^\infty(\mathbf{R}^n)$ *vérifient* $\mathcal{N}_p(f) \leq \delta_0$ *pour* $p = 0, \ldots, n$*, la solution existe globalement et verifie en outre, en notant* u^T *la trace de la solution sur l'hyperplan* $\{t = T\}$

$$\sup_T \sup_{0 \leq p \leq n} \mathcal{N}_p(u^T) \leq C \sup_{0 \leq p \leq n} \mathcal{N}_p(f) \tag{11}$$

et, pour tout p-plan caractéristique Π,

$$\int_\Pi |Q_i(u)| \leq C \left(\sup_{0 \leq p \leq n} \mathcal{N}_p(f) \right)^2 . \tag{12}$$

Nous renvoyons à [4] pour la démonstration, dont le point essentiel est l'inégalité suivante

$$\sup_\Pi \sup_{i \neq j} \int_{\Pi \cap [0,T] \times \mathbf{R}^n} |u_i(P) u_j(P)| \; d^p P \leq$$

$$\mathrm{C^{te}} \left(\sup_{0 \leq p \leq n} \mathcal{N}_p(f) + \sup_\Pi \sup_{i \neq j} \int_{\Pi \cap [0,T] \times \mathbf{R}^n} |u_i(P) u_j(P)| \; d^p P \right)^2 , \tag{13}$$

où Π parcourt la famille des p-plans caractéristiques, et où la constante C est indépendante de T.

3 Diffusion à données petites

Une conséquence facile du théorème d'existence globale est l'existence d'un groupe à un paramètre décrivant l'évolution du système.

Proposition 3.1 *Il existe un voisinage ouvert* Ω *de l'origine dans* $\mathcal{E}$ *et un groupe à un paramètre* G_t *d'applications continues de* Ω *dans lui-même tels que, pour* $f \in \Omega$*, la solution du problème de Cauchy (2) (7) (6) soit donnée par* $u(t, x) = G_t f(x)$*.*

En outre, en posant $\Omega_0 = \Omega \cap \mathcal{E}_0$*, les opérateurs* G_t *appliquent continûment* Ω_0 *dans lui-même.*

Une propriété importante est le caractère lipschitzien *uniforme en t* de la famille des opérateurs G_t.

Théorème 3.2 *Il existe* $\delta > 0$ *et* $C > 0$ *tels que, pour f et g vérifiant* $\|f\|_\mathcal{E} \leq \delta$ *et* $\|g\|_\mathcal{E} \leq \delta$*, on ait*

$$\|G_t f - G_t g\|_\mathcal{E} \leq C \, \|f - g\|_\mathcal{E} \tag{14}$$

Nous ne pouvons que renvoyer à [5] pour la démonstration dont le point important, en notant u et v les solutions associées à f et g, est le contrôle des intégrales des différences des termes d'interaction $|Q_i(u) - Q_i(v)|$ sur tous les p-plans caractéristiques.

L'objectif de la théorie de la diffusion (scattering) est de comparer, asymptotiquement pour $t \to \pm\infty$, l'évolution étudiée jusqu'à présent (et son groupe d'évolution G_t) avec l'évolution libre correspondant au cas $Q_i = 0$ (dont on notera G_t^0 le groupe associé). Cette dernière est bien sûr facile à décrire : pour des données de Cauchy f, la solution u est donnée par

$$G_t^0 f = u(t,\cdot) \quad \text{avec} \quad u_i(t,x) = f_i(x - tC_i) .$$

Quitte à réduire l'ouvert $\Omega_0 \subset \mathcal{E}_0$, on a le résultat suivant.

Théorème 3.3 *Nous supposerons l'hypothèse (9) réalisée.*
(a) *Pour $f \in \overline{\Omega_0}$, les $G_t G_{-t}^0 f$ convergent en norme dans $\mathcal{E}_0$, pour $t \to \pm\infty$, vers un élément de $\overline{\Omega_0}$ noté $W^\pm f$.*
(b) *Les opérateurs $W^\pm$ sont des bijections bilipschitziennes de $\overline{\Omega_0}$ dans lui-même, d'inverses notés $(W^\pm)^{-1}$.*
(c) *Pour $f \in \overline{\Omega_0}$, les $G_{-t}^0 G_t f$ convergent en norme dans $\mathcal{E}_0$, pour $t \to \pm\infty$, vers $(W^\mp)^{-1} f \in \overline{\Omega_0}$.*

Il est assez facile, et classique, de démontrer l'existence de $W^\pm f$ lorsque f est à support compact à partir du théorème d'existence globale. Par contre, le prolongement de ces opérateurs dans tout Ω_0 utilise de manière cruciale l'équicontinuité uniforme des opérateurs G_t énoncée dans le théorème 3.2. De même, à partir de l'estimation des termes d'interaction, il est facile de montrer que

$$\left(G_{-t}^0 G_t f\right)_i \longrightarrow f_i(x) + \int_0^{+\infty} (Q_i(u))(s, x + C_i s)\, ds$$

où la convergence a lieu dans L^1, mais pour démontrer la convergence forte dans $\mathcal{E}_0$ (et notamment la convergence uniforme), il est encore nécessaire d'utiliser l'uniformité en t de l'estimation (14).

Pour $f^0, f^+, f^- \in \overline{\Omega_0}$, l'équation $f = W^+ f^-$ signifie que l'unique évolution asymptote dans le passé à $f_i^-(x - C_i t)$ se trouve dans l'état f^0 à l'instant 0. De même, l'équation $f^+ = (W^-)^{-1} f_0$ signifie que cette même évolution est asymptote, dans l'avenir, à $f_i^+(x - C_i t)$. Nous pouvons donc maintenant définir l'opérateur de diffusion S, reliant les évolutions libres auxquelles une évolution perturbée est asymptote pour $t \to \pm\infty$.

Définition 3.4 *On note S l'opérateur $(W^-)^{-1} W^+$. Il s'agit d'une bijection bilipschitzienne de $\overline{\Omega_0}$ sur lui-même.*

Dans le cas des vrais modèles de cinétique discrète (termes d'interaction du type (3)), les opérateurs $(W^-)^{-1}$, W^+ et S conservent l'ensemble des $f \in \Omega_\delta$ vérifiant $f_i \geq 0$ pour tout i.

4 Existence globale et diffusion en dimension 1

Si on considère les solutions de (2) (3) ne dépendant que d'une des variables (que l'on notera dorénavant x) de l'espace $\mathbf{R}^n$, on est ramené à étudier le système

$$\frac{\partial u_i}{\partial t} + c_i \frac{\partial u_i}{\partial x} = Q_i(u) , \tag{15}$$

pour des fonctions $u_i(t,x)$ définies dans $\mathbf{R}_t \times \mathbf{R}_x$, les scalaires c_i désignant les composantes des C_i selon l'axe des x. Une différence importante avec le cas précédemment traité est que *les c_i ne sont pas nécessairement distincts*.

Les termes d'interaction sont toujours donnés par (3) avec les conditions (4) et (5). Nous supposerons de plus que l'on a conservation de la quantité de mouvement

$$A_{ij}^{kl} \neq 0 \Rightarrow c_i + c_j = c_k + c_l.$$

Enfin, nous étudierons toujours le problème de Cauchy défini par (6), les f_i étant maintenant définies sur $\mathbf{R}$. Nous noterons μ la masse totale

$$\mu = \sum_{i \in I} \int f_i(x)\, dx \ , \tag{16}$$

qui est conservée au cours de l'évolution.

Théorème 4.1 *Pour des données f_i positives sommables et bornées, la solution de (15) (6) est définie et globalement bornée dans $\mathbf{R} \times [0, \infty[$. On a de plus l'estimation*

$$\sup_{i \in I; t \in [0,\infty[; x \in \mathbf{R}} u_i(t,x) \leq C(\mu) \sup_{i \in I; x \in \mathbf{R}} f_i(x) \ ,$$

où la constante $C(\mu)$ ne dépend que de la masse totale (on peut prendre $C(\mu) = \exp(C_1 + C_2\mu^2 \log \mu)$).

Compte tenu de la propagation à vitesse finie, le corollaire suivant en résulte immédiatement.

Corollaire 4.2 *Si les f_i sont positives et localement bornées, le problème (15) (6) admet une unique solution définie globalement dans $\mathbf{R} \times [0, \infty[$.*

Rappelons brièvement le principe de la démonstration (voir [3]). On introduit les deux sous-ensembles de $I \times I$ définis comme suit

$$\mathcal{A} = \{(i,j) \in I \times I \mid c_i \neq c_j\} \tag{17}$$

$$\mathcal{B} = \left\{(i,j) \in I \times I \;\middle|\; \exists (k,l) \in \mathcal{A};\ A_{kl}^{ij} \neq 0\right\} . \tag{18}$$

Pour $0 \leq t_1 \leq t_2$, la solution étant supposée exister jusqu'au temps t_2, on définit la quantité

$$\Delta(t_1, t_2) = \sum_{(i,j) \in \mathcal{A} \cup \mathcal{B}} \iint_{[t_1,t_2] \times \mathbf{R}} u_i(t,x) u_j(t,x)\, dt\, dx \ .$$

L'estimation suivante, qui se démontre par intégration le long des caractéristiques, joue un rôle fondamental. On y désigne par K_1 et K_2 des constantes indépendantes de t_1, t_2 et des conditions initiales, et par μ la masse totale définie par (16).

$$\Big(\sup_{i,x} \sup_{t \leq t_2} u_i(t,x)\Big) \leq (1 + K_1\mu) \Big(\sup_{i,x} \sup t \leq t_1 u_i(t,x)\Big)$$
$$+ K_2 \Delta(\tau_1, t_2) \Big(\sup_{i,x} \sup t \leq t_2 u_i(t,x)\Big) \ .$$

Une telle estimation, jointe au théorème classique d'existence locale en temps, entraîne l'existence globale (avec contrôle de la norme uniforme de la solution) dès lors que l'on connaît une borne a priori, indépendante de t_1 et t_2, de $\Delta(t_1, t_2)$. Le lemme suivant constitue le point clef de la démonstration.

Lemme 4.3 *Supposons la solution u de (15) (6) définie sur* $[0,T] \times \mathbf{R}$. *Il existe alors une constante K, indépendante de T et des données de Cauchy, telle que l'on ait*

$$\sum_{(i,j)\in\mathcal{A}\cup\mathcal{B}} \iint_{[t_1,t_2]\times\mathbf{R}} u_i(t,x)u_j(t,x)\,dt\,dx \leq K(\mu+\mu^2)\ . \tag{19}$$

(a) Le cas des couples $(i,j) \in \mathcal{A}$ est le plus facile, et provient de la décroissance de la quantité suivante.

$$L(t) = \sum_{(i,j)\in I\times I} \iint (c_i - c_j)\mathrm{Sgn}(y-x)u_i(t,x)u_j(t,x)\,dx\,dy\ .$$

Il s'agit de l'intégrale d'une quantité conservative, qui peut s'exprimer uniquement en fonction de $\sum u_i$ et de $\sum c_i u_i$ aux points x et y, et qui n'est donc pas modifiée par les interactions. D'autre part, $L(t)$ représente la somme, pour tous les couples possibles de molécules, des quantités égales à la différence de leurs vitesses (en valeur absolue) affectées du signe plus si ces molécules se rapprochent et du signe moins dans le cas contraire. La seule modification qui puisse avoir lieu est le passage du signe plus au signe moins, lors de leur croisement.

On a en fait

$$L'(t) = -2\sum (c_i - c_j)^2 \int u_i(t,x)u_j(t,x)\,dx\ . \tag{20}$$

Comme $|L(t)|$ reste majoré par $\mathrm{C^{te}}\mu^2$, il suffit d'intégrer (20) entre 0 et T pour obtenir l'estimation (19) pour les couples $(i,j) \in \mathcal{A}$.

(b) Nous avons donné dans [3] une démonstration par récurrence sur l'ensemble des vitesses, valable en toute généralité, de l'estimation (19) pour les $(i,j) \in \mathcal{B}$. Nous allons en donner ici une preuve nettement plus simple mais valable uniquement sous l'hypothèse de microréversibilité

$$A_{ij}^{kl} = A_{kl}^{ij}\ .$$

et en supposant que $\int f_i \left|\log f_i\right| dx < \infty$.

Il est alors possible d'utiliser la décroissance de la fonction de Boltzmann $H(t) = \sum_i \int u_i \log u_i\,dx$

$$H'(t) = -\frac{1}{4}\int A_{ij}^{kl}(u_k u_l - u_i u_j)\log\frac{u_k u_l}{u_i u_j}\,dx$$

et d'après une inégalité classique (voir [9] p.87), on a

$$H'(t) \leq -\frac{1}{2}\int A_{ij}^{kl}(\sqrt{u_k u_l} - \sqrt{u_i u_j})^2\,dx\ .$$

En intégrant cette relation entre 0 et T, on obtient une borne indépendante de T des expressions

$$\iint_{[0,T]\times\mathbf{R}} (\sqrt{u_k u_l} - \sqrt{u_i u_j})^2\,dt\,dx$$

pourvu que A_{ij}^{kl} soit non nul. Or, pour $(i,j) \in \mathcal{B}$, on peut par définition trouver de tels k et l avec $(k,l) \in \mathcal{A}$. L'estimation voulue résulte alors de

$$\iint u_i u_j\,dt\,dx \leq 2\iint u_k u_l\,dt\,dx + 2\iint (\sqrt{u_k u_l} - \sqrt{u_i u_j})^2\,dt\,dx$$

et de la première partie de la démonstration.

Remarque **4.4** Les résultats que l'on peut obtenir sur le comportement asymptotique sont moins complets que dans le cas de la section 3. Si on introduit pour chaque i l'ensemble $J(i) = \{j \mid c_j = c_i\}$ et la fonction $\widetilde{u_i} = \sum_{j \in J(i)} u_j$, on obtient facilement le résultat suivant :

Il existe des fonctions $\varphi_i \in L^1 \cap L^\infty(\mathbf{R})$ *telles que*

$$\|\widetilde{u_i}(t, \cdot) - \varphi_i(\cdot - c_i t)\|_{L^1(\mathbf{R})} \to 0 .$$

Le fait que les u_i elles-mêmes soient asymptotes à une évolution libre est valable pour de très nombreux modèles, mais reste un problème ouvert en général. En outre, nous avons pu obtenir un exemple où, contrairement à ce qui se produit pour des données petites, l'évolution asymptotique dépend de manière discontinue des données de Cauchy.

Bibliographie

[1] J. T. Beale. Large-time behavior of the Broadwell model of a discrete velocity gas *Comm. Math. Phys.* **102** (1985) 217–235

[2] J. T. Beale. Large-time behavior of discrete velocity Boltzmann equations *Comm. Math. Phys.* **106** (1986) 659–678

[3] J.-M. Bony. Solutions globales bornées pour les modèles discrets de l'équation de Boltzmann en dimension 1 d'espace *Actes Journées E.D.P. St. Jean de Monts* (1987) n°XVI

[4] J.-M. Bony. Existence globales à données de Cauchy petites pour les modèles discrets de l'équation de Boltzmann. *soumis à Comm. P.D.E.* (1990)

[5] J.-M. Bony. Problème de Cauchy et diffusion à données petites pour les modèles discrets de la cinétique des gaz. *Actes Journées E.D.P. St. Jean de Monts* (1990) n°I

[6] H. Cabannes. Solution globale du problème de Cauchy en théorie cinétique discrète *J. de Mecan.* **17** (1978) 1-22

[7] H. Cabannes. The discrete Boltzmann equation (Theory and applications) *Lecture notes, University of California, Berkeley* (1980)

[8] H. Cabannes. On the initial-value problem in discrete kinetic theory *à paraître* (1990)

[9] R. Gatignol. Théorie cinétique des gaz à répartition discrète de vitesses *Lecture Notes in Physics* **36** Springer Verlag (1975)

[10] K. Hamdache. Existence globale et comportement asymptotique pour l'équation de Boltzmann à répartition discrète des vitesses *J. de Mecan. Th. Appl.* **3**, 5 (1984) 761–785

[11] R. Illner. Global existence results for discrete velocity models of the Boltzmann equation in several dimension *J. de Mecan. Th. Appl.* **1**, 4 (1982) 611–622

[12] R. Illner. Examples of non-bounded solutions in discrete kinetic theory *J. de Mecan. Th. Appl.* **5**, 4 (1986) 561–571

[13] S. Kawashima. Global solution of the initial value problem for a discrete velocity model of the Boltzmann equation *Proc. Japan Acad.* **57** (1981) 19–24

[14] M. Mimura et T. Nishida. On the Broadwell's model for a simple discrete velocity gas *Proc. Japan Acad.* **50** (1974) 812–817

[15] L. Tartar. Existence globale pour un système hyperbolique semi-linéaire de la théorie cinétique des gaz *Séminaire Goulaouic-Schwartz, Ec. Polytechnique* (1975–76) n°1

[16] L. Tartar. Some existence theorem for semilinear hyperbolic systems in one space variable *Technical Summary Report, Univ. Wisconsin* (1980)

On the Cauchy Problem for the Semidiscrete Enskog Equation

G. Toscani[1], *G. Borgioli*[1], *and A. Pulvirenti*[2]

[1]Dipartimento di Matematica, Università di Ferrara,
Via Machiavelli 35, I-44100 Ferrara, Italy
[2]Dipartimento di Matematica, Politecnico di Torino,
Corso degli Abruzzi 24, I-10129 Torino, Italy

Abstract : We prove that the semidiscrete Enskog equation, an analog of the semidiscrete Boltzmann equation introduced by H.Cabannes [1], has a global mild solution when the initial data are such that $[1 + |\vec{x}|^{\alpha} + \text{Log}\ \varphi]\ \varphi \in L_1$.

1. INTRODUCTION

In 1980, H.Cabannes [1] introduced a kinetic model, from him called semidiscrete Boltzmann equation, obtained in a formal limit procedure from the 2r plane velocity model by R. Gatignol [2], simply going with r to infinity. He obtained in this way a model where the velocities are discretized only in modulus, whenever they can attain any direction in the plane.

In spite of its apparent simplicity, with respect to the complete Boltzmann equation, the Cauchy problem for the semidiscrete model has not been solved, except in particular cases. We quote here the contributions in this direction given by N. Bellomo, R. Illner and G. Toscani [3], G. Toscani [4] and R. Monaco and G. Toscani [5].

Recently, a great interest has been devoted to the study of mathematical problems related to the Enskog equation [6]. This model was proposed by Enskog to take account explicitly of the finite diameter of molecules. In addition, to simulate the effect of a dense gas, the collision frequency is increased by a factor which represents an influence of the pair correlation function. Particular choices of this factor give rise to important differences between the resulting Enskog models.

The Cauchy problem for the Enskog equation has been extensively studied from some years now. A first group of results refers to small initial data or local in time existence results.

The papers by M. Lachowicz [7], G. Toscani and N. Bellomo [8], J. Polewczak [9] and C. Cercignani [10] fall in this class. For large initial data C. Cercignani [11] obtained global in time solutions in the case of one space dimension and geometric factor Y equal to 1. L. Arkeryd ([12], [13]) obtained results in more dimensions when Y is equal to 1.

Recently J. Polewczak [14] proposed a form of the geometric factor, which allows to prove the existence of a Liapunov functional for the Enskog model. In this way, he extended the existence theory by R. L. DiPerna and P. L. Lions [15] to the Enskog equation.

Having in mind the semidiscrete Boltzmann equation, in [16] we constructed a kinetic model with only one velocity modulus, but with the essential features of the Enskog equation. The reason of such an interest is essentially due to the fact that the mathematical structure of these models is relatively simple with respect to the full equations and suitable to obtain quantitative results in fluid dynamics of rarefied gases. Nevertheless, in spite of their relative simplicity in the collision rules, the macroscopic behaviour of the gas represented by these models is rather accurate and in general fits experimental results.

The plane model we proposed, called semidiscrete Enskog equation, simulates the evolution of a fictitious gas whose particles are represented by spheres of finite radius σ, whose centers of mass move in the plane with velocity $\vec{v}(\theta) = v \cos\theta \,\vec{i} + v \sin\theta \,\vec{j}$ with $v = \text{const}$ and $\theta \in [0,2\pi)$. Binary collisions between molecules are possible when the centers are at a distance 2σ, so that if $\vec{\nu}$ is a unit vector, the collision:

$$(\vec{x},\vec{v}(\theta_1)),\ (\vec{x} + 2\sigma\vec{\nu},\vec{v}(\theta)) \text{ in } (\vec{x},\vec{v}(\theta)),\ (\vec{x} + 2\sigma\vec{\nu},\vec{v}(\theta_1))$$

is realized. In such a collision particles simply exchange velocities, but, due to the finite radius, the incoming and outcoming fluxes are modified. (This collision is trivial in the Boltzmann framework).

Moreover, to each collision of the type described above, we associate a collision frequency given by

$$\frac{\sigma}{\pi}\left[\left(v(\theta_1) - \vec{v}(\theta)\right)\cdot \vec{\nu}\right]^+$$

where

$$[y]^+ = y \quad \text{if} \quad y \geqslant 0\ ;\ [y]^+ = 0 \quad \text{if} \quad y < 0\ .$$

If we denote by $f = f(\vec{x},\theta,t)$ the density at time t of the molecules with velocity $\vec{v}(\theta)$ in the position $\vec{x} \in \mathbb{R}^2$, the semidiscrete Enskog equation is written as:

$$\frac{\partial f}{\partial t} + \vec{v}(\theta)\ \vec{\nabla} f = \frac{1}{2\pi}\int_0^{2\pi} d\theta_1 \int_0^{2\pi} d\psi\ Y^+(t,\vec{x},\psi)\ f(\vec{x},\theta_1,t)\ f(\vec{x}+2\sigma\vec{\nu},\theta,t)\ (\vec{q}\cdot 2\sigma\vec{\nu})^+$$

$$- \frac{1}{2\pi} f(\vec{x},\theta,t) \int_0^{2\pi} d\theta_1 \int_0^{2\pi} d\psi\ Y^-(t,\vec{x},\psi)\ f(\vec{x}-2\sigma\vec{\nu},\theta_1,t)\ (\vec{q}\cdot 2\sigma\vec{\nu})^+ \tag{1}$$

where q is the relative velocity, and $\vec{\nu} = \cos\psi\ \vec{i} + \sin\psi\ \vec{j}$.

We suppose that the geometric factor Y reduces to a symmetric function of the local density at $\vec{x}$ and $\vec{x} + 2\sigma\vec{\nu}$, that is

$$Y^{\pm}(t,\vec{x},\psi) = Y(n(\vec{x},t),\ n(\vec{x}\pm 2\sigma\vec{\nu},t)) \tag{2}$$

where

$$n(\vec{x},t) = \int_0^{2\pi} d\theta\ f(\vec{x},\theta,t)\ .$$

With these hypotheses, if Φ is a measurable function on $\mathbb{R}^2 \oplus [0,2\pi)$, and $f \in C_0[\mathbb{R}^2 \oplus [0,2\pi)]$, we have:

$$\int_{\mathbb{R}^2} d\vec{x} \int_0^{2\pi} d\theta\ \Phi(\vec{x},\theta)\ E(f) =$$

$$\frac{1}{\pi}\int_{\mathbb{R}^2} d\vec{x} \int_0^{2\pi} d\theta \int_0^{2\pi} d\theta_1 \int_0^{2\pi} d\psi\ Y^-(t,\vec{x},\psi)\ f(\vec{x},\theta,t)\ f(\vec{x}-2\sigma\vec{\nu},\theta_1,t)$$

$$\cdot \left[\ \Phi(\vec{x},\theta_1) + \Phi(\vec{x}-2\sigma\vec{\nu},\theta) - \Phi(\vec{x},\theta) - \Phi(\vec{x}-2\sigma\vec{\nu},\theta_1)\ \right].[\vec{q}\cdot 2\sigma\vec{\nu}]^+ \tag{3}$$

where, as usual

$$E(f) = E^+(f) - E^-(f) \tag{4}$$

denotes the right-hand side of Eq. (1).

For f a nonnegative solution of (1), and ignoring at this stage any integrability conditions , define :

$$\Gamma(t) = \int_{\mathbb{R}^2} d\vec{x} \int_0^{2\pi} d\theta\ f(\vec{x},\theta,t)\ \mathrm{Log}\ f(\vec{x},\theta,t) - \int_0^t I(s)ds \tag{5}$$

where

$$I(t) = \frac{1}{4\pi} \int_{\mathbb{R}^2} d\vec{x} \int_0^{2\pi} d\theta \int_0^{2\pi} d\theta_1 \int_0^{2\pi} d\psi \Big(f(\vec{x}+2\sigma\vec{\nu},\theta_1,t)\, Y^+(t,\vec{x},\psi)$$

$$- f(\vec{x}-2\sigma\vec{\nu},\theta_1,t)\, Y^-(t,\vec{x},\psi) \Big) f(\vec{x},\theta,t)\, [\vec{q}\cdot 2\sigma\vec{\nu}]^+ . \tag{6}$$

Then, by means of (4), it is a simple matter to prove that $\Gamma(t)$ is monotonically decreasing with time.

This is the analog of the Boltzmann H-theorem for the semidiscrete Enskog equation.

Let us finally remark that from (4) one recovers the usual collisional invariants of mass, momentum and energy.

2. LOCAL EXISTENCE

In this section, by means of the classical contraction mapping principle, we will study the local existence of solutions to the Cauchy problem for the semidiscrete Enskog equation.

After integration along the characteristics, we shall consider the weaker version (mild formulation):

$$\begin{cases} \dfrac{df^{\#}}{dt} = E^{\#}(f) \\ f^{\#}(\vec{x},\theta,t=0) = \varphi(\vec{x},\theta) \end{cases} \tag{7}$$

where, for a given function $g(\vec{x},\theta,t)$ we denoted :

$$g^{\#}(\vec{x},\theta,t) = g(\vec{x}+\vec{v}(\theta)t,\theta,t) .$$

In what follows, we suppose that Y satisfies the following :

Hypothesis : The geometric factor Y is such that :

$$\operatorname*{ess\,sup}_{\vec{x},\psi} Y(t,\vec{x},\psi) \leqslant 1 + K \int_{\mathbb{R}^2} n(\vec{x},t)\, d\vec{x}$$

for some constant K.

We will start to solve (7) with a few definitions. For each constant $\alpha \geqslant 0$, let $L_{1,\alpha}$ denote :

$$L_{1,\alpha} = \Big\{ f(\vec{x},\theta) \;:\; (1+|\vec{x}|^{\alpha})\, f(\vec{x},\theta) \in L_1(\mathbb{R}^2\times[0,2\pi]) \Big\}$$

and, for any $T > 0$, let $\mathcal{B}_\alpha$ be the Banach space :

$$\mathcal{B}_\alpha = \left\{ f(\vec{x},\theta,t) \; : \; f^\#(\vec{x},\theta,t) \in C_0([0,T];L_{1,\alpha}) \right\}$$

endowed with the norm :

$$\|f\|_\alpha = \int_{\mathbb{R}^2} d\vec{x} \int_0^{2\pi} d\theta \; (1+|\vec{x}|^\alpha) \sup_{t\leq T} |f^\#(\vec{x},\theta,t)| \; .$$

Let us introduce the operator $\mathcal{A}$, acting on $\mathcal{B}_\alpha$, defined by

$$(\mathcal{A} f)^\#(\vec{x},\theta,t) = \varphi(\vec{x},\theta) + \int_0^t E^\#(f)(\vec{x},\theta,s)\, ds \tag{8}$$

We obtain :

<u>Lemma 1</u> : Let $\varphi(\vec{x},\theta) \in L_{1,\alpha}$. Then $\mathcal{A}$ maps $\mathcal{B}_\alpha$ into $\mathcal{B}_\alpha$ for every $T > 0$.

<u>Proof</u> : The proof is straightforward. If we define

$$F(\vec{x},\theta) = \sup_{t\leq T} |f^\#(\vec{x},\theta,t)|$$

then, for given $f \in \mathcal{B}_\alpha$ and $T > 0$,

$$\|\mathcal{A} f\|_\alpha \leq \|\varphi\|_\alpha + \frac{1}{2\pi} (1+k\|f\|_0) \; (I_1+I_2)$$

where

$$I_1 = \int_{\mathbb{R}^2} d\vec{x} \int_0^{2\pi} d\theta \int_0^T ds \int_0^{2\pi} d\theta_1 \int_0^{2\pi} d\psi \; (1+|\vec{x}|^\alpha) \; F(\vec{x}+2\sigma\vec{\nu},\theta) \; F(\vec{x}-\vec{q}s,\theta_1) \; [\vec{q} \cdot 2\sigma\vec{\nu}]^+$$

Since

$$(1 + |\vec{x}|^\alpha) \leq K_\alpha \; (1 + |\vec{x}+2\sigma\vec{\nu}|^\alpha)$$

for $K_\alpha < \infty$, and the transformation of variables

$$\begin{cases} \vec{x} + 2\sigma\vec{\nu} = \vec{x}' \\ \theta = \theta' \\ \vec{x} - \vec{q}s = \vec{x}'' \\ \theta_1 = \theta'' \end{cases} \tag{9}$$

has Jacobian $J = \vec{q} \cdot 2\sigma\vec{\nu}$, one recovers : $I_1 \leq K_\alpha \|f\|_\alpha \|f\|_0$.

The same bound holds for I_2. This ends the proof of lemma 1.

If μ is the usual Lebesgue measure on $\mathbb{R}^2\oplus[0,2\pi)$, let $\mathcal{L}$ denote the class of all Borel sets of $\mathbb{R}^2\oplus[0,2\pi)$, and, for $g \in \mathcal{B}_\alpha$:

$$\lambda_\alpha(g)(\beta) = \sup_{B\in\mathcal{L},\ \mu(B)=\beta} \left\{ \iiint_B d\vec{x}\ d\theta\ (1 + |\vec{x}|^\alpha)\ \sup_{t\leq T} |g^\#(\vec{x},\theta,t)| \right\} \tag{10}$$

Finally, let us define the following closex convex subset of $\mathcal{B}_\alpha$:

$$\mathcal{D}_\alpha = \{ f \in \mathcal{B}_\alpha \ : \ \lambda_\alpha(f)(\beta) \leq 2\lambda_\alpha(\varphi)(\beta),\ \ \forall\, \beta \leq T \} \tag{11}$$

Then we prove :

<u>Lemma 2</u> : Let $\varphi \in L_{1,\alpha}$. Then there exits a time T_α such that, for $T \leq T_\alpha$, $\mathcal{A}$ is a contraction mapping from $\mathcal{D}_\alpha$ into $\mathcal{D}_\alpha$.

<u>Proof</u> :

The proof follows from a estimate of a six-fold integral over a small domain. We have in fact to evaluate, for every $\mathcal{B} \in \mathcal{L}$ of measure β, the integral :

$$I_1(B) = \iiint_B d\vec{x}\ d\theta \int_0^T ds \int_0^{2\pi} d\theta_1 \int_0^{2\pi} d\psi\ (1+|\vec{x}|^\alpha)\ F(\vec{x}+2\sigma\vec{\nu},\theta).F(\vec{x}-\vec{q}s,\theta_1).[\vec{q}.2\sigma\vec{\nu}]^+$$

Let at first B be a parallelepiped with the axes parallel to the reference frame given by $(\vec{x},\theta)$. Then, by the transformation (9) one proves that the transformed volume Ω of $\mathbb{R}^6$ can be covered by a union of parallelepipeds in $\mathbb{R}^3_1\oplus\mathbb{R}^3_2$ whose projection into $\mathbb{R}^3_1$ has a volume smaller than $\mu(B)$. On the other hand, from (9) one gets that the projection of Ω into $\mathbb{R}^3_2$ is contained into a sphere of diameter less than T. From this follows :

$$I_1(B) \leq 8\lambda(\varphi)(\beta).\lambda(\varphi)(T)$$

The proof can be easily extended to the case when B is a compact set, and finally to every $B \in \mathcal{L}$.

The evaluation of $I_2(B)$ is immediate.

Lemma 2 gives us a local existence theorem. It remains to prove that positive initial values generate positive solutions.

To obtain this result, we need a preleminary lemma :

<u>Lemma 3</u> : Let $\varphi \in L_{1,\alpha}$ for some $\alpha > 0$, and, in addition $|\varphi(\vec{x},\theta)| \leq c\ (1+x^2)^{-2}$ for some constant c. Then, in $[0,T_\alpha]$:

$$|f^\#(\vec{x},\theta,t)| \leq K_c(1+x^2)^{-2} \tag{12}$$

Proof :

This result follows from a standard domain of dependence argument. For sufficiently great ρ, out of a circle of radius ρ one obtains [9] :

$$|f^{\#}(\vec{x},\theta,t)| \leq 2\ (1+x^2)^{-2}$$

If $|\vec{x}| \leq \rho$, and $|f^{\#}(\vec{x},\theta,t)| \geq 3c\left[1 + (\rho+2vT_\alpha)^2\right]^{-2}$ we are in contradiction with the result of Lemma 2, since $f \in \mathfrak{D}_\alpha$.

Positivity follows easily from lemma 3 and from the classical Carleman argument.

3. GLOBAL EXISTENCE

Global existence follow if we are able to prove that Γ functional is bounded, from below and above. In this case, the classical Tartar's method can be applied, and one can extend the local existence theorem to any finite T. It is clear, from the estimation of lemma 1, that $\left|\int_0^t I(s)\,ds\right| \leq C_T \|\varphi\|_\alpha^2$ for some constant C_T. Then, proceeding as in [17], we prove that H is bounded.

Finally, the following holds :

Theorem : Let $0 \leq \varphi \in L_{1,\alpha}$ for some $\alpha > 0$, and let Y satisfy the hypothesis. Then, if $\varphi\,\mathrm{Log}\varphi \in L_1(\mathbb{R}^2 \oplus [0,2\pi))$, the Cauchy problem (7) has a unique nonnegative global mild solution.

Remark : The detailed calculations will be published elsewhere [18]. We remark here that the crucial point besides in proving a local existence theorem in $\mathfrak{D}_\alpha$. For this reason, we sketched only the proof of this result.

Acknowledgment : This research has been partially supported by the project MMAI of the Gruppo Nationale per la Fisica Matematica of National Council for Research (CNR). One of the authors (G.T.) wishes to express his thanks for the ospitality of the "Laboratoire de Modelisation en Mécanique" of Paris VI, where part of this paper has been written.

REFERENCES

[1] H. Cabannes, "The discrete Boltzmann Equation (Theory and applications)", Lecture Notes at the Univ. of California, Berkeley, (1980).

[2] R. Gatignol, "Théorie cinétique des gaz à répartition discrète de vitesses" Lecture Notes in Physics, Vol 36, Springer Verlag, Heidelberg (1975).

[3] N. Bellomo, R. Illner and G. Toscani, "Sur le problème de Cauchy pour l'équation de Boltzmann semi-discrète", C. R. Acad. Sc., Paris, **239**, (1984), p. 835.

[4] G. Toscani, "The semidiscrete Boltzmann equation for hard spheres", Meccanica, **20**, (1985), p. 249.

[5] R. Monaco and G. Toscani, "New results on the semidiscrete Boltzmann equation for a binary gas mixture", Meccanica, **22**, (1987), p. 179.

[6] D. Enskog, "Kinetische Theorie", Kungl.Svenska Vetenskaps Akademiens Handl., **63** n.4, (1921); English transl. in S.Brush, Kinetic Theory, Vol. 3, Pergamon Press, New York, (1972).

[7] M. Lachowicz, "On the local existence and uniqueness of solution of initial value problem for the Enskog equation", Bull. Polish Acad. of Sci., **31**, (1983), p. 89.

[8] G. Toscani and N. Bellomo, "The Enskog-Boltzmann equation in the whole space : Some global existence, uniqueness and stability results", Comput. Math. Applic., **13**, (1987), p. 851.

[9] J. Polewczak, "Global existence and asymptotic behavior for the nonlinear Enskog equation", SIAM J. Appl. Math., **49**, (1989), p. 952.

[10] C. Cercignani, "Small data existence for the Enskog equation in L_1", J. Stat. Phys., **51**, (1988), p. 291.

[11] C. Cercignani, "Existence of global solutions for the space inhomogeneous Enskog equation", Trans. Th. Stat. Phys., **16**, (1987), p. 213.

[12] L. Arkeryd, "On the Enskog equation in two space variables", Trans.-Th. Stat. Phys., **15**, (1986), p. 673.

[13] L. Arkeryd, "On the Enskog equation with large initial data", preprint, Dept. Mathematics University of Goeteborg, (1988).

[14] J. Polewczak, "Global existence in L_1 for the generalized Enskog equation", J. Stat. Phys., **59**, (1990), p. 461.

[15] R. L. DiPerna and P. L. Lions, "On the Cauchy problem for the Boltzmann Equation : Global existence and weak stability", Ann. of Math., **130**, (1989), p. 321.

[16] G. Borgioli, R. Monaco and G. Toscani, "The semidiscrete Enskog equation", Proceedings of "Waves and stability in continuous media", S. Rionero ed., World Scientific, Singapore, (1990).

[17] G. Toscani, "On the Cauchy problem for the discrete Boltzmann equation with initial values in $L_1^+(\mathbb{R})$", Commun. Math. Phys., **121**, (1989), p. 121.

[18] G. Borgioli, A. Pulvirenti and G. Toscani, "Existence and uniqueness for the semidiscrete Enskog equation", Preprint, (1990).

On Uniform Boundedness of Solutions to Discrete Velocity Models in Several Dimensions

R. Illner

Department of Mathematics and Statistics, University of Victoria, Victoria, B.C. V8W 3P4, Canada

1. INTRODUCTION.

The global existence question for discrete velocity models in more than one space dimension remains unsolved, except for initial values which are small in some sense ([2],[6],[7]). The crucial difficulty is that we do not seem to have the tools to obtain uniform L^∞ - bounds in time on the local solution in terms of the initial values (this, of course, would entail global existence of a mild solution). The purpose of this article is to compare the situation with the better understood one-dimensional case, spell out some crucial differences, and point out a possible way to progress.

2. GROWTH RESULTS FOR THE 4-VELOCITY BROADWELL MODEL.

The initial value problem for discrete velocity models in one space dimension,

$$\frac{\partial}{\partial t} f_i + \xi_i \frac{\partial}{\partial x} f_i = Q_i(f,f) , \qquad f_i(0,x) = f_{i,o}(x) , \tag{1}$$

$$i = 1 , \ldots , n$$

is well understood if $f_{i,o} \in L^1_+ \cap L^\infty(\mathbb{R})$. In fact, Beale [1] and later, with a more elegant method, Bony [3], proved that (1) admits a global, uniformly bounded solution, and were also able to show that the asymptotic behaviour of this solution is given by free streaming. Cabannes and Kawashima [4] obtained the global existence via the older methods pioneered by Nishida and Mimura [10] and Crandall and Tartar [12]. However, this method only proves global existence, not uniform boundedness.

If $f_{i,o} \in L^{\infty}_{+}(\mathbb{R})$ (but $\notin L^1_+(\mathbb{R})$), we can use the strict hyperbolicity of equations (1) to conclude that the solution to the Cauchy problem will still exist globally. What the methods from [1], [3], [4], do not show is uniform boundedness of this solution, though it is hard to imagine how the solution could grow indefinitely - we expect uniform boundedness ! How difficult this problem is can be seen from the many, as yet unsuccessful attempts to solve the one-dimensional Broadwell model :

$$\partial_t v + \partial_x v = z^2 - vw$$

$$\partial_t w - \partial_x w = z^2 - vw \tag{2}$$

$$\partial_t z = \frac{1}{2}(vw - z^2)$$

with periodic boundary conditions :

$$v(t,0) = w(t,0) \ , \qquad v(t,1) = w(t,1) \ ,$$

and smooth data on $[0,1]$ such that :

$$v_0(0) = w_0(0) \ , \qquad v_0(1) = w_0(1) \ .$$

This problem, as is well known, can be recast as a pure Cauchy problem for periodic initial values. The big prize is to show that, as $t \to \infty$, there is a constant $a > 0$ such that :

$$\lim_{t \to \infty} (v(t,x) \ , w(t,x) \ , z(t,x)) = (a \ , a \ , a)$$

uniformly in x, and the big hurdle to this end is the lack of global L^{∞}- bounds for the solution. Recently, M. Slemrod [11] has proved a result on the asymptotic behaviour (more precisely, the orbital stability) of solutions to this initial boundary value problem, but in spite of skillful use of modern methods (like compensated compactness), he could not prove uniform boundedness. As a consequence, he could also not establish that the asymptotic state is a constant vector ; the method only shows that v, w and z approach, in the weak topology, solutions to the collision-free system, i. e. waves travelling without interaction.

In more than one dimension, <u>we know</u> that $f_{i,o} \in L^1_+ \cap L^{\infty}$ is not sufficient for uniform boundedness, because these are counterexamples (see [8]). The initial values from [8] which lead to unlimited growth of the L^{∞}-norm of the solution are characteristic functions and therefore discontinuous, but the geometric idea behind these examples can be applied to prove the following type of growth result for smooth data.

We are concerned with the solutions of the standard 4-velocity model in two space dimensions :

$$
\begin{aligned}
(\partial_t + \partial_x)\, f_1 &= Q(f,f) \\
(\partial_t - \partial_x)\, f_2 &= Q(f,f) \\
(\partial_t + \partial_y)\, f_3 &= -\, Q(f,f) \\
(\partial_t - \partial_y)\, f_4 &= -\, Q(f,f)
\end{aligned}
\qquad (3)
$$

with $Q(f,f) = f_3 f_4 - f_1 f_2$.

Theorem : *For any two constants* $\delta > 0$ *and* $C > 0$, *there are continuous initial values* $f_{i,o}$ *for (3),* $i = 1,\dots,4$, *such that :*

$$0 \leqslant f_{i,o}(x,y) \leqslant \delta \quad \text{for all } x,\ y \text{ and } i, \quad \text{but} \quad \sup_{t,(x,y),i} f_i(t,x,y) \geqslant C.$$

In addition, the L^1*- norm of the initial values can be chosen arbitrarily small.*

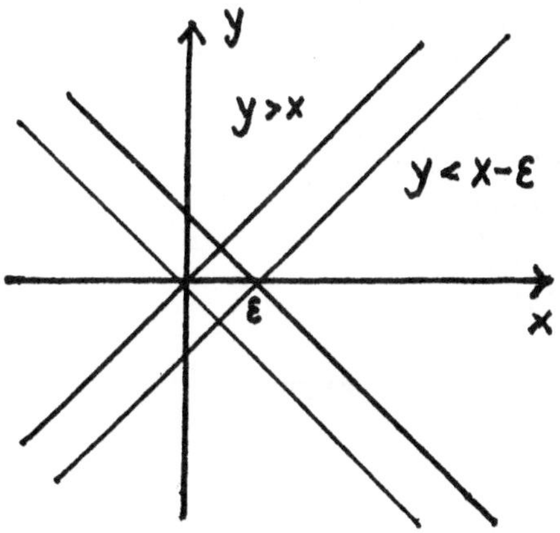

$$f_{4,o}(x,y) = \begin{cases} \delta & \text{for } y > x \\ 0 & \text{for } y < x - \varepsilon \\ \text{linear} & \text{for } x - \varepsilon \leqslant y \leqslant x \end{cases}$$

$$f_{3,o}(x,y) = \begin{cases} \delta & \text{for } y < -x \\ 0 & \text{for } y > -x + \varepsilon \\ \text{linear} & \text{for } -x \leqslant y \leqslant -x + \varepsilon \end{cases}$$

Figure 1

Proof : Consider the initial values $f_{1,0} = f_{2,0} = 0$, and construct $f_{3,0}$ and $f_{4,0}$ as indicated in Fig. 1 . ε is a nonnegative parameter to be chosen later. Fix $\delta > 0$, and suppose that there is a constant $C > 0$ such that, for any $\varepsilon > 0$, the solution to (3) with these initial values will remain bounded by C . Then, consider the solution at time t at the point $x = t$, $y = 0$: We have :

$$f_3(t;t,0) = f_3(t-\varepsilon;t,-\varepsilon) + \int_0^\varepsilon (f_1 f_2 - f_3 f_4)(t-\varepsilon+\tau;t,-\varepsilon+\tau)\, d\tau \qquad (4)$$

For $\sigma \in [0, t-\varepsilon]$, f_3 is constant along the (backward) characteristic connecting $(t-\varepsilon;t,-\varepsilon)$ and $(0;t,-t)$, because f_1 , f_2 and f_4 are all identically zero on that characteristic. This follows from the geometric setup : f_1 and f_2 , being zero initially, can only become nonzero by interaction between f_3 and f_4 . Such interaction has, until time σ , only occurred in the shaded region in Fig. 2. Clearly, f_3 and f_4 do not interact along the mentioned characteristic until $\sigma = t-\varepsilon$.

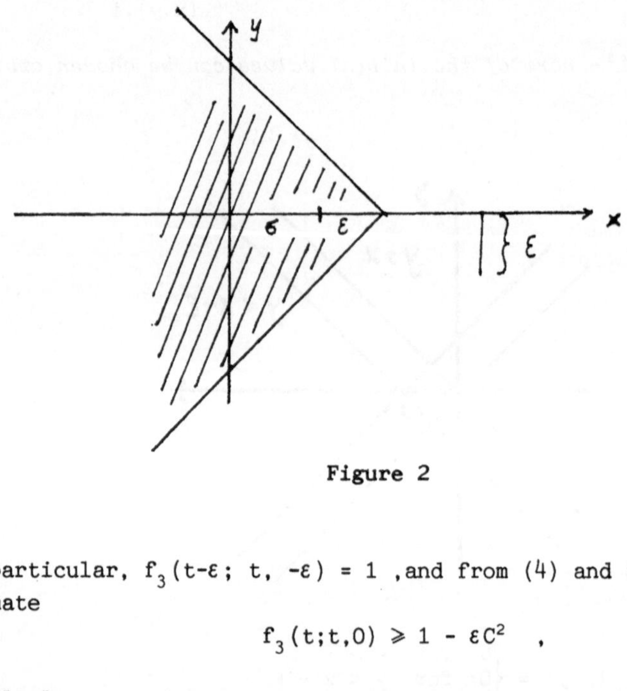

Figure 2

In particular, $f_3(t-\varepsilon;\ t,\ -\varepsilon) = 1$,and from (4) and the assumed bounds, we estimate

$$f_3(t;t,0) \geqslant 1 - \varepsilon C^2 \quad ,$$

and similarly

$$f_4(t;t,0) \geqslant 1 - \varepsilon C^2 \quad .$$

Choose ε such that $1 - \varepsilon C^2 \geqslant \frac{1}{2}$, then $f_3(t;t,0).f_4(t;t,0) \geqslant \frac{1}{4}$.

For $f_2(t;t,0)$, we have the simple estimate :

$$f_2(t;t,0) \leqslant f_2(t-\varepsilon;t+\varepsilon,0) + \int_0^{\varepsilon} f_3 f_4(t-\varepsilon+\tau,t+\varepsilon-\tau,0)\, d\tau \leqslant \varepsilon.C^2 \quad .$$

Now, consider the equation for f_1 in mild form :

$$f_1(t;t,0) = f_{1,o}(0,0) + \int_0^t [f_3 f_4 - f_1 f_2](\tau;\tau,0)\, d\tau \quad .$$

By the estimate from below of $f_3.f_4$ and the (assumed) estimate from above on $f_1.f_2$ by $\varepsilon.C^3$, we get :

$$f_1(t;t,0) \geqslant 0 + \frac{t}{4} - t.\varepsilon C^3 \quad .$$

By choosing ε so small that $\varepsilon.C^3 < \frac{1}{4}$, we could conclude that f_1 would grow indefinitely along the characteristic $(t;t,0)$. This contradicts our assumption on boundedness. The proof is complete.

Remark : The principle of this proof can certainly be applied to many other discrete velocity models. For discontinuous data, the underlying geometric idea can also be used to show that the semi-discrete velocity models suggested by Cabannes [5] admit solutions which grow without bounds. The data constructed above have infinite L^1- norm, but the proof is easily modified to show that they can actually be chosen to have arbitrarily small L^1- norm.

3. TENTATIVE STEPS TOWARDS GROWTH CONTROL.

The data constructed in Section 2 are, of course, very pathological. The unlimited growth is predetermined by the fact that the integrals $\int_{L_1} f_{4,o}$ and $\int_{L_2} f_{3,o}$, where L_1 and L_2 are the lines given by $y = x$ and $y = -x$ respectively, are divergent. In [7], it was shown that, if $\sup_{L,i} \int_L f_{i,o}$ is small enough, where $i = 1,\dots,4$ and L is any line parallel to L_1 or L_2 , then there is indeed a uniformly bounded global solution.

Remark : This result is one example of a general global existence result proved in [7]. While all the examples from [7] remain of interest, Bony [2] has recently pointed out that the conditions imposed on the initial values are only true for the zero function (unless the admitted velocities satisfy restrictive conditions). This flaw was removed by Bony in [2] ; his method does not use the conservation laws intrinsic to the model, but rather the convolution structure forced into the collision terms by the flow terms.

We return to the 4-velocity model. It is unknown whether the solution to the Cauchy problem even exists globally if $\sup_{L,i} \int_L f_{i,o}$ is finite (but large). Our growth result shows, however, that control of mixed norms of the above type is essential to obtain L^∞- control of the solution.

This is an important observation with respect to a possible generalization of the "potential for interaction" which Bony introduced in [5] for the one-dimensional case. For any one-dimensional discrete velocity model with mass and momentum conservation and for which $\xi_i \neq \xi_j$ if $i \neq j$, let

$$B[f] := \sum_{i,j} \iint_{y<x} (\xi_i - \xi_j)\, f_i(t,x)\, f_j(t,y)\, dx\, dy \quad .$$

$B[f]$ is clearly bounded in terms of the largest x-component of the velocities and $\max_i \int f_{i,o}(x)\, dx$, and

$$\frac{d}{dt} B[f] = - \sum_{i,j} (\xi_i - \xi_j)^2 \int f_i f_j(t,x)\, dx \quad .$$

This last identity is the key to very efficient control of the L^∞- norm of the solution in terms of the initial "mass" $\sum_i \int f_{i,o}(x)\, dx$. $B[f]$ has become known as "potential for interaction".

Suppose we call a potential for interaction in higher dimensions any functional of the system state which will yield L^∞- control of the solution. The above growth results show that such a functional could not be bounded in terms of the total mass. It can, at least for the 4-velocity model, at best be bounded in terms of integrals $\int_L f_{i,o}$.

Finally, we point out that the integrals $\int_L f_{i,o}$ arise, for the 4-velocity model, quite naturally from simple cancellations of the collision terms : It is easily checked that, if $f_1, \ldots, f_4$ is a solution to (3), and if $L_{1,+}(t)$ is any line parallel to $y = x$ and moving with velocity (1,0) (or, equivalently, with (0,1)), then :

$$\frac{d}{dt} \int_{L_{1,+}(t)} (f_1 + f_3)(t,.) = 0 \quad . \tag{5}$$

Similarly, if $L_{2,-}(t)$ is a line parallel to L_2 and moving with velocity (-1,0) or (0,1) , then :

$$\frac{d}{dt} \int_{L_{2,-}(t)} (f_2 + f_3)(t,.) = 0 \quad .$$

$f_1 + f_4$ and $f_2 + f_4$ also satisfy such conservation laws. These conservations are quite clear from a mechanical point of view, given the underlying particle model. Equivalent conservation laws exist for the corresponding lattice gas model ; as was pointed out to me at the Symposium, this fact is known to at least some of the experts in cellular automata theory ([9]).

The conservation law (5) is an example for a general principle, which we now explain. Let $u_1, \dots, u_n \in \mathbb{R}^3$ be the admissible velocities for a certain discrete velocity model :

$$\frac{\partial f_i}{\partial t} + u_i . \nabla_x f_i = Q_i(f,f) \quad ,$$

and suppose that $M \subset \{1,\dots,n\}$ is an index set such that there are real numbers α_i , $i \in M$, with

$$\sum_{i \in M} \alpha_i \, Q_i(f,f) = 0$$

(mass, momentum and energy conservation are special cases with $M = \{1,\dots,n\}$).

Now, suppose that L is a linear submanifold of $\mathbb{R}^3$ such that the sets :

$$L_i(t) = \{x \in \mathbb{R}^3 \; ; \; x \in L + tu_i\}$$

are all identical (say $L_i(t) = L(t)$). Then, clearly :

$$\frac{d}{dt} \sum_{i \in M} \alpha_i \int_{L(t)} f_i(t,x) \, dx = 0 \quad . \tag{6}$$

It is easily checked that (5) is just a special case of (6). We suggest that a potential of interaction, if it exists in 2 or 3 dimensions, would have to involve integrals like the ones in (6) in some way, with submanifolds of dimensions 1 , 2 , and 3.

If u_1, $u_2 \in \mathbb{R}^2$ are linearly independent, then there is exactly one line L through the origin such that $L + tu_1$ and $L + tu_2$ are identical for all t (Figure 3).

Similarly, if u_1, u_2 and $u_3 \in \mathbb{R}^3$ are linearly independent then there is exactly one plane P through the origin such that $P + tu_i$ is the same set for $i = 1, 2$ and 3.

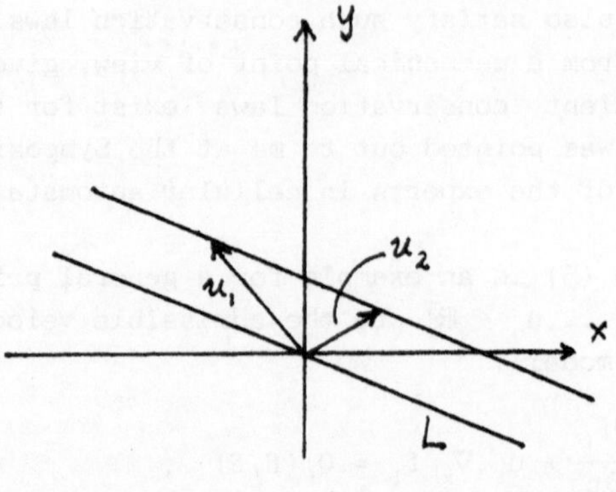

Figure 3

Acknowledgement.

I would like to express my deepest gratitude to Henri Cabannes, who, through his kindness and his continued support and appreciation of my work, has greatly contributed to my scientific career.

REFERENCES

[1] J. T. BEALE, "Large-time behavior of discrete velocity Boltzmann equations", Comm. Math. Phys., 106, 1986, p. 659-678.

[2] J. M. BONY, "Existence globale à données de Cauchy petites pour les modèles discrets de l'équation de Boltzmann", Ecole Polytechnique, Palaiseau, 1990, (preprint).

[3] J. M. BONY, "Solutions globales bornées pour les modèles discrets de l'équation de Boltzmann en dimension 1 d'espace", Actes Journées E.D.P., St Jean de Monts, 1987, N° XVI.

[4] H. CABANNES, S. KAWASHIMA, "Le problème aux valeurs initiales en théorie cinétique discrète", C. R. Acad. Sci. Paris, 241, I, 1988.

[5] H. CABANNES, in *"Mathematical methods in the kinetic theory of gases"* D. Pack and H. Neunzert Ed., Verlag D. Lang, Frankfurt, 1980.

[6] K. HAMDACHE, "Existence globale et comportement asymptotique pour l'équation de Boltzmann à répartition discrète des vitesses", J. de Mécan. Th. Appl., 3, 1984, p. 761-785.

[7] R. ILLNER, "Global existence results for discrete velocity models of the Boltzmann equation in several dimensions", J. Mécan. Th. Appl., 1, 1982, p. 611-622.

[8] R. ILLNER, "Examples on non-bounded solutions in discrete kinetic theory", J. Mécan. Th. Appl., 5, 1986, p. 561-571.

[9] D. LEVERMORE, Personal communication.

[10] M. MIMURA, T. NISHIDA, "On the Broadwell's model for a simple discrete velocity gas", Proc. Japan Acad., 50, 1974, p. 812-817.

[11] M. SLEMROD, "Large time behavior of the Broadwell model of a discrete velocity gas with specularly reflective boundary conditions", Archive Rat. Mech. Anal. (to appear).

[12] L. TARTAR, "Existence globale pour un système hyperbolique semi-linéaire de la théorie cinétique des gaz", Séminaire Goulaouic-Schwartz, Ecole Polytechnique, 1975-76, n° 1.

Temperature and Local Entropy Overshoots for the Second Fourteen-Velocity Cabannes Model

H. Cornille

Service de Physique Théorique, CEN-Saclay,
F-91191 Gif-sur-Yvette Cedex, France

ABSTRACT

For the Cabannes 14 velocities model with speeds $\sqrt{3}, 2$, we construct different classes of similarity shock waves solutions and study temperature and local entropy overshoots. We observe temperature overshoots only when the shock front speed is higher than the speed of the slow particles. On the contrary the local entropy overshoot can always be present when the scaling parameter of the solutions is close to a critical value which limits the local entropy increase.

I. INTRODUCTION

For the discrete velocity Boltzmann models as well as for the associated lattice gas models, the velocity can only take discrete values[(1)] : $\vec{v}_i, i = 1, p$. However the drawback of many discrete models is that they only have one speed $|\vec{v}_i| = \text{const}$ so that the temperature is ill-defined. Recently[(2)], from a study of shock waves for non trivial temperature models with two and three speeds in R($4\vec{v}_i$ model[(3)]), R^2($8\vec{v}_i$ and $9\vec{v}_i$ models[(4)]) and R^3 (Cabannes $14\vec{v}_i$ model[(5)]), then temperature[(2)] and local entropy overshoots[(6−2)] across the shock have been observed.

Here we go on to study these overshoots effects by investigating the second Cabannes model[(5)]. For the Cabannes models[(5)] eight of the velocities $\vec{u}_i$, $|\vec{u}_i| = \sqrt{3}$ are along the diagonals of a cube with sides 2 while six $\vec{v}_i$, $|\vec{v}_i|$ being 1 or 2 are perpendicular to the faces. For the first model with $|\vec{v}_i| = 1$, Cabannes[(5)] had previously studied shock waves. For the same model temperature and local entropy overshoots were observed[(2)], but the $|\vec{v}_i| = 2$ model was not investigated .

For the Cabannes models the coordinates in the x_1, x_2, x_3 space are $(-1,1,1)$, $(1,1,1)$, $(-1,-1,1)$, $(1,-1,1)$ for $\vec{u}_i$ $i = 1,2,3,4$, $|\vec{u}_i| = \sqrt{3}$, $\vec{u}_i + \vec{u}_{9-i} = 0$ and $(v,0,0), (0,v,0)$, $(0,0,v)$ for $\vec{v}_i$, $i = 1,2,3$ and $v = 1$ or 2, $\vec{v}_i + \vec{v}_{i+3} = 0$. To $\vec{u}_i, \vec{v}_i$ we associate the densities N_i, M_i and for the solutions in one spatial dimension $x = x_1$ we have $N_j = N_{2i+j}$, $j = 1,2$, $i = 1,2,3$, $M_6 = M_5 = M_3 = M_2$. There remains five independent densities N_1, N_2, M_1, M_4, M_2 satisfying five equations but three of them are equivalent to the linear conservation laws of mass, momentum and energy. The main difference between the $v = 1$ and 2 models is that the projections of the velocities along the x_1−axis have only one modulus 1 for $v = 1$ and two 1,2 for $v = 2$. Consequently for the $v = 1$ model we cannot really, on the x_1−axis, distinguish between the slow and the fast particles while this is possible for $v = 2$.

In Section 2 we give a general explanation of the local entropy overshoot, first discussed by Platkowski[6], for the similarity shock waves $N_i = n_{0i} + n_i/D$, $D = 1 + e^{\gamma\eta}$, $\eta = x - \xi t$ solutions of the discrete models equations in one dimension. We can always define a scaling parameter such that macroscopic quantities like the mass, the momentum or the energy are proportional to it. Further the temperature is independent of this scaling parameter but this not true for the local entropy which contains $N_i \log N_i$ terms. There exists a critical value of this scaling parameter which limits the local entropy increase, overshoots occur for values close to the critical value and disappear for very large values (rarefactive shocks) or very small values (compressive shocks).

In Section 3 we study the shock waves solutions of the second Cabannes model. Two types of collision between particles occur: either with different speeds $\sqrt{3}, 2$ or with the same speed 2 for which, generalising slightly the Cabannes model, we assume a cross-section $d > 0$ arbitrary. Two different methods exist for the study of shock waves.

In the first approach we assume traveling waves and the existence of two Maxwellian states Ma_0, Ma_s associated to N_i, M_i when $|\eta| \longrightarrow \infty$

$$N_i(\eta), M_i(\eta), \eta = x - \xi t; Ma_0 : n_{0i}, m_{0i}; Ma_s : s_{0i} = n_{0i} + n_i, p_{0i} = m_{0i} + m_i \quad (1.1)$$

Assuming both the Rankine-Hugoniot relations (conservation laws and that Ma_0, Ma_s correspond to the vanishing of the collision terms, the solutions can be constructed from three arbitrary parameters $n_{01}, \bar{m}_1 = m_1/n_1, \xi$. We prove the positivity of Ma_0, Ma_s for three classes with different intervals for the shock front velocity ξ : class I, $0 < \xi < 1$, class II, $-1 < \xi < 0$, class III, $-2 < \xi < -1$. The drawback of this method is that the solutions are compatible with positive as well as negative cross-sections.

In the second approach we start with similarity shock waves

$$N_i = n_{0i} + n_i/D, M_i = m_{0i} + m_i/D, D = 1 + e^{\gamma\eta}, \eta = x - \xi t, s_{0i} = n_{0i} + n_i, p_{0i} = m_{0i} + m_i \quad (1.2)$$

and, in addition to the previous relations, find two new relations linking d, γ to the arbitrary parameters. Then we know the solutions with $d > 0$ and the Maxwellian states when $\eta \longrightarrow \pm\infty$. We prove that our solutions of class II violate $d > 0$ and obtain for classes I, III the subclasses for which $d > 0$. We determine the characteristic velocities ξ_0, ξ_s associated to Ma_0, Ma_s. For this we apply the Lax-Whitham weak shock theory, introduced in discrete kinetic theory by Broadwell and Gatignol[7]. Class I corresponds to rarefactive shocks, class II to compressive ones and the supersonic and subsonic inequalities are satisfied. We introduce the temperature $\mathcal{T}(\eta)$, the local entropy $-\mathcal{H}(\eta)$ and a criterion for a temperature overshoot is that $\mathcal{T}(\eta = 0)$ is larger than both the values at the upstream and downstream states. We assume that the local entropy value at the downstream state is larger than the one at the upstream state and introduce a scaling parameter n_{01} with critical value $n_{01}^{(c)}$. Applying the Section 2 results we foresee local entropy overshoots for n_{01} values close to $n_{01}^{(c)}$ which disappear for n_{01} large for class I and n_{01} small for class III. All these theoretical previsions are verified numerically. For class I with shock front speed $|\xi| < 1$, no temperature overshoots are observed, contrary to class III with $|\xi| > 1$.

II. LOCAL ENTROPY OVERSHOOT

We assume similarity shock waves for the discrete Boltzmann equation[1]

$$N_i = n_{0i} + n_i/D, D = 1 + e^{\vec{\gamma}.\vec{\eta}}, \vec{\gamma}.\vec{\eta} = \Sigma\gamma_m\eta_m, \vec{\eta} = \vec{x} - \vec{\xi}t, \eta_m = x_m - \xi_m t \quad (2.1)$$

$$\ell_i N_i = \sum_{j=1}^{p} \sum_{k\ell} A_{kl}^{ij}(N_k N_\ell - N_i N_j), \quad \ell_i = \partial_t + \Sigma a_{im}\partial_{x_m} \quad (2.2)$$

To $|\vec{\gamma}.\vec{\eta}| \longrightarrow \infty$ we associate the Maxwellian states Ma_0, Ma_s defined by the parameters $n_{0i}, s_{0i} = n_{0i} + n_i$ with masses $M_0 = \Sigma n_{0i}$, $M_s = \Sigma s_{0i}$ and local entropies $-H_0, H_0 = \Sigma n_{0i}\log n_{0i}$, $-H_s, H_s = \Sigma s_{0i}\log s_{0i}$.

Lemma 1: Among the parameters (n_{0i}, n_i, γ_i), we can always choose a scaling parameter $z = n_{0i}$ or n_i or γ_i such that the scaled parameters $\bar{n}_{0i} = n_{0i}/z, \bar{n}_i = n_i/z, \bar{s}_{0i} = \bar{n}_{0i} + \bar{n}_i$, $\bar{\gamma}_i = \gamma_i/z$ as well as $\bar{M}_0 = M_0/z$, $\bar{M}_s/z$ are z-independent. We get :

$$\sum A_{k\ell}^{ij}(n_{0k}n_{0\ell} - n_{0i}n_{0j}) = 0$$

$$n_i \sum \gamma_m(a_{im} - \xi_m) = \sum A_{k\ell}^{ij}(n_k n_\ell - n_i n_j) = -\sum A_{k\ell}^{ij}(n_{0k}n_\ell + n_{0\ell}n_k - n_{0i}n_j - n_{0j}n_i) \quad (2.3)$$

These relations are quadratic in $n_{0k}n_{0\ell}$, $n_i\gamma_m, n_k n_\ell$, multiplying (2.3) by z^{-2} we find relations between the scaled parameters, which are z-independent. Consequently the scaled masses $\bar{M}_0, \bar{M}_s$ and $\tilde{H}_0 = \Sigma\bar{n}_{0i}\log\bar{n}_{0i}$, $\tilde{H}_s = \Sigma\bar{s}_{0i}\log\bar{s}_{0i}$, are z-independent.

Lemma 2: For the scaling parameter z there exists a critical value $z^{(c)}$ for which $(H_0 - H_s)(M_0 - M_s) \gtrless 0$ for $z \gtrless z^{(c)}$. We obtain from the definitions of M_0, M_s, H_0, H_s :

$$(\bar{M}_0 - \bar{M}_s)(H_0 - H_s)/z = (\bar{M}_0 - \bar{M}_s)^2\log z + (\bar{M}_0 - \bar{M}_s)(\tilde{H}_0 - \tilde{H}_s)$$

Both sides are positive for $z \gg 1$, negative for $z \ll 1$ and have only one zero for $z = z^{(c)}$ such that $\log z^{(c)} = (\tilde{H}_s - \tilde{H}_0)/(\bar{M}_0 - \bar{M}_s)$ and $H_0 = H_s$.

First we assume $M_0 > M_s$ and compressive shocks (Ma_s, Ma_0 as the upstream and downstream states). Further we require that the local entropy increases $H_s > H_0$ then $z < z^{(c)}$, while $z > z^{(c)}$ violates this condition. Second we assume $M_0 > M_s$, rarefactive shock and an increase of the local entropy $H_0 > H_s$. Then $z > z^{(c)}$ while $z < z^{(c)}$ violates this condition.

These similarity solutions being in fact unidimensional let us assume, for simplicity, solutions in one spatial coordinate $\vec{\gamma}.\vec{\eta} = \gamma\eta = \gamma(x - \xi t) = \bar{\gamma}\eta z$, $\bar{\gamma}$ being z-independent. We introduce the macroscopic η-dependent functions: mass $\mathcal{M} = M_0 + (M_s - M_0)/D$, momentum $\mathcal{J} = J_0 + (J_s - J_0)/D$, velocity $\mathcal{U} = \mathcal{J}/\mathcal{M}$, energy $\mathcal{E} = E_0 + (E_s - E_0)/D$, temperature $3\mathcal{T} = 2\mathcal{E}/\mathcal{M} - \mathcal{U}^2$, pressure $\mathcal{P} = \mathcal{M}\mathcal{T}$ where J_0, E_0, J_s, E_s associated to Ma_0, Ma_s are linear combination of the n_{0i}, n_i. The functions $\mathcal{M}/M_0, \mathcal{P}/P_0$ normalised at one Maxwellian state as well as $\mathcal{U}, \mathcal{T}$ and the normalised $\mathcal{U}/U_0$, $\mathcal{T}/T_0$, plotted as functions of ηn_{01} are z-independent. On the contrary the local entropy $-\mathcal{H} = -\Sigma N_i\log N_i$, $-\mathcal{H}/H_0$, (normalised or nor at one Maxwellian state) plotted as function of ηz is z-dependent. When z crosses $z^{(c)}$two different regimes exist: either the local entropy increases or not. At the critical value, the two limits $-H_0, -H_s$ are the same but $-\mathcal{H}(\eta)$ being not a constant, bumps or dips are present which still exist for z values not too far from $z^{(c)}$ but disappear for $z \gg 1$ or $z \ll 1$.

III. SHOCK WAVES SOLUTIONS FOR THE 14$\vec{v}_i$ CABANNES MODEL WITH TWO SPEEDS $\sqrt{3}, 2$

In one spatial coordinate the five independent N_i, M_i satisfy[5] :

$$\begin{aligned}&\Delta_0 M_2 = D - C_1 + C_2, D_- N_1 = C_1 + C_2, D_+ N_2 = -C_1 - C_2, \Delta_+ M_1 = 4C_1 - 2D,\\ &\Delta_- M_4 = -4C_2 - 2D, D = 2d(M_1 M_4 - M_2^2), d > 0, C_1 = \sqrt{11}(N_2 M_2 - N_1 M_1),\\ &C_2 = \sqrt{11}(N_2 M_4 - N_1 M_2), \Delta_0 = \partial_t, D_\pm = \partial_t \pm \partial_x, \Delta_\pm = \partial_t \pm 2\partial_x\end{aligned} \tag{3.1}$$

$C_i, i = 1,2$ and D are the collision terms for the particles with speeds $2, \sqrt{3}$ and $2, 2$. We deduce three linear relations equivalent to the Rankine-Hugoniot relations.

$$D_- N_1 + D_+ N_2 = \Delta_+ M_1 + \Delta_- M_4 + 4\Delta_0 M_2 = \Delta_+ M_1 - \Delta_- M_4 - 4D_- N_1 = 0 \tag{3.2}$$

Two methods exist : (i) We assume (1.1), $C_1 = D = 0$ for Ma_0, Ma_s

$$m_{01} m_{04} = m_{02}^2, n_{02} m_{02} = n_{01} m_{01}, p_{01} p_{04} = p_{02}^2, s_{02} p_{02} = s_{01} p_{01} \tag{3.3}$$

$C_2 = 0$ or $n_{02} m_{04} = n_{01} m_{02}$, $s_{02} p_{04} = s_{01} p_{02}$ being not new relations and (3.2) gives

$$m_1(2-\xi) = m_4(2+\xi) + 4\xi m_2, n_1(\xi+1) = n_2(1-\xi), m_1(2-\xi) + m_4(2+\xi) + 4n_1(1+\xi) = 0 \tag{3.4}$$

(ii) We assume (1.2) and in addition to (3.3.4) obtain both γ, d

$$\begin{aligned}&-\xi\gamma m_2 = 2d(m_1 m_4 - m_2^2) + \sqrt{11}(n_1 m_1 - n_2 m_2 + n_2 m_4 - n_1 m_2)\\ &-(\xi+1) n_1 \gamma = \sqrt{11}(n_2(m_2 + m_4) - n_1(m_2 + m_1)), \quad d > 0\end{aligned} \tag{3.5}$$

For (3.1.2) there exist invariance relations coming from the transform $\mathcal{T}$.

$$\mathcal{T} : N_1(x,t) \longleftrightarrow N_2(-x,t), M_1(x,t) \longleftrightarrow M_4(-x,t), M_2(x,t) \longleftrightarrow M_2(-x,t) \tag{3.6}$$

and giving: $D(x) \longrightarrow D(-x), C_i \longrightarrow -C_j(-x), i \neq j$. Applying $\mathcal{T}$ to (3.1.2) some equations are exchanged $D_- N_1 = C_1 + C_2 \longleftrightarrow D_+ N_2 = -C_1 - C_2$ while others are invariant $D_- N_1 + D_+ N_2 = 0$. The $-\xi$ solutions are deduced from the ξ ones

$$\begin{aligned}\xi \longleftrightarrow -\xi, \gamma \longleftrightarrow -\gamma, n_1 \longleftrightarrow n_2, m_1 \longleftrightarrow m_4, m_2 \longleftrightarrow m_2, n_{01} \longleftrightarrow n_{02},\\ m_{01} \longleftrightarrow m_{04}, m_{02} \longleftrightarrow m_{02}\end{aligned} \tag{3.6'}$$

III.1 Nonnegative $\mathbf{Ma_0, Ma_s}$, taking into account the R-H- relations

Here for the 11 parameters ξ, n_{0i}, m_{0i} and s_{0i}, p_{0i} (or n_i, m_i) we want to construct $n_{0i} \geq 0$, $s_{0i} \geq 0$, $p_{0i} \geq 0$ satisfying (3.3.4). We define $n_2 = \bar{n}_2 n_1$, $m_i = \bar{m}_i n_1$, $n_{02} = \bar{n}_{02} n_{01}$, $m_{0i} = \bar{m}_{0i} n_{01}$, $\bar{D}, \bar{C}_i, \bar{G}, \bar{F}$ and rewrite (3.3.4):

$$\begin{aligned}&\bar{n}_2 = (1+\xi)/(1-\xi), \bar{m}_2 = (2-\xi)\bar{m}_1/2\xi + (1+\xi)/\xi, \bar{m}_4(2+\xi) = (\xi-2)\bar{m}_1 - 4(1+\xi)\\ &\bar{m}_{01} = \bar{n}_{02}^2 \bar{m}_{04}, \bar{m}_{02} = \bar{n}_{02}\bar{m}_{04}, p_{01}\bar{D} = (\bar{m}_1\sqrt{\bar{m}_{04}} - \bar{m}_2\sqrt{\bar{m}_{01}})^2,\\ &p_{04}\bar{D} = [\bar{m}_4\sqrt{\bar{m}_{01}} - \bar{m}_2\sqrt{\bar{m}_{04}}]^2,\\ &\bar{D} = \bar{m}_2^2 - \bar{m}_1\bar{m}_4, p_{02}\bar{D}/m_{04} = \bar{G} = (\bar{m}_1 - \bar{n}_{02}\bar{m}_2)(\bar{m}_2 - \bar{n}_{02}\bar{m}_4),\end{aligned} \tag{3.4'}$$

$$s_{01} = n_{01}(\bar{n}_{02} - \bar{n}_2)(\bar{m}_{02i} - \bar{m}_{2i})/\bar{C}_i, \bar{C}_i = \bar{n}_2\bar{m}_{2i} - \bar{m}_i, i = 1,2$$
$$n_1\bar{D}/m_{04} = \bar{F} = \bar{m}_1 + \bar{m}_4\bar{n}_{02}^2 - 2\bar{m}_2\bar{n}_{02} \tag{3.3'}$$

To illustrate this method we explain how to obtain the p_{01} relation. We start with $m_2^2 - m_1m_4 = m_{01}m_4 + m_{04}m_1 - 2m_2m_{02}$ found from (3.3), deduce $n_1\bar{D} = m_{01}\bar{m}_4 + \bar{m}_{04}\bar{m}_1 - 2\sqrt{m_{01}m_{04}}\ \bar{m}_2$ and substitute into $p_{01} = m_{01} + m_1 = m_{01} + n_1\bar{m}_1$.

III.1.1 Construction of the solutions

From (3.3') two general classes of solutions are possible.

$$(\bar{n}_{02} - \bar{n}_2)((\bar{m}_{02} - \bar{m}_2)\bar{C}_2 - (\bar{m}_{04} - \bar{m}_4)\bar{C}_1) = 0, \tag{3.7}$$

With this additional relation remains three arbitrary parameters chosen to be

$$n_{01} > 0, \xi, \bar{m}_1 = m_1/n_1 \tag{3.8}$$

(i) From (3.7) one class is defined by $\bar{n}_{02} = \bar{n}_2$. From (3.3'.4') we get $s_{01} = n_{01} + n_1 = 0$, $n_{02} = -n_2 = n_{01}(1+\xi)/(1-\xi) > 0$ if $|\xi| < 1$, $s_{02} = 0, m_i = \bar{m}_i n_{01}$, $\bar{m}_{0i}$, $m_{0i} = n_{01}\bar{m}_{0i}$ and for $p_{01} > 0$, $\bar{m}_{04} > 0$, $p_{02} > 0$ we have the conditions

$$\bar{D} > 0, \quad \bar{F} < 0, \quad \bar{G} > 0, \qquad \text{Classe I} : 0 < \xi < 1. \tag{3.9}$$

(ii) From (3.7) another class is defined by $(\bar{m}_{02} - \bar{m}_2)\bar{C}_2 = (\bar{m}_{04} - \bar{m}_4)\bar{C}_1$. We choose $\bar{m}_{02} = \bar{m}_2$, $\bar{m}_{04} = \bar{m}_4$ and get $s_{01} = p_{02} = p_{04} = 0$, $s_{02} \neq 0$, $p_{01} \neq 0$, $n_2 = \bar{n}_2 n_1$, $m_i = \bar{m}_i n_1$, $m_{02} = \bar{m}_2 n_{01}$, $m_{04} = \bar{m}_4 n_{01}$, $m_{01} = m_{02}^2/m_{04}$.

Lemma 3: For the solutions $\bar{m}_{0i} = \bar{m}_i$, $i = 2,4$ we have $-2 < \xi < 0$ and $m_{0i} > 0$, $i = 2,4$ if $-((2-\xi)\bar{m}_1 + 4(1+\xi))/(2+\xi) > 0$, $((2-\xi)\bar{m}_1/2 + 1 + \xi)/\xi > 0$.

First if $\xi > 2$ we get $4 < \bar{m}_1(\xi-2)/(1+\xi) < 2$. Second if either $\xi < -2$ or $0 < \xi < 2$ we get $-4 > \bar{m}_1(2-\xi)/(1+\xi) > -2$. There remains $-2 < \xi < 0$ for which $m_{0i} > 0$ gives $\bar{m}_1 < \inf(-4(1+\xi)/(2-\xi), -2(1+\xi)/(2-\xi))$ and we define two classes.

$$\begin{aligned} &\text{Class II} : \ -1 < \xi < 0, \bar{m}_1 < -4(1+\xi)/(2-\xi) < 0, \ n_{01} > 0 \\ &\text{Class III} : \ -2 < \xi < -1, \bar{m}_1 < -2(1+\xi)/(2-\xi), \ n_{01} > 0 \end{aligned} \tag{3.10}$$

For nonnegative Maxwellians it remains to find the restrictions on the arbitrary parameters leading to (3.9) for class I and $p_{01} > 0$, $s_{02} > 0$ for classes II, III.

$$\begin{aligned} &4\bar{D}\xi^2(2+\xi)/(4+3\xi^2)(2-\xi) = (\bar{m}_1 - \bar{m}_D^+)(\bar{m}_1 - \bar{m}_D^-), \\ &\bar{m}_D^\pm = 2(1+\xi)(1 \pm 2\xi/\sqrt{4+3\xi^2})/(\xi-2) \\ &-4\bar{G}\xi^2(1-\xi)^2 = (4+3\xi^2+\xi^4)(\bar{m}_1 - \bar{m}_G^+)(\bar{m}_1 - \bar{m}_G^-), \ \bar{m}_G^+(2-\xi+\xi^2) = -2(1+\xi)^2 \\ &\bar{m}_G^-(2-\xi)(2+\xi+\xi^2) = -2(1+\xi)(2+3(\xi+\xi^2)) \\ &\bar{F}\xi(1-\xi)^2(2+\xi)/(4+\xi^2-\xi^4) = \bar{m}_F - \bar{m}_1, \bar{m}_F = -2(1+\xi)^2(2+\xi+\xi^2)/(4+\xi^2-\xi^4) \end{aligned} \tag{3.11}$$

III.1.2 Construction of three nonnegative Maxwellian classes

Theorem 1A: For class I, the Maxwellians are non negative if

$$0 < \xi < 1, n_{01} > 0 \text{ and } \ \bar{m}_F < \bar{m}_1 < \bar{m}_D^+ < 0 \tag{3.12}$$

First $\bar{D} > 0$, $\bar{F} < 0$, $\bar{G} > 0$ if $\bar{m}_1 \in (\bar{m}_D^+, \bar{m}_D^-)$, $\bar{m}_1 \in (\bar{m}_G^-, \bar{m}_G^+)$, $\bar{m}_1 > \bar{m}_F$. Second $\bar{m}_G^+ < \bar{m}_D^-$, $\bar{m}_D^+ < \bar{m}_G^+$ or $(2-\xi)(1+\xi^2) > 0$, $\bar{m}_F < \bar{m}_D^+$ or $4 + 7\xi^2 + 2\xi^2 + 2\xi^4 - \xi^6 > 0$, $\bar{m}_G^- < \bar{m}_F$ or $(1-\xi)(1+\xi^2)(2+\xi) > 0$.

For classes II, III, we get $p_{01} > 0$ if $\bar{m}_1 \notin (\bar{m}_D^-, \bar{m}_D^+)$ and $s_{02}/n_{01} = -\bar{n}_2 + \bar{m}_{02}/\bar{m}_{04}$

$$m_{04}s_{02}(2+\xi)(1-\xi)2\xi/n_{01}^2(2-\xi)(2+\xi+\xi^2) = \bar{m}_1 - \bar{m}_s, \bar{m}_s = \frac{-2(1+\xi)(2+3\xi+3\xi^2)}{(2-\xi)(2+\xi+\xi^2)} \tag{3.13}$$

Theorem 2A: <u>For class II, (3.10) the Maxwellians are non negative.</u>

First we recall (Lemma 3) that $m_{0i} > 0$ if $\bar{m}_1 < -4(1+\xi)/(2-\xi)$, from (3.13) $s_{02} > 0$ if $\bar{m}_1 < \bar{m}_s < 0$ and $p_{01} > 0$ if $\bar{m}_1 \notin (\bar{m}_D^-, \bar{m}_D^+)$ with $\bar{m}_D^- < \bar{m}_D^+ < 0$. Second we prove that $-4(1+\xi)/(2-\xi) < \bar{m}_D^-$ or $-1 < 2\xi/\sqrt{4+3\xi^2}$ or $\xi^2 < 1$. Third we prove that $-4(1+\xi)/(2-\xi) < \bar{m}_s$ or $2 > (2+3\xi+\xi^2)/(2+\xi+\xi^2)$ or $2-\xi-\xi^2 > 0$.

Theorem 3A: <u>For class III, the Maxwellians are non negative</u> if $n_{01} > 0, -2 < \xi < -1$, $\bar{m}_1 < \bar{m}_D^+$ with $\bar{m}_D^+ > 0$ in (3.11). First (Lemma 3) $m_{0i} > 0$ if $\bar{m}_1 < -2(1+\xi)/(2-\xi)$, from (3.13) $s_{02} > 0$ if $\bar{m}_1 < \bar{m}_s$, $\bar{m}_s > 0$ and $p_{01} > 0$ if $m \notin [\bar{m}_D^+, \bar{m}_D^-]$ with $0 < \bar{m}_D^+ < \bar{m}_D^-$, second $\bar{m}_D^+ < -2(1+\xi)/(2-\xi)$ or $1 + 2\xi/\sqrt{4+3\xi^2} < 1$ or $\xi < 0$, and third $-2(1+\xi)/(2-\xi) < \bar{m}_s$ or $\xi(1+\xi) > 0$.

From these classes of solution, applying (3.6') with $n_{01} \longrightarrow n_{02}$, $\bar{m}_1 = m_1/n_1 \longrightarrow m_4/n_2$, $\xi \longrightarrow -\xi$ we get other classes. In this method, the nonlinear part of the equations has not been fully taken into account and we must check whether $d > 0$.

III.2 Nonnegative similarity shock waves solutions

For the similarity (1.2) solutions, in addition to (3.3'.4') we have (3.5) or

$$-\gamma/n_1 = (\bar{C}_1 + \bar{C}_2)\sqrt{11}/(\xi+1) = [-2d\bar{D} + \sqrt{11}(\bar{C}_1 - \bar{C}_2)]/\xi\bar{m}_2 \tag{3.5'}$$

with $\bar{D}, \bar{C}_1, \bar{C}_2$ written down in (3.3'). One relation gives γ and the other d :

$$\begin{aligned} &d = -\sqrt{11}\bar{E}/2\bar{D}, \bar{E} = \xi\bar{m}_2(\bar{C}_1 + \bar{C}_2)/(\xi+1) + \bar{C}_1 - \bar{C}_2 \\ &2\bar{E}(1-\xi^2)(2+\xi)/\xi^2(2-\xi) = (\bar{m}_1 - \bar{m}_E^+)(\bar{m}_1 - \bar{m}_E^-) \\ &\bar{m}_E^{\pm} = (-4 - 2\xi^3 + \xi^2 \pm \sqrt{\Delta})(1+\xi)/\xi^3(2-\xi), \Delta = 16 - 15\xi^4 + 8\xi^6 - 8\xi^2 \end{aligned} \tag{3.14}$$

Recalling that $\bar{D} > 0$ for $p_{01} > 0$ we see the new constraint $\bar{E} < 0$ for $d > 0$.

<u>III.2.1 Positivity of the density and of the cross-section d</u>

Theorem 1B: For the class I, with $0 < \xi < 1$, $n_{01} > 0$, $s_{01} = s_{02} = 0$ the densities N_i, M_i and d are positive if $\bar{m}_F < \bar{m}_1 < \bar{m}_E^+ < 0$ with $\bar{m}_F, \bar{m}_E^+$ written down in (3.12.14). First $\bar{E} < 0$ if $\bar{m}_1 \in [\bar{m}_E^-, \bar{m}_E^+]$ with $\bar{m}_E^- < \bar{m}_E^+ < 0$ and $\bar{m}_1 < 0$, second from Theorem 1A we must have $[\bar{m}_E^-, \bar{m}_E^+] \cap [\bar{m}_F, \bar{m}_D^+] \neq 0$ or $\bar{m}_F < \bar{m}_E^+ < 0$, or $|\bar{m}_F| > |\bar{m}_E^+|$ or $(4+\xi^2-\xi^4)\sqrt{\Delta} - 16 + 13\xi^4 - \xi^6 > 0$ or $\xi^8(1-\xi^4)(4-\xi^2) > 0$.

Theorem 2B: For the class II with $-1 < \xi < 0$, $s_{01} = p_{02} = p_{04} = 0$, we find $d < 0$ and the similarity solutions violate positivity for the cross-section. First $d > 0$ or $\bar{E} < 0$ requires $\bar{m}_1 \in [\bar{m}_E^+, \bar{m}_E^-]$ with $\bar{m}_E^+ < 0$, $\bar{m}_E^- > 0$. Second from Theorem 2A $\bar{m}_1 < -4(1+\xi)/(2-\xi) < 0 < \bar{m}_E^+$ or $\sqrt{\Delta} < 4 - \xi^2 - 2\xi^3$ or $\xi^3(1-\xi)(4-\xi^2) < 0$. Consequently $\bar{m}_1 > \bar{m}_E^+$ is impossible as well as $d > 0$.

We study the last class III with $-2 < \xi < 1$ and we recall that $\bar{m}_1 < \bar{m}_D^+$ from Theorem 3A. $\bar{E} < 0$ means no restriction if $\bar{m}_E^\pm$ are not real or if $\Delta < 0$.

Lemma 4: $\Delta > 0$ either if $-\sqrt{2} > \xi > -2$ or if $-1 < \xi < -2/(15)^{1/4} = 1.016$. In (3.14) $\Delta = \xi^4 + 8(\xi^4 - 1)(\xi^2 - 2) > 0$ if $|\xi|^2 > 2$ or if $\xi^4 < 16/15$. Numerically we get

$$\Delta < 0 \text{ if } -\xi \in]1.04, 1.34[, \Delta > 0 \text{ if } -\xi \notin]1.04, 1.34[\quad (3.15)$$

If $\Delta < 0$ then $\bar{m}_E^\pm$ are not real and $d > 0$ for all positive Maxwellians (Theorem 3A).

Lemma 5: If $-\xi \in]1.34, 2[$ then $0 < \bar{m}_D^+ < \bar{m}_E^- < \bar{m}_E^+$ and $d > 0$ if $\bar{m}_1 < \bar{m}_D^+$. First $4 + 2\xi^3 - \xi^2 < 0$ and $0 < \bar{m}_E^- < \bar{m}_E^+$. Second $\bar{m}_D^+ < \bar{m}_E^-$ is equivalent to $\sqrt{\Delta} < \xi^2 - 4 + 4\xi^4/\sqrt{4 + 3\xi^2}$ or $\xi^2 + 2 > \sqrt{4 + 3\xi^2}$. Third $d > 0$ if $\bar{m}_1 \notin (\bar{m}_E^-, \bar{m}_E^+)$.

Lemma 6: If $-\xi \in]1, 1.04[$ then $\bar{m}_E^- < \bar{m}_E^+ < 0$. We notice that $4 + 2\xi^3 - \xi^2 > 0$. Due to $\bar{m}_1 < \bar{m}_D^+$, $\bar{m}_D^+ > 0$ for $N_i > 0$, $M_i > 0$ and $\bar{m}_1 \notin [\bar{m}_E^-, \bar{m}_E^+]$ for $d > 0$ the two possibilities are either $\bar{m}_1 < \bar{m}_E^- < 0$ or $\bar{m}_E^+ < \bar{m}_1 < \bar{m}_D^+$.

Theorem 3B: For the class III with $-2 < \xi < -1$, $s_{01} = p_{02} = p_{04} = 0$, the densities N_i, M_i and the cross-section d are positive if: (i) $-2 < \xi < -1.04$ and $\bar{m}_1 < \bar{m}_D^+$, (ii) $-1.04 < \xi < -1$ and either $\bar{m}_1 < \bar{m}_E^-$ or $m_E^+ < \bar{m}_1 < \bar{m}_D^+$.

Only $\bar{m}_1$ and ξ enter into the discussion of positivity but not $n_{01} > 0$ which is a scaling variable. For ξ fixed and $\bar{m}_1$ varying in Fig. 1a, $0 < \xi < 1$ class I and in Fig. 2a, $-2 < -\xi < -1$ class III, we plot the $d > 0$ domains. For class I this domain grows with ξ increasing but $d < 1$ while for class III, solutions $d = 1$, $d > 1$ or $d < 1$ exist.

III.2.2 Further properties for the similarity shock waves solutions

We begin with the class I solutions satisfying Theorem 1B.

Lemma 7: For class I a sufficient condition for $p_{01} > p_{04}$ or $p_{02} > p_{04}$ is $\bar{m}_1 < \bar{m}_{ss} < 0$ with $\bar{m}_{ss} = -2(1+\xi)(5\xi+2)/(2-\xi)(2+3\xi)$. In fact $p_{02}/p_{04} > 1$ is equivalent to $\bar{m}_4 - \bar{m}_2 > \bar{m}_{04} - \bar{m}_{02} = \bar{m}_{02}(-1 + n_{01}/n_{02}) = -2\xi\bar{m}_{02}/(1+\xi)$ which is negative. Consequently it is sufficient that $\bar{m}_4 - \bar{m}_2 > 0$ or equivalently $\bar{m}_1 < \bar{m}_{ss} < 0$.

Lemma 8: $\bar{m}_E^+ < \bar{m}_{ss} < 0$ or $|\bar{m}_E^+| > |\bar{m}_{ss}|$. This result is equivalent to $\sqrt{\Delta}(2+3\xi) < -2\xi^3(2+5\xi) + (2+3\xi)(4+2\xi^3-\xi^2)$ or $12 + 16\xi - 9\xi^2 - 7\xi^3 > 0$ which is true.

Lemma 9: For class I: $\sqrt{p_{01}/p_{04}} = p_{02}/p_{04} < (1+\xi)/(1-\xi)$ if $\bar{m}_1 > \bar{m}_s$ with $\bar{m}_s = -2(1+\xi)(2+3\xi+3\xi^2)/(2-\xi)(2+\xi+\xi^2) < 0$. This is equivalent to $-\bar{m}_2 + \bar{m}_{02} < \bar{n}_{02}(-\bar{m}_4 + \bar{m}_{04})$ or to $\bar{m}_2 - \bar{m}_4\bar{n}_{02} > 0$ or to $\bar{m}_1 > \bar{m}_s$.

Lemma 10: $\bar{m}_s < \bar{m}_F$ or $|\bar{m}_s| > |\bar{m}_F|$. This result is equivalent to $(2+\xi-\xi^2)(2+\xi+\xi^2)^2 < (4+\xi^2-\xi^4)(2+3\xi+3\xi^2)$ or to $2+\xi+\xi^2+\xi^3 > 0$.

Theorem 4: For the class I satisfying Theorem 1B we have $1 < \sqrt{p_{01}/p_{04}} < (1+\xi)/(1-\xi)$. We recall (Theorem 1B) that $\bar{m}_F < \bar{m}_1 < \bar{m}_E^+ < 0$ and apply Lemmas 8-10.

We go on with class III:. $-2 < \xi < -1$, $\bar{m}_{0i} = \bar{m}_i$, $i = 2, 4$ and apply (3.4').

Lemma 11: For class III we find: $m_{02}/m_{04} < (2+\xi)/(-2\xi) < \sqrt{(2+\xi)/(2-\xi)} < 1$, $\bar{n}_{02} < 1$, $m_{01}/m_{04} < (2+\xi)/(2-\xi)$. These results come from $2\xi\bar{m}_{02} + (2+\xi)\bar{m}_{04} = -2(1+\xi) > 0$, $4 < 5\xi^2$, $\xi < 0$, $\bar{n}_{02} = m_{02}/m_{04}$, $m_{01}/m_{04} = (m_{02}/m_{04})^2$.

Lemma 12: For class III and $\bar{m}_1 < 0$ we find $\bar{m}_{02}/\bar{m}_{04} > -(2+\xi)/4\xi$, $4\xi^2\bar{m}_{01} > -\bar{m}_1(4-\xi^2)/4 - (2+\xi)(1+\xi)$. For the results we use the identity $4\xi\bar{m}_{02} + (2+\xi)\bar{m}_{04} = (2-\xi)\bar{m}_1 < 0$, that we report into $\bar{m}_{01} = \bar{m}_1^2/m_{04}$ for the second one $-4\xi\bar{m}_{01} =$

$\bar{m}_{02}(2+\xi)-(2-\xi)\bar{m}_1\bar{m}_{02}/\bar{m}_{04} > \bar{m}_{02}(2+\xi)+(4-\xi^2)\bar{m}_1/4\xi$. Finally we find the second result with (3.4') for $\bar{m}_{02}=\bar{m}_2$.

III.3 Weak shocks

We apply the Lax-Whitham weak shock theory[7]. Let ξ_0, ξ_s be the characteristic velocities, for weak shocks, associated to Ma_0, Ma_s. Two methods exist: (i) we determine from the R-H relations and the Maxwellian states relations, the $\xi=\xi_0$ such that $s_{0i}=n_{0i}$, $p_{0i}=m_{0i}$ or equivalently such that $n_i=m_i=0$, (ii) we linearize the equations (3.1.2) around the Maxwellian states.

III.3.1 Characteristic velocities ξ_0, ξ_s

In the first method we put in (3.3') $n_1=0, n_2=\bar{n}_2 n_1=m_i=\bar{m}_i n_1=0$, obtain $\bar{C}_i-(\bar{n}_{02}-\bar{n}_{02})(\bar{m}_{0i}-\bar{m}_i)=0$, $i=1,2$ and

$$\begin{aligned}(1-\xi)\bar{m}_1 &= (\xi(1-\xi)m_{01}-\xi(1+\xi)m_{02}-(1-\xi^2)n_{02})/(-\xi n_{01}+(2-\xi)n_{02}/2)\\ &= ((\xi(2+\xi)((1+\xi)m_{04}-(1-\xi)m_{02})-4n_{02}\xi(1-\xi^2)-n_{01}(1-\xi^2)(2+\xi))\\ &\quad /(2-\xi)(\xi n_{02}+(2+\xi)n_{01}/2)\end{aligned}$$

We get cubic polynomials $\hat{n}(\xi_0)=0$, $\hat{s}(\xi_0)=0$ associated to Ma_0, Ma_s

$$\begin{aligned}\hat{n} &= (-1+\xi^2)[(2-\xi)n_{02}^2-n_{01}^2(2+\xi)-4n_{02}n_{01}\xi]+m_{04}(2+\xi)\\ &\quad [-n_{01}\xi(1+\xi)+2n_{02}(1+\xi-\xi^2)]-m_{01}(2-\xi)[n_{02}\xi(1-\xi)+2n_{01}(1-\xi-\xi^2)]\end{aligned} \tag{3.16}$$

$$\begin{aligned}\hat{s} &= (-1+\xi^2)[(2-\xi)s_{02}^2-s_{01}^2(2+\xi)-4s_{02}s_{01}\xi]+p_{04}(2+\xi)\\ &\quad [-s_{01}\xi(1+\xi)+2s_{02}(1+\xi-\xi^2)]-p_{01}(2-\xi)[s_{02}\xi(1-\xi)+2s_{01}(1-\xi-\xi^2)]\end{aligned} \tag{3.17}$$

In the second method we linearize (3.1.2) around Ma_0 with $N_i=n_{0i}(1+X_i(\eta_0))$, $M_i=m_{0i}(1+Y_i(\eta_0))$, $\eta_0=x-\xi_0 t$, define $d_-=D_-n_{01}$, $d_+=D_+n_{02}$, $\delta_+=\Delta_+m_{01}$, $\delta_-=\Delta_-m_{04}$, $\delta_0=\Delta_0 m_{02}$, $\bar{e}_i=n_{01}m_{0i}\sqrt{11}$, $i=1,2$, $\bar{d}=2dm_{01}m_{04}$. We find $\Delta\hat{Z}(\eta_0)=0$, $\hat{Z}$ a column vector with components X_1, X_2, Y_1, Y_4, Y_2. We get for $\hat{\Delta}=\det\Delta$

$$\hat{\Delta}=\begin{vmatrix} d_- & d_+ & 0 & 0 & 0\\ 0 & 0 & \delta_+ & \delta_- & 4\delta_0\\ -4d_- & 0 & \delta_+ & -\delta_- & 0\\ d_-+\bar{e}_1+\bar{e}_2 & -\bar{e}_1-\bar{e}_2 & \bar{e}_1 & -\bar{e}_2 & -\bar{e}_1+\bar{e}_2\\ 4\bar{e}_1 & -4\bar{e}_1 & \delta_++4\bar{e}_1+2\bar{d} & 2\bar{d} & -4\bar{e}_1-4\bar{d}\end{vmatrix}$$

an operator which is the sum of a fifth $(-4d_+d_-\delta_+\delta_-\delta_0)$, fourth and third order differential terms. The lowest third order differential operator is proportional to $\tilde{\Delta}_3$

$$\tilde{\Delta}_3=d_-d_+(\delta_-+\delta_++4\delta_0)+(d_-+d_+)(\delta_+\delta_-+\delta_0(\delta_-+\delta_+)) \tag{3.18}$$

It is easily verified that $\hat{\Delta}_3$ applied to an $\eta_0=x-\xi_0 t$ dependent function gives $m_{02}\hat{n}(\xi_0)=0$ where $\hat{n}$ is the previous cubic polynomial (3.16). Similarly if we linearize around Ma_s, we go back to $\hat{s}(\xi_s)$ written down in (3.17).

We notice that under the transform (3.6'): $\xi \longrightarrow -\xi, \ldots$ then $\hat{n}(\xi) \longrightarrow -\hat{n}(\xi)$. We determine $\hat{n}$ for $\xi_0 = 2, 1, 0$ and apply this transform: $\hat{n}(2) \equiv -12n_{01}(n_{01} + 2n_{02}) - 8m_{04}(3n_{01} + n_{02}) < 0$, $\hat{n}(-2) > 0$, $\hat{n}(1)/2 = 3m_{04}(n_{02} - n_{01}) + m_{01}n_{01} > 0$ if $n_{02}/n_{01} > 1$, $\hat{n}(-1) < 0$ if $n_{01}/n_{02} > 1$, $\hat{n}(0) = -2(n_{02} - n_{01})(n_{02} + n_{01} + 2m_{02}) \gtrless 0$ if $n_{02}/n_{01} \lessgtr 1$. The three ξ_0 roots are real and belong to the following intervals

$$n_{02}/n_{01} > 1 \longrightarrow (-2,0), (0,1), (1,2); n_{02}/n_{01} < 1 \longrightarrow (-2,-1), (-1,0), (0,2) \quad (3.19)$$

and s_{02}/s_{01}, instead of n_{02}/n_{01} for $\hat{s}(\xi_s)$. We note that $|\xi|$, $|\xi_0|$, $|\xi_s|$ are < 2.

III.3.2 $\xi_s < \xi < \xi_0$ inequality for the class I satisfying Theorem 1B

Recalling $0 < \xi < 1$, $n_{02}/n_{01} = (1+\xi)/(1-\xi) > 1$ then $\xi_0 \in]0,1[$ from (3.19). Further $m_{04}(1+\xi)^2 = m_{01}(1-\xi)^2$ and substituting into (3.16) we get $\hat{n}(\xi_0 = \xi) = -2\xi(1+\xi^2) < 0$. It follows that in (0,1): $\hat{n}(0) < 0$, $\hat{n}(\xi) < 0$, $\hat{n}(1) > 0$, $\xi < \xi_0$. For class I $s_{01} = s_{02} = 0$ so that $\hat{s} \equiv 0$ in (3.17). We use a limiting procedure $s_{01} = \varepsilon$, $\varepsilon \longrightarrow 0$, define $q = \sqrt{p_{01}/p_{04}}$ recall $s_{02} = s_{01}\sqrt{p_{01}/p_{04}} = \varepsilon q$, substitute into (3.17) and get

$$\hat{\hat{s}} = \hat{s}/\varepsilon p_{04} = (2+\xi)[-\xi(1+\xi) + 2q(1+\xi-\xi^2)] - \xi^2(2-\xi)[q\xi(1-\xi) + 2(1-\xi-\xi^2)]$$

$$\hat{\hat{s}} = [q - 1 - \xi(q+1)][q(\xi^2-4) + \xi(\xi+2) + q^2\xi(\xi-2)] \quad (3.17')$$

The quadratic polynomial has two real roots and $\hat{\hat{s}}$ three real ξ_s roots, which $q > 1$ and (3.17') belong to (-2,0), (0,1), (1,2) and consequently here $\xi_s \in [0,1]$. Recalling that $q < (1+\xi)/(1-\xi)$ we get finally

$$\text{Class I}: q = \sqrt{p_{01}/p_{04}},\ \xi_s = (q-1)/(q+1) < \xi < \xi_0 \quad (3.20)$$

III.3.3 $\xi_0 < \xi < \xi_s$ inequality for the class III satisfying Theorem 3B

We recall : $-2 < \xi < -1$, $s_{01} = p_{02} = p_{04} = 0$, $s_{02} > 0$, $p_{01} > 0$ and (3.17) gives

$$\hat{s} = s_{02}(2-\xi)(\xi-1)(s_{02}+p_{01})((\xi+1)s_{02} + \xi p_{01}), \xi < -1 < \xi_s = -s_{02}/(s_{02}+p_{01}) < 0 \quad (3.17'')$$

We go on with ξ_0. From Lemma 11 and (3.19) we find $\bar{n}_{02} < 1$ and $\xi_0 \in]-2,-1[$. For the inequality $\xi_0 < \xi$ we must prove $\hat{n}(\xi_0 = \xi) < 0$. This result is numerically verified for all $\xi, \bar{m}_1$ satisfying Theorem 3B but the analytical proof is tedious. We rewrite $\hat{n}$ coming from $\hat{\Delta}_3$ in (3.18) and define, $X = (1-\xi)\bar{m}_{01}((2-\xi)\bar{m}_1 + 3(1+\xi))$, $Y = (1+\xi)[(2+\xi)\bar{m}_{04}\bar{m}_{02} + (1-\xi)\bar{m}_1]$, $Z = (\bar{n}_{02}(1-\xi) - 1 - \xi)(\bar{m}_{04}(2+\xi) + \bar{m}_{01}(\xi-2))\xi\bar{m}_{02} < 0$

$$\hat{n}m_{02}/n_{01}^3 = \bar{n}_{02}(2-\xi)(X+Y) + Z \quad (3.21)$$

with $Z < 0$ due to Lemma 11. Then $\hat{n} < 0$ if either $X < 0$, $Y < 0$ or $X + Y < 0$.

Lemma 12: For class III $X < 0$ and further $\hat{n} < 0$ if $\bar{m}_1 > 0$. We get $X < 0$ if $\bar{m}_1(2-\xi) < -3(1+\xi)$ and recall (Theorem 3B) that $\bar{m}_1 < \bar{m}_D^+$. We get $\bar{m}_D^+(2-\xi) = -2(1+\xi)(1+2\xi/\sqrt{4+3\xi^2}) < -3(1+\xi)$ and $X < 0$. Assuming $\bar{m}_1 > 0$ we get $Y < 0$, $\hat{n} < 0$. There remains the case $\bar{m}_1 < 0$.

Lemma 13: $Y < 0$ and $\hat{n} < 0$ if $-2 < \xi < (1 - \sqrt{4\sqrt{2}+5})/2 = -1.13$. We find

$$(-\xi)Y = (1+\xi)[a\bar{m}_1^2 + b\bar{m}_1 + c], a = (2-\xi)^2/2, b = 6 + 2\xi - 2\xi^2, c = 4(1+\xi^2) \quad (3.22)$$

$Y < 0$ if the two roots are not real or $b^2 < 4ac$ or $2\xi < 1 - \sqrt{4\sqrt{2}+5}$. With $\bar{m}_1 < 0$ we can include a part of $X < 0$ into Y. We get $X < 3(1-\xi^2)\bar{m}_{01} = X_1$ and use Lemma 12: $X + Y < X_1 + Y < (1+\xi)[a\bar{m}_1^2 + b'\bar{m}_1 + c']/(-\xi), b' = b + 3(1-\xi)(4-\xi^2)/16\xi < b$, $c' = c + 3(1-\xi^2)(2+\xi)/4\xi > c$, $b'^2 < 4ac'$ for $-2 < \xi < -1.02$.

$$\text{Class III}: -2 < \xi_0 < \xi < -1 < \xi_s < 0 \text{ for } -2 < \xi < -1.02 \tag{3.23}$$

III.4 Supersonic and subsonic inequalities for the similarity shock waves

We introduce the η-dependent macroscopic quantities: mass $\mathcal{M} = 4(N_1 + N_2 + M_2) + M_1 + M_4 = M_0 + (M_s - M_0)/D$, $D = 1 + e^{\gamma\eta}$, momentum $\mathcal{J} = 2(M_1 - M_4) + 4(N_2 - N_1) = J_0 + (J_s - J_0)/D$, velocity $\mathcal{U} = \mathcal{J}/\mathcal{M}$. To the Maxwellian states Ma_0, Ma_s, in addition to M_0, M_s, J_0, J_s, we define the shock velocities $V_0 = J_0/M_0 - \xi$, $V_s = J_s/M_s - \xi$, and the sound-waves velocities $W_0 = V_0 + \xi - \xi_0$, $W_s = V_s + \xi - \xi_s$.

III.4.1 Classe I: $s_{01} = s_{02} = 0$, $0 < \xi_s < \xi < \xi_0 < 1$, $n_{02}/n_{01} = (1+\xi)/(1-\xi)$ (Theorem 1B). We define $\bar{M}_0 = M_0/n_{01}$, $\bar{M}_s = M_s/n_{01}$, $\bar{J}_0 = J_0/n_{01}$, $\bar{\gamma} = \gamma/n_{01}$, recall $\bar{m}_{0i} = m_{0i}/n_{01}$, $\bar{m}_i = m_i/n_1 = -m_i/n_{01}$ and find:

$$\begin{aligned}
\bar{M}_0 &= \bar{m}_{01} + \bar{m}_{04} + 4(1 + \bar{n}_{02} + \bar{m}_{02}) = 8/(1-\xi) + 2\bar{m}_{01}(3-\xi^2)/(1+\xi)^2 > 0\\
\bar{J}_0 &= 2(\bar{m}_{01} - \bar{m}_{04}) + 4(-1 + \bar{n}_{02}) = 8\xi(1/(1-\xi) + \bar{m}_{01}/(1+\xi)^2) > 0\\
\bar{\gamma}(\xi+1) &= \sqrt{11}(\bar{C}_1 + \bar{C}_2), (\bar{C}_1 + \bar{C}_2)(1-\xi)(2+\xi) = \bar{m}_1\xi^2 - 2\xi(1+\xi) \qquad (3.24)\\
\bar{M}_0 - \bar{M}_s &= 4(1 + \bar{n}_{02}) + \bar{m}_1 + \bar{m}_4 + 4\bar{m}_2 = (8/\xi(2+\xi))(\bar{m}_1 + (2\xi+1)/(1-\xi))
\end{aligned}$$

Theorem 5: For class I the following inequalities hold: $V_0 > 0$, $V_s > 0$, $\gamma < 0$, $W_0 < V_0$, $W_s > V_s > 0$, $M_0 > M_s$. We find first $M_0 V_0 = 2m_{01}\xi(1+\xi^2)/(1+\xi)^2 > 0$, second $V_s = M_0V_0/M_s > 0$, $\bar{C}_1 + \bar{C}_2 < 0$, $W_0 - V_0 = \xi - \xi_0 < 0$, $W_s - V_s = \xi - \xi_s > 0$ and for the last result $(2\xi+1)/(1-\xi) > -\bar{m}_F$ or $2\xi + \xi^2 + 6\xi^3 + 3\xi^4 > 0$.

Ma_0, Ma_s are associated to the limits $\eta \longrightarrow -\infty, +\infty$, further, Ma_0 is the upstream state, Ma_s the downstream one and due to $M_0 > M_s$ the shock is rarefactive.

III.4.2 Class III: $s_{01} = p_{02} = p_{04} = 0$, $-2 < \xi_0 < \xi < -1 < \xi_s < 0$, $\bar{n}_2 = (1+\xi)/(1-\xi)$, $\bar{m}_2 = \bar{m}_{02}$, $\bar{m}_4 = \bar{m}_{04}$ (Theorem 3B). We get $M_s = 4s_{02} + p_{01}$, $J_s = 2p_{01} + 4s_{02} > 0$

$$\bar{M}_0 - \bar{M}_s = 4(1 + \bar{n}_2) + 4\bar{m}_2 + \bar{m}_4 + \bar{m}_1 = (8/\xi(2+\xi))(\bar{m}_1 + (2\xi+1)/(1-\xi)) \tag{3.25}$$

Theorem 6: The solutions of class III satisfy $V_s > 0$, $W_s > 0$, $V_0 > 0$, $W_s < V_s$, $W_0 > V_0 > 0$, $\gamma > 0$, $M_0 > M_s$. We get $J_s > 0$, $-\xi M_s > 0$, $-\xi_s M_s > 0$, $V_0 = V_s M_s/M_0$, $M_s(W_s - V_s) = \xi - \xi_s < 0$, $M_0(W_0 - V_0) = \xi - \xi_0 > 0$. γ has the same expression (3.24) in both classes I, III and $\gamma > 0$ if $\bar{C}_1 + \bar{C}_2 < 0$ or $\bar{m}_1 < 2(1+\xi)/\xi$ or if $\bar{m}_D^+ < 2(1+\xi)/\xi$ or $-\xi^2/\sqrt{4+3\xi^2} < 1$. For $M_0 - M_s$ we note the same $\xi, \bar{m}_1$ expression for classes I, III. $M_0 > M_s$ if $\bar{m}_1 < -(2\xi+1)/(1-\xi)$ or with $\gamma > 0$, $2(1+\xi)/\xi < -(2\xi+1)/(1-\xi)$ or $\xi > -2$. Ma_0, Ma_s correspond $\eta \longrightarrow \infty, -\infty$, Ma_0 is the downstream state, Ma_s the upstream one and the shock is compressive.

III.5 Temperature and local entropy for the similarity shock waves

We introduce the remaining macroscopic quantities: energy $\mathcal{E}(\eta) = 2(M_1 + M_4 + 4M_2) + 6(N_1 + N_2)$, temperature $3\mathcal{T}(\eta) = 2\mathcal{E}/\mathcal{M} - \mathcal{U}^2$, pressure $\mathcal{P}(\eta) = \mathcal{T}(\eta)\mathcal{M}(\eta)$

and local entropy $-\mathcal{H}(\eta), \mathcal{H}(\eta) = \Sigma N_i \log N_i + M_i \log M_i$. For Ma_0, Ma_s, we associate: $E_0, E_s, T_0, T_s, P_0, P_s, -H_0, -H_s$. For the rarefactive and compressive shocks, respectively of classes I, III, with $M_0 > M_s$, we retain in both cases the solutions for which $T_0 > T_s$ so that $P_0 > P_s$. Consequently from the upstream state to the downstream one, the pressure decreases for the rarefactive shocks of class I and increases for the compressive ones of class III. We assume that the local entropy increases between the two limits at the upstream state and at the downstream one or $-H_0 < -H_s$ for class I and $-H_s < -H_0$ for class III.

Although the mass $\mathcal{M}(\eta) = M_0 + (M_s - M_0)/(1 + e^{\gamma\eta})$ is necessarily either decreasing, $\partial_\eta \mathcal{M} < 0$, for rarefactive shocks of class I or increasing, $\partial_\eta \mathcal{M} > 0$, for compressive shocks of class III this is not true for $\mathcal{T}(\eta)$ and $-\mathcal{H}(\eta)$. It can happen that $\partial_\eta \mathcal{T}$ or $\partial_\eta \mathcal{H}$ have a change of sign across the shock. A criterion for the presence of a bump or a dip across the shock is that $\mathcal{T}(\eta = 0)$ or $\mathcal{H}(\eta = 0)$ is either larger or smaller than both limits $|\eta| \longrightarrow \infty$. We find $3\mathcal{T}(0) = 2(E_0 + E_s)/(M_0 + m_s) - (J_0 + J_s)^2/(M_0 + M_s)^2$ and if $\mathcal{T}(0) > \sup(T_0, T_s)$, then an overshoot across the shock will occur.

For the local entropy overshoots we apply the Section 2 results. For classes I and III the scaling parameter is n_{01} while the scaled quantities $\bar{M}_0, \bar{M}_s, \bar{\gamma}$ (3.25) both n_{01}-independent, are $\xi, \bar{m}_1$-dependent. It is sufficient to recall that $\bar{m}_{04}$ defined in (3.9.11) for class I and by $\bar{m}_{04} = \bar{m}_4(3.3')$ for class III is only $\xi, \bar{m}_1$-dependent. For both classes with $\xi, \bar{m}_1$ (or d) fixed and n_{01} varying the corresponding family of solutions has a critical value $n_{01}^{(c)}$ for which $H_0 = H_s$ and overshoots, for n_{01} close to $n_{01}^{(c)}$, which disappear for n_{01} small or large.

III.6 Numerical calculations

The similarity solutions depend on three arbitrary parameters $n_{01}, \xi, \bar{m}_1$. However, due to (3.14): $\bar{E} + 2\bar{D}d/\sqrt{11} = 0$, with $\bar{E}, \bar{D}$ quadradic $\bar{m}_1$ polynomials it is equivalent to choose $\bar{m}_1$ or d (and verify that $\bar{m}_1$ satisfies the positivity conditions of Theorem 1B, 3B) as arbitrary. Due to the scaling parameter n_{01}, both $\mathcal{M}/M_0$, $\mathcal{P}/P_0, \mathcal{T}$ plotted as functions of ηn_{01} are n_{01}-independent but not $\mathcal{H}/H_0$.

A useful test for shock waves is the ratio of the thickness $w = 4/\mid\gamma\mid$ by an effective mean free path $\lambda_{\text{ef}} = \Sigma \mid \vec{v}_i \mid /\Sigma\nu_i$. For the collision frequency $\nu_i = \sum_j \sum A_{k\ell}^{ij} N_j$ associated to N_i we sum up only the N_j for which collisions $N_i N_j$ occur in R^3.

In Fig.1 we choose the class I of rarefactive similarity shock waves. In Fig.1a we plot the ξ, d positivity domain and note that $d = 1$ is excluded. In Fig.1-b-c we represent the curves for $\xi = 0.96$, $\bar{m}_1 = -7.27$ or $d = 0.5$ fixed and n_{01} varying. We note that both $\mathcal{M}, \mathcal{P}, \mathcal{T}$ decrease continuously across the shock and that there is no temperature overshoot. We have verified for all $\xi \in]0, 1[$ or $|\xi| < 1$, that no temperature overshoot occurs. In Fig.1b we present also the local entropy $\mathcal{H}/H_0$ for $n_{01} > n_{01}^{(c)} = 2.22\ 10^{-2}$ with an increase between the two limits satisfied. We see a large dip for $n_{01} = 2.27\ 10^{-2}$ close to $n_{01}^{(c)}$ which has disappeared for a large $n_{01} = 10$ value. We determine the mean free path at the Maxwellian Ma_s for which $M_s = \Sigma p_{0i}$. We find $w/\lambda_{\text{ef}} = 4(d + \sqrt{11})M_s/|\gamma|(3 + 2\sqrt{3})$ with a value 2.89 for the present example.

In Fig.2-3 we choose the class III, $-2 < \xi < -1$ of compressive similarity shock waves. In Fig.2-a we plot the ξ, d positivity domain and note that $d = 1$ and $d \gtrless 1$ are allowed. In Fig.2-b-c we choose a cross-section $d = 1$ or $\bar{m}_1 = -0.071$ and $\xi =$

-1.23 fixed while n_{01} is varying. The local entropy increase is satisfied in Fig.2-b for $n_{01} < n_{01}^{(c)} = 1.11$ and not in Fig. 2-c for $n_{01} > n_{01}^{(c)}$. We observe overshoots both for the temperature and the local entropy. The local entropy bump becomes more and more important when n_{01} reaches $n_{01}^{(c)}$ and disappears when $n_{01} = 0.1$ is small. In Fig.2-c we observe for $\mathcal{H}(\eta)/H_0$ three different regimes when $n_{01} > n_{01}^{(c)}$ increases: first an increase for the bump, second a dip and finally for $n = 100$ a monotonic curve. In Fig.3-a-b-c we present the curves for different $\xi = -1.06, -1.54, -1.57$, $d = 0.5, 3, 2$ or $\bar{m}_1 = -98\ 10^{-4}, 0.0109, -54\ 10^{-5}$ values but always $n_{01} < n_{01}^{(c)} = 1.26, 0.573, 0.51$ with a local entropy increase satisfied. We observe bumps for $\mathcal{T}$ which increase slowly with $|\ \xi\ |$. For $\mathcal{H}$ the largest bumps are still for n_{01} close to $n_{01}^{(c)}$ and disappear $n_{01} \ll 1$.

REFERENCES

(1) Gatignol R., Lectures Notes in Physics Vol. 36, (Springer, Berlin 1975); Frisch et al. Complex Systems 1, 649 (1987); Platkowski T. and Illner R., SIAM Review 30 (1988) 213; d'Humières, Proceedings in Phys. 46, Ed. Manneville P., Springer Verlag (1989); Cornille H. "Partially Integrable Evolution Eq. in Physics" Ed. Conte R., Kluwer Ac.: Press (1990) p.39.

(2) Cornille H. and Qian Y.H., CRAS 309, 1883 (1989), J. Stat. Phys. (1990); Cornille H. "Inverse Methods in Action" Ed. Sabatier, Springer-Verlag (1990) p.414, Saclay SPhT/90-047, 067.

(3) Qian Y.H., d'Humières D. and Lallemand P., J. Stat. Phys. (1990).

(4) d'Humières D., Lallemand P. and Frisch U., Europh. Let. 2, 291 (1986); Nadiga B.T., Broadwell J.E. and Sturtevant B. "R.G.D." Astro. Aero. 16, Ed. Muntz p.155 (1988); Chen S., Lee M., Zhaok H. and Doolen G.D., Physica D37, 42 (1989).

(5) Cabannes H., Journ. Mécan. 14, 703 (1975), Journ. Fluid Mech. 76, 273 (1976), Lect. Notes at Berkeley (1980), Mech. Res. Commun. 12, 289 (1985); Cabannes H. and Tiem D.M., CRAS, 304, 29 (1987), Comp. Syst. 1, 574 (1987).

(6) Platkowski T., Transp. Theo. Stat. Phys. 18, 222 (1989) "Discrete Kinetic Theory, Lattice Gas Dynamics" Ed. Monaco R., ISI, World Scientific (1988), p.248.

(7) Broadwell J.E., Phys. Fluids 7, 1243 (1964); Gatignol R., Phys. Fluids 18, 153 (1987).

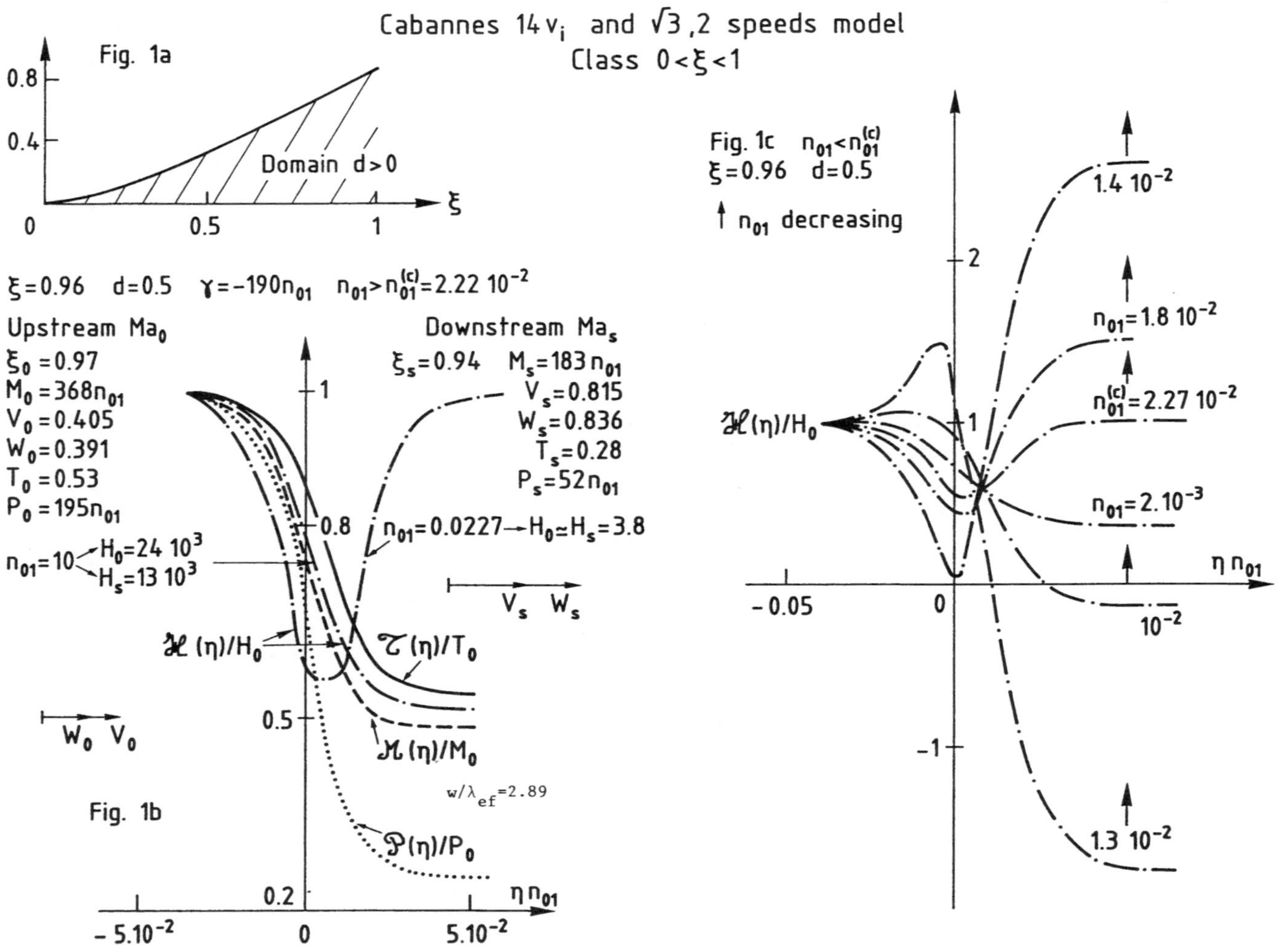

Cabannes 14 v_i and $\sqrt{3}$,2 speeds model
Class $0<\xi<1$
Fig. 1a
0.8
0.4
0
0.5
1
ξ
Domain $d>0$
$\xi=0.96$ $d=0.5$ $\gamma=-190n_{01}$ $n_{01}>n_{01}^{(c)}=2.22\ 10^{-2}$
Upstream Ma_0
$\xi_0=0.97$
$M_0=368n_{01}$
$V_0=0.405$
$W_0=0.391$
$T_0=0.53$
$P_0=195n_{01}$
$n_{01}=10$ $H_0=24\ 10^3$ $H_s=13\ 10^3$
Downstream Ma_s
$\xi_s=0.94$ $M_s=183n_{01}$
$V_s=0.815$
$W_s=0.836$
$T_s=0.28$
$P_s=52n_{01}$
$n_{01}=0.0227 \rightarrow H_0 \simeq H_s=3.8$
V_s W_s
W_0 V_0
$\mathcal{H}(\eta)/H_0$
$\mathcal{T}(\eta)/T_0$
$\mathcal{M}(\eta)/M_0$
$\mathcal{P}(\eta)/P_0$
$w/\lambda_{ef}=2.89$
Fig. 1b
1
0.8
0.5
0.2
-5.10^{-2}
0
5.10^{-2}
ηn_{01}
Fig. 1c $n_{01}<n_{01}^{(c)}$
$\xi=0.96$ $d=0.5$
n_{01} decreasing
$\mathcal{H}(\eta)/H_0$
2
1
-1
-0.05
0
ηn_{01}
$1.4\ 10^{-2}$
$n_{01}=1.8\ 10^{-2}$
$n_{01}^{(c)}=2.27\ 10^{-2}$
$n_{01}=2.10^{-3}$
10^{-2}
$1.3\ 10^{-2}$

Cabannes 14 v_i and speeds $\sqrt{3}$, 2 model

Class $-2<\xi<-1$

Fig. 2a

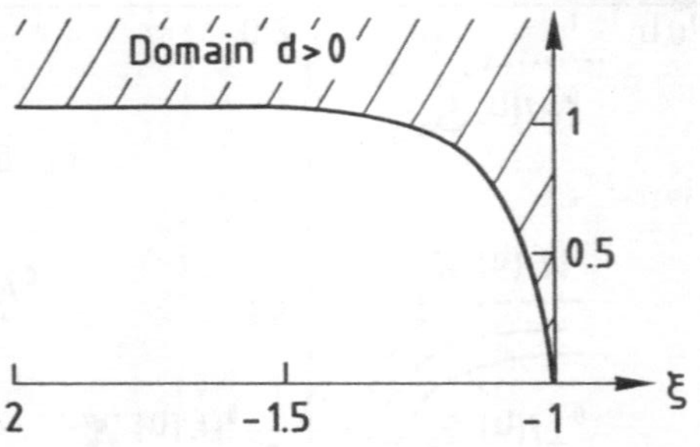

Fig. 2b $\xi=-1.23$ $d=1$ $\gamma=5.7n_{01}$ $n_{01}<n_{01}^{(c)}=1.11$

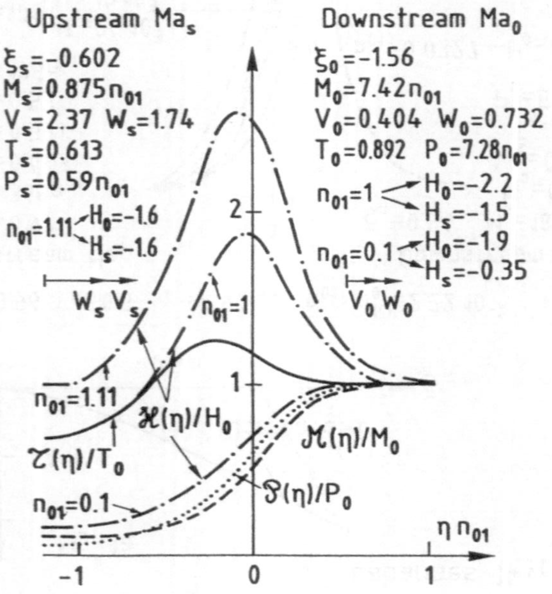

Fig. 2c

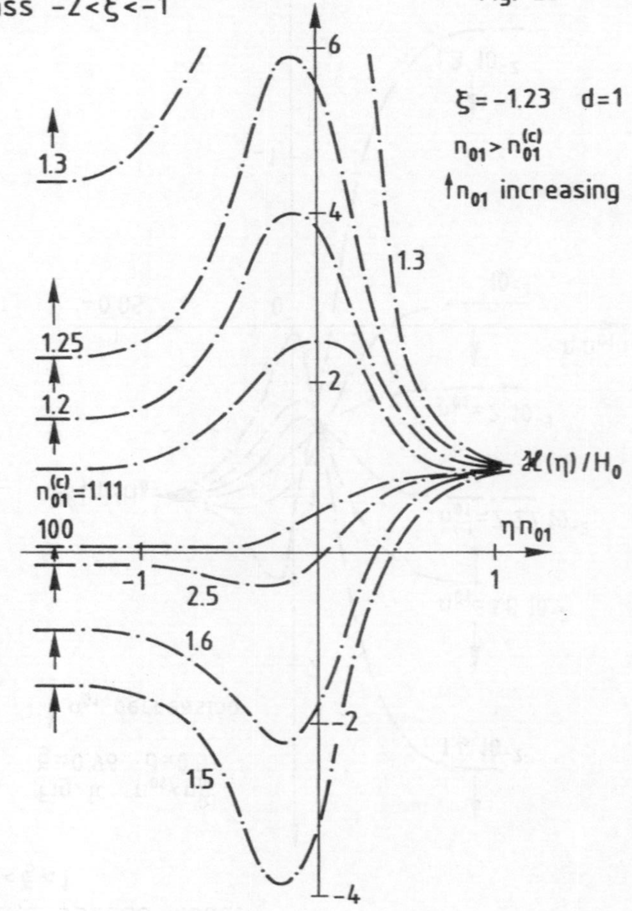

Fig- 3a - Cabannes $14 \vec{v_i}$ and speeds $\sqrt{3}, 2$ model
$$\xi = -1.06 \qquad d = 0.5 \qquad \gamma = 3.9\, n_{01} \qquad n_{01} < n_{01}^{(c)} = 1.26$$

Upstream Ma_s

$\xi_s = -0.9$

$M_s = 1.01\, n_{01}$

$V_s = 2.1$

$W_s = 1.93$

$T_s = 0.657$

$P_s = 0.666\, n_{01}$

$n_{01} = 1 \quad \left< \begin{array}{l} H_0 = -2.6 \\ H_s = -1.5 \end{array} \right.$

$n_{01} = 0.21 \quad \left< \begin{array}{l} H_0 = -2.4 \\ H_s = -0.64 \end{array} \right.$

Downstream Ma_0

$\xi_0 = -1.22$

$M_0 = 5.57\, n_{01}$

$V_0 = 0.42$

$W_0 = 0.58$

$T_0 = 0.98$

$P_0 = 5.01$

$n_{01} = 1.26 \quad \left< \begin{array}{l} H_0 = -1.6 \\ H_s = -1.6 \end{array} \right.$

$W_s \quad V_s$

$\mathcal{H}(\eta)/H_0$

$V_0 \quad W_0$

$n_{01} = 1.26$

$n_{01} = 1$

$n_{01} = 0.21$

$\mathcal{M}(\eta)/M_0$

$\mathcal{P}(\eta)/P_0$

$\eta\, n_{01}$

$-1 \qquad 0 \qquad 1$

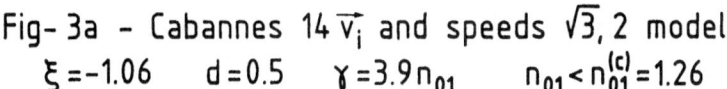

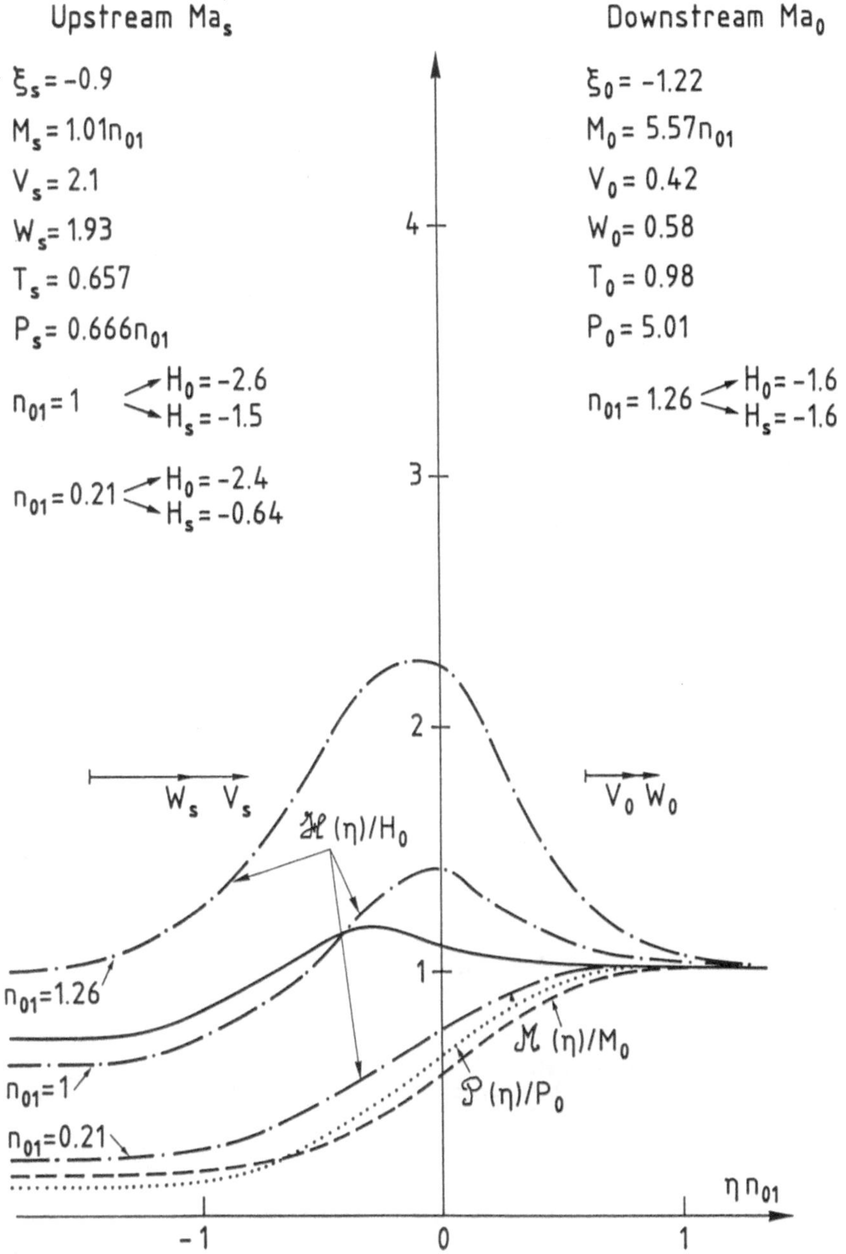

Fig- 3b - Cabannes 14 $\vec{v_i}$ and speeds $\sqrt{3}, 2$ model

$\xi = -1.54$ $d = 3$ $\gamma = 8.6 n_{01}$ $n_{01} < n_{01}^{(c)} = 0.573$

Upstream Ma_s

$\xi_s = -0.84$

$M_s = 0.307 n_{01}$

$V_s = 2.59$

$W_s = 1.89$

$T_s = 0.651$

$P_s = 0.23 n_{01}$

$n_{01} = 0.5$ $\begin{array}{l} H_0 = -1.2 \\ H_s = -0.52 \end{array}$

$n_{01} = 0.1$ $\begin{array}{l} H_0 = -1.9 \\ H_s = -1.5 \end{array}$

Downstream Ma_0

$\xi_0 = -1.88$

$M_0 = 10.28 n_{01}$

$V_0 = 0.287$

$W_0 = 0.625$

$T_0 = 0.72$

$P_0 = 6.9 n_{01}$

$n_{01} = 0.573$ $\begin{array}{l} H_0 = -0.57 \\ H_s = -0.57 \end{array}$

W_s V_s

V_0 W_0

$n_{01} = 0.573$

$\mathcal{H}(\eta)/H_0$

$\mathcal{T}(\eta)T_0$

$\mathcal{M}(\eta)/M_0$

$\mathcal{P}(\eta)/P_0$

$n_{01} = 0.5$

$n_{01} = 0.1$

$\eta\, n_{01}$

-1

0

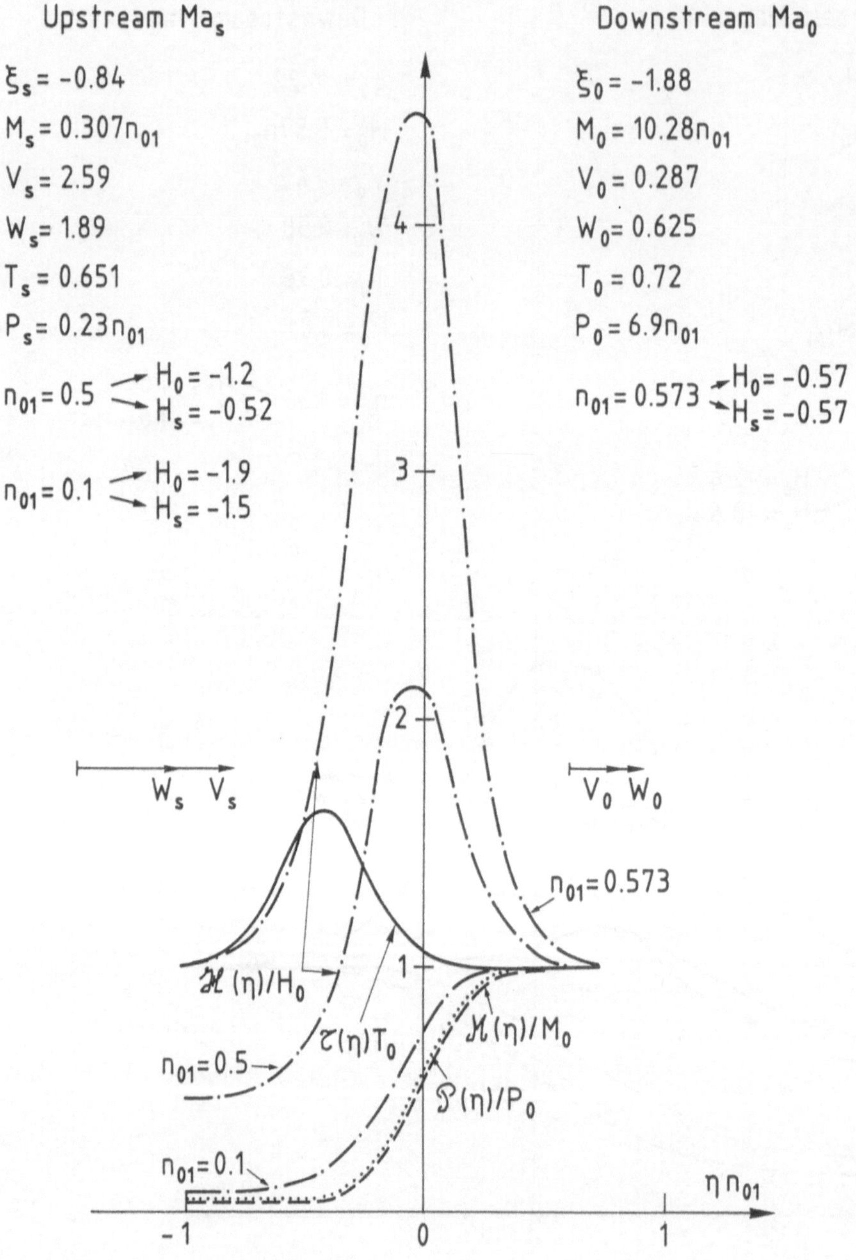

Fig- 3c - Cabannes $14\,\vec{v_i}$ and speeds $\sqrt{3}, 2$ model

$\xi = -1.57$ $d = 2$ $\gamma = 2.35n_{01}$ $n_{01} < n_{01}^{(c)} = 0.51$

Upstream Ma_s

$\xi_s = -0.73$

$M_s = 0.3n_{01}$

$V_s = 2.66$

$W_s = 1.81$

$T_s = 0.636$

$P_s = 0.19n_{01}$

$n_{01} = 0.5$ $\begin{cases} H_0 = -0.55 \\ H_s = -0.52 \end{cases}$

$n_{01} = 0.24$ $\begin{cases} H_0 = -2.2 \\ H_s = -0.3 \end{cases}$

$n_{01} = 0.06$ $\begin{cases} H_0 = -0.07 \\ H_s = -0.003 \end{cases}$

Downstream Ma_0

$\xi_0 = -1.89$

$M_0 = 11n_{01}$

$V_0 = 0.278$

$W_0 = 0.603$

$T_0 = 0.648$

$P_0 = 5.16n_{01}$

$\overrightarrow{W_s\ V_s}$

$\overrightarrow{W_0\ V_0}$

$\mathcal{T}(\eta)/T_0$

$\mathcal{H}(\eta)/H_0$

$\mathcal{M}(\eta)/M_0$

$\mathcal{P}(\eta)/P_0$

$n_{01} = 0.5$

$n_{01} = 0.06$

$n_{01} = 0.24$

$\eta\, n_{01}$

-1 0 1

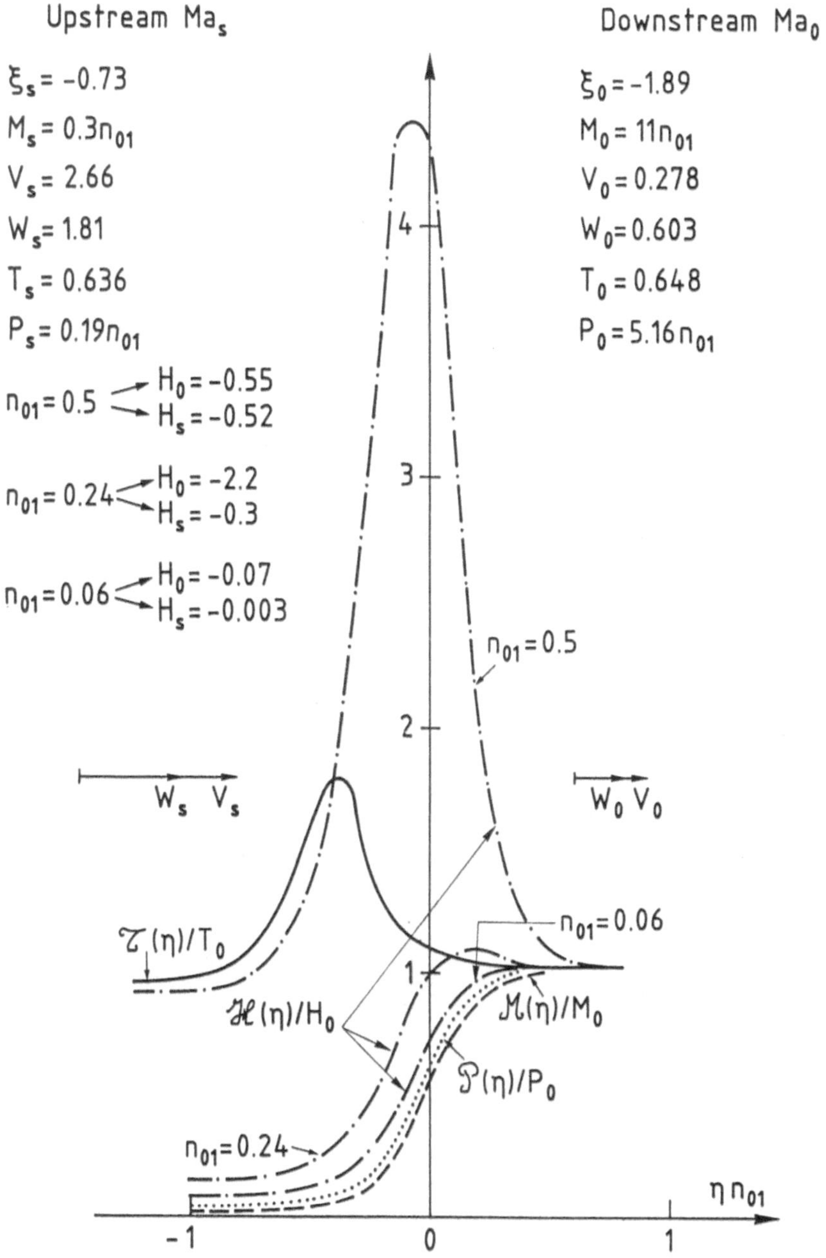

Fig. 3c – Cabannes $\Delta\nu$ and speeds v_3^2 model
$\xi=-15.7$ $d=2$ $\gamma=2.35\, n_f$ $n_{01}<n_{01}^{(l)}=0.51$

A One Dimensional Lattice Boltzmann Equation with Galilean Invariance

Yue-Hong Qian[1], *D. d'Humières*[2], *and P. Lallemand*[2]

[1]Renault–Direction des Recherches, 9–11, avenue du 18 Juin 1940, F-92500 Rueil-Malmaison, France

[2]Laboratoire de Physique Statistique, Ecole Normale Supérieure, 24, rue Lhomond, F-75231 Paris Cedex 05, France

Abstract: A three-velocity one-dimensional Lattice Boltzmann Equation is presented. The stability of the equilibrium distribution is studied and the dynamical equations are given. It is shown that Galilean invariance can be recovered for the proper choice of the collision operator. This particular model is then used to compare the results of numerical simulations to the behavior of a shock tube.

1 INTRODUCTION.

Since the work of Frisch, Hasslacher and Pomeau [1], a lot of interest has been devoted to lattice gas models [2]. These models deal with Boolean particles moving synchronously on a regular lattice with local collision rules conserving some physical quantities (number of particles, momentum or energy) on each node of the lattice. Despite very encouraging results, these models suffer some drawbacks, particularly for those having engineering applications in mind. One of these limitations comes from the intrinsic noise occuring in the measurement of the macroscopic quantities through the averaging of Boolean fields. As in every Monte-Carlo processes, the cost of a simulation increases as the square of the signal to noise ratio, consequently limiting the practical accuracy to about two digits. A second limitation is the lack of Galilean invariance of the dynamical equations governing the lattice gas evolution, restricting the use of lattice gases to low Mach numbers (incompressible limit). This problem is severe for the simulation of multiphase flows, for which the vorticity and concentration fields may be advected at different speeds, and a method has been derived to recover Galilean invariance, at least for a small range of density [3,4,5]. Finally, a last limitation comes from the existence of spurious conserved quantities [6,7], leading to new and nonphysical dynamical equations and modifications of the physical ones.

---

*associated to C.N.R.S. and to the Universities Paris 6 and Paris 7

Different approaches have been used to remove the noise limitation. All of them are based on the Lattice Boltzmann Equation where the Boolean variables of the lattice gases are replaced by floating point quantities. These quantities represent the probability to find a particle moving in a given direction and are advected as their Boolean counterparts. These models differ by the implementation of the collision terms. The first attempt was done by McNamara and Zanetti [8], with a collision operator directly derived from the Boolean one, the collision polynomials being evaluated in terms of the occupation probabilities. In this case, the probabilities are bounded by 0 and 1 and the scheme is stable, numerical problems having some likelihood only when a probability goes very close to one of the bounds. This technique is very convenient to study the noise and correlation effects in lattice gases: they are directly exhibited by the differences between simulations of the two algorithms in identical physical situations [9]. However, the complexity of the collision polynomial increases very rapidly with the number of velocities and the collision efficiency, limiting the practical use to the four- or six-bit models or to highly viscous ones. A more practical approach was proposed by Higuera and coworkers [10,11,12]. In this case, the collision operator is approximated by the product of the linearized collision operator, evaluated at the given density, times the difference between the current probabilities and those computed from the local equilibrium. With this approximation, the most complex part of the algorithm is the product of a $b \times b$ matrix by a b vector, where b is the number of velocities ($b = 18$ in three dimensions). In addition, there is no need for keeping a close relationship with lattice gases but for the symmetry requirements. This argument has been used to derive collision matrices giving very small viscosity or to replace the basic Fermi-Dirac equilibrium of lattice gases by other equilibrium distributions, for instance Bose-Einstein (boson-like particles) [13]. Although there is no guaranty of numerical stability for this scheme, the simulations show that it is well behaved, at least for not too small viscosities. While solving the noise problem, none of these techniques have yet made significant contributions to solve at the same time one of the two other limitations mentionned above.

In his thesis [14], Qian has introduced a new class of models curing at the same time both the noise and the Galilean invariance. These models are lattice versions of the well known discrete velocity gases [15] with the addition of zero velocity particles. Using the same trick as in lattice gases, Galilean invariance can be recovered, provided the proper choice of the collision rules has been made. In addition, the viscosity of these models can be set to an arbitrary small value, limited only by stability criteria. In this paper, we shall describe in detail a very simple three-velocity one-dimensional model for which we shall study both the stability and the effects of the spurious invariant. These theoretical results will be illustrated by numerical simulations of shock profiles.

2 THEORETICAL RESULTS.

2.1 The Dynamical Equations.

Let us consider three kinds of particles moving on a periodic one-dimensional lattice of length L with constant velocities chosen from the set $\{0, 1, -1\}$ and masses from $\{2, 1, 1\}$. The lattice sites are represented by $x_\star$, with $1 \leq x_\star \leq L$. The time evolution of the average occupation numbers on each site is computed at discrete time steps $t_\star$ according to the following lattice Boltzmann equation:

$$\begin{aligned} N_0(x_\star, t_\star + 1) &= N_0(x_\star, t_\star) + \Delta(x_\star, t_\star) , \\ N_1(x_\star + 1, t_\star + 1) &= N_1(x_\star, t_\star) - \Delta(x_\star, t_\star) , \\ N_2(x_\star - 1, t_\star + 1) &= N_2(x_\star, t_\star) - \Delta(x_\star, t_\star) , \end{aligned} \tag{1}$$

where the collision term Δ is

$$\Delta(x_\star, t_\star) = \kappa \left(f^2 N_1(x_\star, t_\star) N_2(x_\star, t_\star) - N_0^2(x_\star, t_\star) \right) . \tag{2}$$

κ is a positive constant giving the collision efficiency, chosen to set the viscosity of the model but also to enforce the numerical stability of equations 1. f is a positive constant tuning the ratio between resting and moving particles. The use of these constants will be clarified later. Each node is then described at each time step by a 3-vector $\mathbf{N}(x_\star, t_\star)$.

For L even, it is easy to check that there exist three global conserved quantities:

$$L\rho = \sum_{x_\star=1}^{L} (2N_0(x_\star, t_\star) + N_1(x_\star, t_\star) + N_2(x_\star, t_\star)) , \tag{3}$$

$$L\rho u = \sum_{x_\star=1}^{L} (N_1(x_\star, t_\star) - N_2(x_\star, t_\star)) , \tag{4}$$

$$L\rho h = \sum_{x_\star=1}^{L} (-1)^{(x_\star + t_\star)} (N_1(x_\star, t_\star) - N_2(x_\star, t_\star)) . \tag{5}$$

The first two equations correspond to the mass and momentum conservation, defining the average density ρ and velocity u. The last equation corresponds to a spurious invariant staggered both in space and time, with an average value h. Using the method of reference [16], it can be shown that there is no additional independent linear combination of the N_i's conserved during the time evolution. When h is non zero, the state $\mathbf{N}$ of a node cannot be a smooth function of space and time, even if ρ, u, and h evolve very slowly in space and time. In order to make possible the Chapman-Enskog expansion, we have to write

$$\mathbf{N}(x_\star, t_\star) = \widetilde{\mathbf{N}}(x_\star, t_\star) + (-1)^{(x_\star + t_\star)} \widehat{\mathbf{N}}(x_\star, t_\star) , \tag{6}$$

where $\widetilde{\mathbf{N}}$ and $\widehat{\mathbf{N}}$ are now smooth functions of space and time. When the conserved quantities are uniform in space, $\widetilde{\mathbf{N}}$ and $\widehat{\mathbf{N}}$ share the same property and are given by

$$\widetilde{\mathbf{N}}(t_\star) = (\mathbf{N}(2x_\star, t_\star) + \mathbf{N}(2x_\star + 1, t_\star))/2 , \tag{7}$$

$$\widehat{\mathbf{N}}(t_\star) = (-1)^{t_\star} (\mathbf{N}(2x_\star, t_\star) - \mathbf{N}(2x_\star + 1, t_\star))/2 . \tag{8}$$

This will not be the case for the nonuniform cases and the derivation will be given in section 2.3.

Separating the slow and the fast evolutions, equations 1 can be written

$$\begin{aligned} \widetilde{N}_0(x_\star, t_\star + 1) &= \widetilde{N}_0(x_\star, t_\star) + \tilde{\Delta}(x_\star, t_\star) , \\ \widetilde{N}_1(x_\star + 1, t_\star + 1) &= \widetilde{N}_1(x_\star, t_\star) - \tilde{\Delta}(x_\star, t_\star) , \\ \widetilde{N}_2(x_\star - 1, t_\star + 1) &= \widetilde{N}_2(x_\star, t_\star) - \tilde{\Delta}(x_\star, t_\star) , \end{aligned} \tag{9}$$

and

$$\begin{aligned}
\widehat{N}_0(x_\star, t_\star + 1) &= -\widehat{N}_0(x_\star, t_\star) - \widehat{\Delta}(x_\star, t_\star), \\
\widehat{N}_1(x_\star + 1, t_\star + 1) &= \widehat{N}_1(x_\star, t_\star) - \widehat{\Delta}(x_\star, t_\star), \\
\widehat{N}_2(x_\star - 1, t_\star + 1) &= \widehat{N}_2(x_\star, t_\star) - \widehat{\Delta}(x_\star, t_\star),
\end{aligned} \tag{10}$$

where

$$\widetilde{\Delta} = \kappa\left(f^2\left(\widetilde{N}_1\widetilde{N}_2 + \widehat{N}_1\widehat{N}_2\right) - \left(\widetilde{N}_0^2 + \widehat{N}_0^2\right)\right), \tag{11}$$

$$\widehat{\Delta} = \kappa\left(f^2\left(\widetilde{N}_1\widehat{N}_2 + \widehat{N}_1\widetilde{N}_2\right) - 2\widetilde{N}_0\widehat{N}_0\right). \tag{12}$$

2.2 Homogoneous Equilibrium and Stability.

Let us assume in this section that ρ, u, and h are uniform. Then the space dependence of equations 9 to 12 can be removed and the conservation relations can be written

$$\rho = 2\widetilde{N}_0(t_\star) + \widetilde{N}_1(t_\star) + \widetilde{N}_2(t_\star), \tag{13}$$

$$\rho u = \widetilde{N}_1(t_\star) - \widetilde{N}_2(t_\star);, \tag{14}$$

$$\rho h = \widehat{N}_1(t_\star) - \widehat{N}_2(t_\star). \tag{15}$$

The equilibrium state is obtained when the quantities $\widetilde{\Delta}$, $\widehat{\Delta}$, and $\widehat{N}_0$ are simultaneously equal to zero, therefore it is completely determined by

$$\widetilde{N}_0 = \rho_0, \qquad \widetilde{N}_1 = \frac{\rho_m + \rho u}{2}, \qquad \widetilde{N}_2 = \frac{\rho_m - \rho u}{2}, \tag{16}$$

$$\widehat{N}_0 = 0, \qquad \widehat{N}_1 = \frac{\rho_m + \rho u}{2\rho_m}\rho h, \qquad \widehat{N}_2 = -\frac{\rho_m - \rho u}{2\rho_m}\rho h, \tag{17}$$

where ρ_0 and $\rho_m = \rho - 2\rho_0$ are the equilibrium densities of zero velocity and moving particles, ρ_m being solution of

$$\rho_m^2(\rho - \rho_m)^2 = f^2\left(\rho_m^2 - \rho^2 u^2\right)\left(\rho_m^2 - \rho^2 h^2\right). \tag{18}$$

This equation has in general four solutions with no obvious useful analytical expression. For $u = h = 0$, the physical solution is $\rho_m = \rho/(1+f)$ and $\rho_0 = f\rho/(2(1+f))$, the three others being either the trivial solution $\rho_m = 0$ or giving a negative value for ρ_0. For small values of u and h, ρ_m can be obtained as an expansion of the solution of equation 18 around $\rho/(1+f)$. Up to the second order in u and h, this expansion leads to

$$\rho_m = \frac{\rho}{2(f+1)}\left(2 + f(f+1)(u^2 + h^2)\right), \tag{19}$$

$$\rho_0 = \frac{f\rho}{4(f+1)}\left(2 - (f+1)(u^2 + h^2)\right). \tag{20}$$

To the same order of approximation, the equilibrium distributions are then given by

$$\begin{aligned}
N_0^{eq} &= \frac{f\rho}{2}\left(\frac{1}{f+1} - \frac{u^2 + h^2}{2}\right), \\
N_1^{eq} &= \frac{\rho}{2}\left(\frac{1}{f+1} + u + \frac{f}{2}(u^2 + h^2) + (-1)^{(x_\star + t_\star)} h(1 + (f+1)u)\right), \\
N_2^{eq} &= \frac{\rho}{2}\left(\frac{1}{f+1} - u + \frac{f}{2}(u^2 + h^2) - (-1)^{(x_\star + t_\star)} h(1 - (f+1)u)\right).
\end{aligned} \tag{21}$$

The linear stability of the different solutions is derived from the linearization of equations 9 and 10 around the equilibrium solutions given by equations 16 to 18. Since there are six variables and three constrains, the stability analysis reduces to a system of three independent variables $\widetilde{N}_0$, $\widehat{N}_0$, and $\widehat{N}_1$, leading to the following linearized matrix

$$\mathcal{L} = \begin{pmatrix} 1-\kappa\left(f^2\rho_m + 2\rho_0\right) & 0 & \dfrac{\kappa f^2\rho^2 hu}{\rho_m} \\ \dfrac{\kappa f^2\rho^2 hu}{\rho_m} & -1+2\kappa\rho_0 & -\kappa f^2\rho_m \\ \dfrac{\kappa f^2\rho^2 hu}{\rho_m} & 2\kappa\rho_0 & 1-\kappa f^2\rho_m \end{pmatrix} . \tag{22}$$

An equilibrium solution is stable if and only if the moduli of the three eigenvalues of $\mathcal{L}$ are all smaller than one. When u or h is equal to zero, one eigenvalue of $\mathcal{L}$ is

$$\lambda_1 = 1-\kappa\left(f^2\rho_m + 2\rho_0\right) , \tag{23}$$

and the two others are solutions of

$$\lambda^2 - (\lambda_1 - 1 + 4\kappa\rho_0)\lambda - \lambda_1 = 0 . \tag{24}$$

Around the physical solution for $u = h = 0$, λ_1 is always smaller than 1 and must be larger than -1 for stability. For $|\lambda_1| < 1$, it can be shown that the modulus of the roots of equation 24 are also smaller than one. In addition, when λ_1 is close to -1, these two roots are complex conjugate and their moduli are equal to $|\lambda_1|$.

When $u = h = 0$, it is easy to check that the stability condition is

$$\kappa f\rho \leq 2 . \tag{25}$$

Similar exact conditions can be obtained if u or h is equal to zero. Otherwise when $f > 1$, the stability domain of the product $\kappa f\rho$ can be bounded from above by the following relation:

$$(f^2-1)\frac{u^2+h^2}{2} \leq \frac{2}{\kappa f\rho} - 1 . \tag{26}$$

2.3 The Macroscopic Transport Equations.

In this section, we assume small variations of the macroscopic variables with very large wavelength. Then the local distributions can be obtained from the values of the local equilibria, derived in the previous section, plus a small perturbation coming from the gradients of the macroscopic quantities. Separating the fast and the slow components, it comes $\widetilde{\mathbf{N}} = \widetilde{\mathbf{N}}^{eq} + \widetilde{\mathbf{N}}^{ne}$ and $\widehat{\mathbf{N}} = \widehat{\mathbf{N}}^{eq} + \widehat{\mathbf{N}}^{ne}$, where $\widetilde{\mathbf{N}}^{eq}$ and $\widehat{\mathbf{N}}^{eq}$ are given by equations 21 and $\widetilde{\mathbf{N}}^{ne}$ and $\widehat{\mathbf{N}}^{ne}$ are related to the previous quantities by equations 9 and 10, after linearization of $\widetilde{\Delta}$ and $\widehat{\Delta}$ around the local equilibrium. Relations 13 to 15 lead to the following constraints:

$$\begin{aligned} 2\widetilde{N}_0^{ne} + \widetilde{N}_1^{ne} + \widetilde{N}_2^{ne} &= 0 , \\ \widetilde{N}_1^{ne} - \widetilde{N}_2^{ne} &= 0 , \\ \widehat{N}_1^{ne} - \widehat{N}_2^{ne} &= 0 , \end{aligned} \tag{27}$$

and the linearization of $\widetilde{\Delta}$ and $\widehat{\Delta}$, around the local equilibrium for $u = h = 0$, gives

$$\widetilde{\Delta} = -\kappa f \rho \widetilde{N_0^{ne}} , \tag{28}$$

$$\widehat{\Delta} = \frac{\kappa f \rho}{f+1}(f\widehat{N_1^{ne}} - \widehat{N_0^{ne}}) = -2\widehat{N_0^{ne}} . \tag{29}$$

Since we have introduced smooth functions, the differences can be replaced by time and space derivatives. Using a Chapman-Enskog expansion, it comes:

$$\widetilde{N_0^{ne}} = -\widetilde{N_1^{ne}} = \frac{1}{2\kappa(f+1)\rho}\partial_x(\rho u) , \tag{30}$$

$$\widehat{N_1^{ne}} = -\frac{1}{2f}\left(\frac{f+1}{\kappa f \rho} - \frac{1}{2}\right)\partial_x(\rho h) . \tag{31}$$

After some algebra, one gets the following macroscopic equations:

$$\partial_t \rho + \partial_x(\rho u) = 0 , \tag{32}$$

$$\partial_t(\rho u) + \frac{f}{2}\partial_x(\rho u^2) = -\frac{1}{f+1}\partial_x\left(\rho + \frac{f(f+1)}{2}\rho h^2\right) + \partial_x\left(\frac{f}{f+1}\left(\frac{1}{\kappa f \rho} - \frac{1}{2}\right)\partial_x(\rho u)\right) , \tag{33}$$

$$\partial_t(\rho h) + (f+1)\partial_x(\rho u h) = \partial_x\left(\frac{f+1}{f}\left(\frac{1}{\kappa f \rho} - \frac{1}{2}\right)\partial_x(\rho h)\right) . \tag{34}$$

Using relations 25 and 26, it can be shown that the kinematic viscosity,

$$\nu = \frac{f}{f+1}\left(\frac{1}{\kappa f \rho} - \frac{1}{2}\right) , \tag{35}$$

is always positive in the stability domain and can be set very close to zero[1].

When $h = 0$ and $f = 2$, the usual one-dimensional Navier-Stokes equation is recovered with Galilean invariance. The pressure P is related to the density by the equation of state $\rho = 3P$ and the speed of sound is then $c_s = 1/\sqrt{3}$. The kinematic viscosity is given by

$$\nu = \frac{1}{3}\left(\frac{1}{\kappa\rho} - 1\right) . \tag{36}$$

Note that, for this particular value of f, one finds the discrete version of the Broadwell model studied by Caflish and Liu [17]. For $h \neq 0$, equation 34 must be taken into account and corresponds to an advection of the spurious invariant with a speed $3u$, thus the transport equation for the spurious invariant is not Galilean invariant. In addition, the spurious invariant gives a nonphysical contribution to the pressure term.

Until now, we have skipped the problem of computing $\widetilde{\mathbf{N}}(x_\star, t_\star)$ and $\widehat{\mathbf{N}}(x_\star, t_\star)$ from the values of $\mathbf{N}(x_\star, t_\star)$ given by the numerical simulations. In fact, this problem has no exact solution and can only be solved to a prescribed order of accuracy in the gradient

[1] Note the term $-\frac{1}{2}$ due to the lattice.

expansion. The simplest solution, accurate up to the second order in gradients, is:

$$\widetilde{\mathrm{N}}(x_\star, t_\star) \approx \frac{1}{4}\left(2\mathrm{N}(x_\star, t_\star) + \mathrm{N}(x_\star + 1, t_\star) + \mathrm{N}(x_\star - 1, t_\star)\right) , \tag{37}$$

$$\widehat{\mathrm{N}}(x_\star, t_\star) \approx \frac{1}{4}\left(2\mathrm{N}(x_\star, t_\star) - \mathrm{N}(x_\star + 1, t_\star) - \mathrm{N}(x_\star - 1, t_\star)\right)(-1)^{(x_\star + t_\star)} . \tag{38}$$

These formulae can be easily extended for higher accuracy taking more points around $x_\star$, the price to pay for this stronger filtering of the fast or slow components is a smoothing of the probability profiles.

3 SIMULATIONS OF A SHOCK TUBE.

3.1 Analytical solutions.

In this section we shall study the time evolution of one-dimensional fronts in a shock tube, filled at time $t_\star = 0$ with a gas at rest with uniform density ρ_- ($u_- = 0$), for $x_\star < 0$, and ρ_+ ($u_+ = 0$), for $x_\star \geq 0$, and we shall assume $\rho_- > \rho_+$. This is a very classical problem in which it appears a compressive shock front, moving in the low density (to the right here), and a rarefaction front, moving in the high density region (to the left). These two fronts leave an intermediate region in the center of the tube, with uniform density ρ_c and velocity u_c. In what follows, the transient will be discarded and the two fronts will be assumed to be independent, ρ_c and u_c (resp. ρ_+ and u_+) being taken as the values at $-\infty$ (resp. $+\infty$) for the compressive shock, and ρ_- and u_- (resp. ρ_c and u_c) at $-\infty$ (resp. $+\infty$) for the rarefaction front.

For the inviscid case $\nu = 0$, the density and velocity profiles present a discontinuity across the shock. This shock moves with a velocity v_s, such that the Navier-Stokes equations are time independent in the frame moving with the shock. The shock speed is given by the Rankine-Hugoniot relations

$$v_s(\rho_+ - \rho_c) + \rho_c u_c = 0 , \tag{39}$$

$$v_s \rho_c u_c - \rho_c u_c^2 = -c_s^2(\rho_+ - \rho_c) . \tag{40}$$

These relations give

$$v_s^2 = r_c c_s^2 , \qquad u_c = \frac{r_c - 1}{\sqrt{r_c}} c_s , \qquad \text{where} \quad r_c = \frac{\rho_c}{\rho_+} . \tag{41}$$

The rarefaction front is obtained as the steady solution of the Navier-Stokes equations, written as a function of the reduced variable x/t. This solution is given by:

$$\begin{array}{lll} u = 0 , & \rho = \rho_- , & \text{for} \quad x \leq -c_s t , \\ u = c_s + \dfrac{x}{t} , & \rho = \rho_- \exp\left(-\dfrac{x + c_s t}{c_s t}\right) , & \text{for} \quad -c_s t \leq x \leq (u_c - c_s)t , \\ u = u_c , & \rho = \rho_c , & \text{for} \quad x \geq (u_c - c_s)t . \end{array} \tag{42}$$

The continuity of the density between the two fronts leads to

$$\ln \rho_c = \ln \rho_- - \frac{u_c}{c_s} , \tag{43}$$

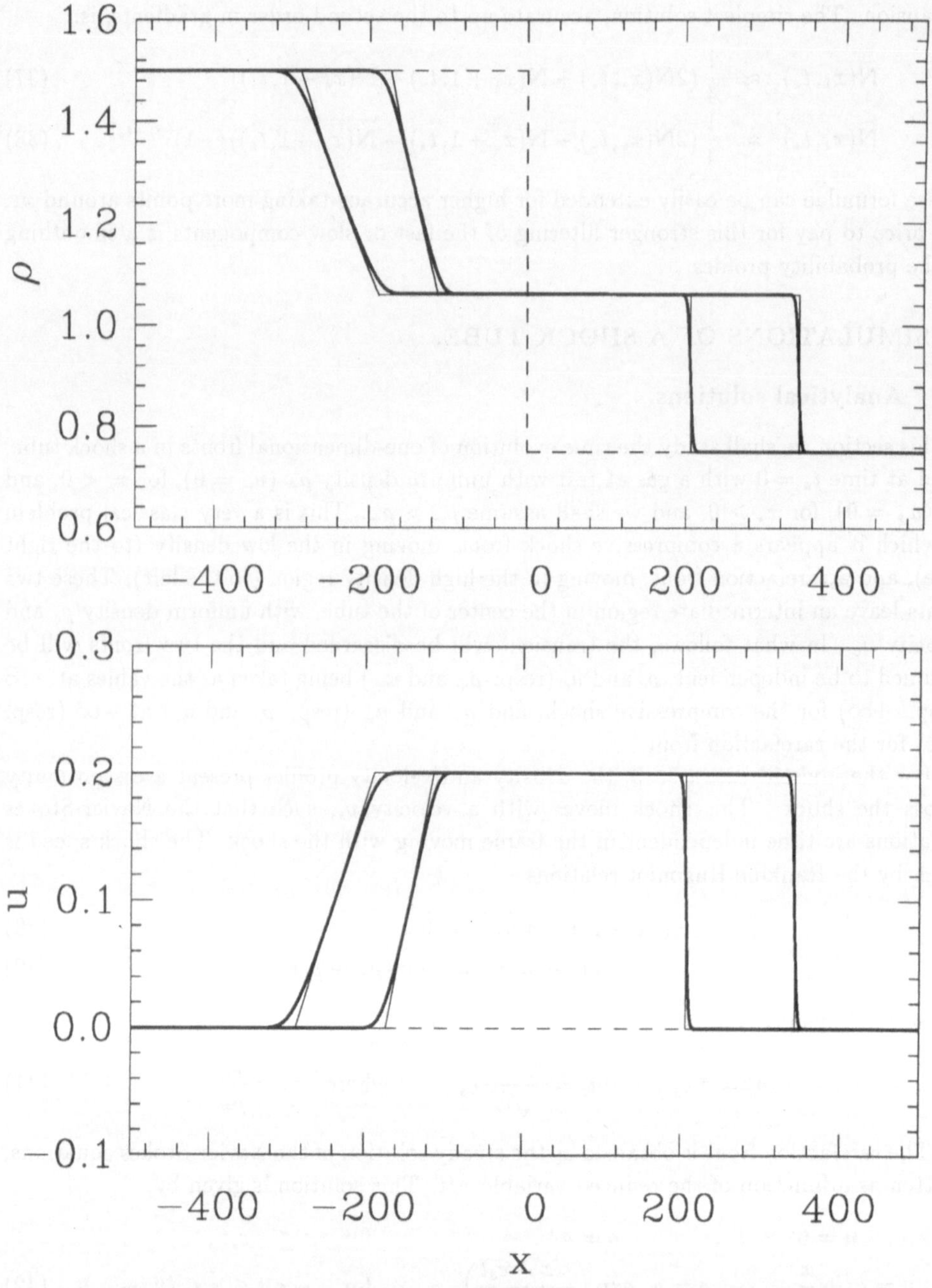

Figure 1: Time evolution of the density and velocity profiles, ρ and u, in a shock tube for $\psi = 2$. The dashed lines give the initial profiles. The thin (bold) curves correspond to the theoretical (numerical) profiles for $t_\star = 300$ and 500.

or

$$\ln r_c + \frac{r_c - 1}{\sqrt{r_c}} = \ln R\,, \tag{44}$$

where $R = \rho_-/\rho_+$. This equation gives r_c as a function of R.

The previous results use the state equation $P = c_s^2\rho$ relating the pressure P to the density. They cannot capture thermal effects occuring in real gases and producing an additional shock for the density profile. Thus the results of the following section must be compared to the velocity and pressure profiles obtained with other methods [18].

3.2 Numerical simulations.

The lattice Boltzmann equations 1 is simulated for the Galilean invariant case $f = 2$, using the standard scheme in which the time evolution is split in two steps. In the first step, the collision operator is computed independently for each node. In order to increase the stability of the scheme and to set a constant viscosity, the coefficient κ is locally adjusted so that $\kappa\rho = 1/\psi$ is constant (with the assumption $h = 0$). In the second step, the field N_1 is shifted one position to the right and the field N_2 to the left. At each end of the chain the particles bounce back, simulating rigid walls.

Figure 1 shows the density and velocity profiles for $R = 2$ ($\rho_- = 1.5$ and $\rho_+ = 0.75$) and $\psi = 2$, for time steps $t_\star = 0$, $t_\star = 300$, and $t_\star = 500$, and the same profiles computed from equations 41 and 42. These curves show a very good agreement between the simulations and the theoretical predictions. The rounding of the simulation curves is due to the non-zero viscosity. The measured values of ρ_c and u_c are respectively: 1.06156 and 0.20042, to be compared to the theoretical values 1.05974 and 0.20059. The agreement is again very good, the discrepancies coming from the higher order terms in the Chapman-Enskog expansion, terms neglected in the theoretical results given here. This agreement becomes even better when R is closer to one.

The value $\psi = 2$ corresponds to the maximum stability of the lattice Boltzmann equations. When ψ becomes close to one, the marginal stability decreases and we find that the system becomes numerically unstable when $\psi < 1.0715$, producing "overflows" problems very rapidly. The theoretical prediction of the first instability is $\psi = 1.0583$ for $u = u_c$ and $h = 0$, thus the system becomes unstable after a shock slightly sooner than expected, probably because the shock produces a non-zero value of the spurious invariant. Figure 2 shows the velocity and spurious invariant profiles for $\psi = 1.0715$, the values of u and h being extracted from the local distribution using equations 37 and 38. Large oscillations can be observed behind the shock with a noticeable value of the local spurious invariant. A strange feature is the spatial period four of the oscillations that is probably related to an instability of the period two spurious invariant.

A shock is a very stringent situation for a system close to instability. The stability of simulations with smaller velocity and density gradients can be better than the stability reported here. For instance, the rarefaction front produces only a small perturbation when the velocity starts to decrease, in addition this perturbation is highly damped.

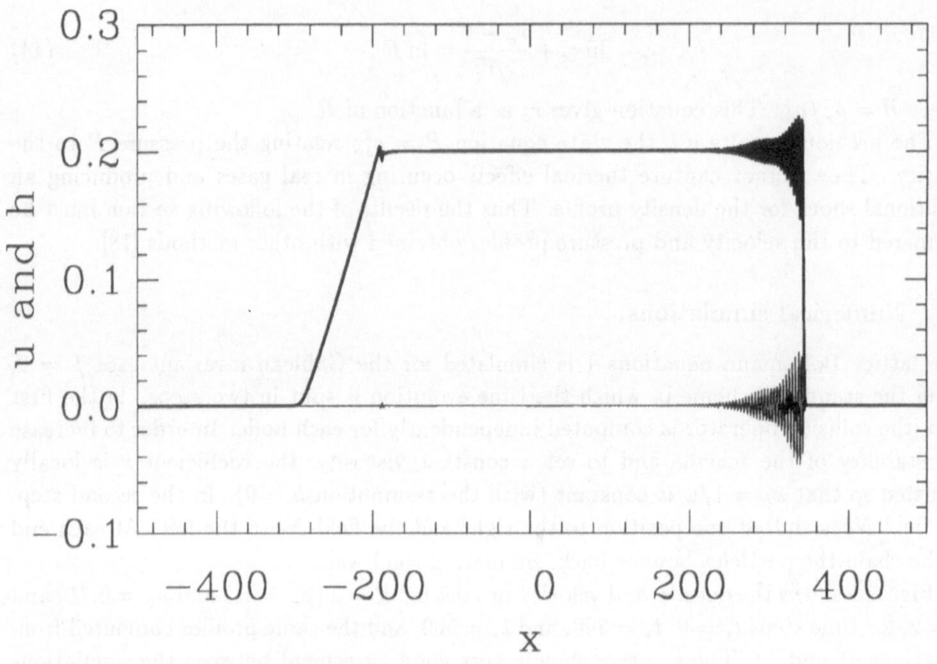

Figure 2: Velocity (bold line) and spurious invariant (thin line) profiles for $\psi = 1.0715$ and $t_\star = 500$.

4 CONCLUSIONS.

The three-velocity lattice Boltzmann equation presented here is a very convenient system for which all the analytical results can be easily worked out. We have been able to study in details the effects of the spurious invariant and the stability of the system. Comparisons between theory and numerical simulations show a very good agreement.

These results can be extended to models with more than three velocities and in higher dimensions. However, in dimension two and more, some care must be exercised with respect to the isotropy of the transport properties. Using the models described in [14], it is possible to use square or cubic lattices with both isotropy and Galilean invariance. Without spurious invariants, the linear stability of these models can be directly derived from the eigenvalues of the linearized collision operator [19]. If λ is any non-zero eigenvalue of this operator, the stability criteria is $|1 + \lambda| < 1$. Since $\lambda < 0$, the system is linearly stable at $\vec{u} = 0$ for $-2 < \lambda$. When the collision operator is constructed and if small viscosities are of interest, it is important to keep close to -1 all the λ but those entering in the viscosity. In addition, the results presented in this paper indicate that the stability problem has to be studied in detail for simulations of low viscosity models: the viscosity cannot be too small and the initial approximation has to be close enough to the solution to stay in its vicinity. Finally, a last problem, no studied here, is the sub-grid modelling arising when the grid size is not small enough to capture the smallest scales of the flow.

References

[1] U. Frisch, B. Hasslacher, and Y. Pomeau, "Lattice-Gas Automata for the Navier-Stokes Equation", *Phys. Rev. Lett.* **56**, 1505-1508 (1986).

[2] An extended bibliography can be found at the end of the Proceedings of the workshop *Lattice Gas Methods for PDE's: Theory, Application, and Hardware*, 6–9 September 1989, Los Alamos, USA, to appear in *Physica* **D 47**, (1991).

[3] C. Burges and S. Zaleski, "Buoyant Mixtures of Cellular Automata Gases", *Complex Sys.* **1**, 31-50 (1987).

[4] D. d'Humières, P. Lallemand, and G. Searby, "Numerical Experiments on Lattice Gases: Mixtures and Galilean Invariance", *Complex Sys.* **1**, 633-647 (1987).

[5] V. Zehnlé and G. Searby, "Lattice Gas Experiments on a Non-Exothermic Diffusion Flame in a Vortex Field", *J. Physique* **50**, 1083-1097 (1989).

[6] G. Zanetti, "The Hydrodynamics of Lattice Gas Automata" *Phys. Rev.* **A40**, 1539-1548 (1989).

[7] D. d'Humières, Y.H. Qian, and P. Lallemand, "Invariants in Lattice Gas Models", in *Discrete Kinematic Theory, Lattice Gas Dynamics, and Foundations of Hydrodynamics*, R. Monaco ed. (World Scient., 1989) pp. 102-113.

[8] G. McNamara and G. Zanetti, "Use of the Boltzmann Equation to Simulate Lattice-Gas Automata", *Phys. Rev. Lett.* **61**, 2332-2335 (1988).

[9] R. Cornubert, D. d'Humières, and D. Levermore, "A Knudsen Layer Theory for Lattice Gases", to appear in *Physica* **D 47**, 241-259 (1991).

[10] F.J. Higuera, "Lattice Gas Simulation Based on the Boltzmann Equation", in *Discrete Kinematic Theory, Lattice Gas Dynamics, and Foundations of Hydrodynamics*, R. Monaco ed. (World Scient., 1989) pp. 329–342.

[11] F.J. Higuera and S. Succi, "Simulating the Flow Around a Circular Cylinder with a Lattice Boltzmann Equation", *Europhys. Lett.* **8**, 517-521 (1989).

[12] F.J. Higuera and J. Jimenez, "Boltzmann Approach to Lattice Gas Simulations", *Europhys. Lett.* **9**, 663-668 (1989).

[13] F.J. Higuera, S. Succi, and R. Benzi, "Lattice Gas Dynamics with Enhanced Collisions", *Europhys. Lett.* **9**, 345-349 (1989).

[14] Y.H. Qian, "Gaz sur Réseaux et Théorie Cinétique sur Réseaux Appliquée à l'Équation de Navier-Stokes", Doctoral thesis, Paris VI (1990).

[15] R. Gatignol, "Théorie Cinétique des Gaz à Répartition Discrète de Vitesses", *Lecture Notes in Physics* **36** (Springer, Berlin,1975).

[16] D. d'Humières, Y.H. Qian, and P. Lallemand, "Finding the Linear Invariants of Lattice Gases", in *Computational Physics and Cellular Automata*, A. Pires, D.P. Landau, and H. Hermann eds. (World Scient., 1990) pp. 97-115.

[17] R.E. Caflish and T.P. Liu, "Stability of Shock Waves for the Broadwell Equations", *Commun. Math. Phys.* **114**, 103-130 (1988).

[18] I.M. Katz and E.J. Shaughnessy, "Computer Aided Analysis of 1-D Compressible Flow Problems in a Lagrangian Particle Description Using the α Method", *Computers & Fluids* **18**, 75-101 (1988).

[19] M. Vergassola, R. Benzi, and S. Succi, "On the Hydrodynamic Behaviour of the Lattice Boltzmann Equation", *Europhys. Lett.* **13**, 411-416 (1990).

The Euler Description for a Class of Discrete Models of Gases with Multiple Collisions

P. Chauvat, F. Coulouvrat, and R. Gatignol

Laboratoire de Modélisation en Mécanique, associé au CNRS,
Université Pierre et Marie Curie, 4, place Jussieu,
F-75252 Paris Cedex 05, France

Abstract : Two difficulties arise in discrete kinetic theory with binary collisions only. The first one is the existence of some macroscopic variables other than mass, momentum and energy, in relation to parasite summational invariants. The second one concerns the anisotropic character generally related to the models. In order to eliminate these difficulties, multiple collisions are introduced, and some symmetry properties on the models are adopted. The Euler equations are then given for discrete models with different moduli.

1. INTRODUCTION

Inthe discrete kinetic theory of gases, the main idea is that the velocities of the particles belong to a *given finite set of vectors*.

The theory for a general model with discrete velocities has been given, for the first time in 1970 [10]. The Boltzmann equation is replaced by a system of partial differential equations with a very interesting mathematical structure. Many papers concern the proof of the existence of the solutions with initial data ([2],[14],[16]), and the trends of kinetic equations to the hydrodynamical equations ([1],[3],[5],[7],[11],[12],[15]). Alike, for particular models some exact solutions have been found [6].

It should be remarked that in discrete kinetic theory, only the velocity space is discretized, the space and time variables being continuous. For a lattice gas ([8],[9]), the space and time variables are also discretized. A main and important consequence is to have one's way to study flow problems by simulation on a computer of cellular automaton type. This aspect is presented in the Reference [9], and many classical problems of fluid dynamics have been studied with this point of view ([9],[17]). For lattice gases, by replacing the discretization in space and time by a continuous approach, it is possible to write the kinetic equations describing the microscopic world. The equations of the lattice gas theory and those of the discrete models of gases are very similar. The difference is due to the exclusion principle used in lattice gases.

In this paper, multiple collisions are introduced (Section 2). After describing the macroscopic state of the gas (Section 3), we adopt some symmetry properties on the models (Section 4). We define "good" models as isotropic models for which the summational invariants are reduced to the physical ones (mass, momentum and energy). It is possible to compare the Euler equations associated with the "good" models and the classical equations of the fluid dynamics (Section 5). A good agreement is obtained between these two descriptions.

2. DISCRETE KINETIC THEORY WITH MULTIPLE COLLISIONS

In previous works [11], we described the general model of a gas with discrete velocities and with binary collisions only. The gas is composed of identical particles of mass m. The velocities of these particles are restricted to a given finite set of p vectors : $\vec{u}_1, \vec{u}_2, \ldots, \vec{u}_p$. We denote by $N_i(\vec{r},t)$ the number density of particles with the velocity $\vec{u}_i$ (called particles "i") at the point $\vec{r}$ and at the time t.

Previous theories with binary collisions only are generalized to multiple collisions, with the aim to reducing the number of summational invariants and, if possible, to obtain the physical invariants (mass, momentum and energy) as the only ones.

By definition, an r-collision involves r particles ($r \geqslant 2$). The velocities of the r particles are $\vec{u}_{i_1}, \ldots, \vec{u}_{i_r}$ before the collision and $\vec{u}_{j_1}, \ldots, \vec{u}_{j_r}$ after the collision. Such a collision is written $I_r \Longrightarrow J_r$ where I_r and J_r denote the r-sets $(i_1, \ldots, i_r)$ and $(j_1, \ldots, j_r)$. We emphasize that the r-set $I_r = (i_1, \ldots, i_r)$ is the set of the r-not arranged numbers $i_1, \ldots, i_r$. In the same way, in the r-set J_r, the r numbers $j_1, \ldots, j_r$ are not arranged. The collision must bear out the conservation of momentum and energy (the conservation of mass being borne out automatically) :

$$\sum_{k=1}^{r} \vec{u}_{i_k} = \sum_{k=1}^{r} \vec{u}_{j_k} \ , \qquad \sum_{k=1}^{r} |\vec{u}_{i_k}|^2 = \sum_{k=1}^{r} |\vec{u}_{i_k}|^2 \ . \tag{1}$$

A "transition probability" denoted by $A_{I_r}^{J_r}$ is associated to each r-collision $I_r \Longrightarrow J_r$. The quantities $A_{I_r}^{J_r}$ are positive or zero. We assume that they do not depend on the order of the r indexes in I_r or in J_r. The number of r-collisions $I_r \Longrightarrow J_r$ per unit volume and unit time is $A_{I_r}^{J_r} N_{I_r}$ where N_{I_r} denotes the product $N_{i_1} N_{i_2} \dots N_{i_r}$. As in the binary collision theory, it is convenient to assign a transition probability equal to zero to an unrealizable collision. So the $A_{I_r}^{J_r}$ coefficients are defined for each pair (I_r, J_r). Finally, we introduce the set $\mathcal{E}_r$ of the r-sets of the r-not arranged numbers taken in the set $\{1,2,\dots,p\}$.

It is assumed that the transition probabilities satisfy the hypothesis of microversibility (I), or more generally, the hypothesis of the semi-detailed balance (II):

$$(I) : A_{I_r}^{J_r} = A_{J_r}^{I_r} \quad ; \quad (II) : \sum_{J_r \in \mathcal{E}_r} \left(A_{I_r}^{J_r} - A_{J_r}^{I_r} \right) = 0 . \tag{2}$$

If we pay attention to the binary collisions (r=2), it is important to remark that the transition probability $A_{I_2}^{J_2}$ introduced in this paper is identical to the transition probability $A_{ij}^{k\ell}$ introduced in the Reference [10] or [11] by putting $I_2 = (i,j)$ and $J_2 = (k,\ell)$.

Now, a balance equation should be written for the number density of particles "k". Through an r-collision, several particles "k" can be created or destroyed. We explain this remark by observing that after the r-collision, the three following situations may occur :

1. after the r-collision, we have 1, 2 or more new particles "k",
2. after the r-collision, 1, 2 or more particles "k" desappear,
3. after the r-collision, zero new particle "k" appear.

These situations are illustrated on the hereafter figure :

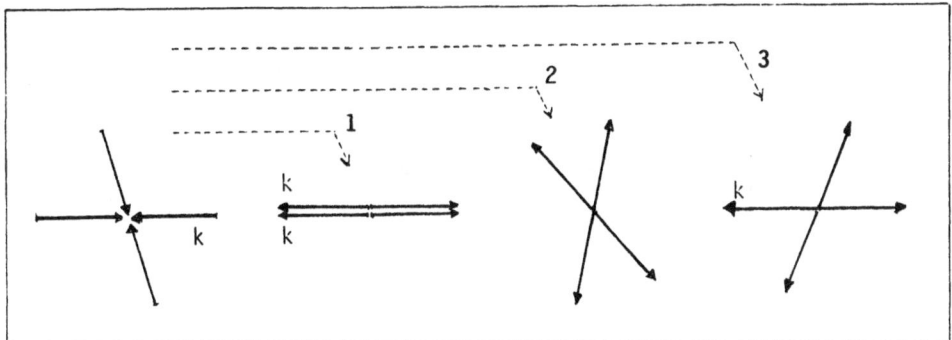

The number of r-collisions between the particles "i_1",...,"i_r", per unit volume and unit time is $\Sigma_{J_r \in \mathcal{E}_r} A_{I_r}^{J_r} N_{I_r}$, and the total number of r-collisions per unit volume and time is $\Sigma_{I_r \in \mathcal{E}_r} \Sigma_{J_r \in \mathcal{E}_r} A_{I_r}^{J_r} N_{I_r}$. We denote by $\delta(k,I_r)$ the number of indices k present in the r-set $I_r = (i_1,\dots,i_r)$. It is clear that $\delta(k,I_r)$ is positive or zero. Now we put $\delta(k,J_r,I_r) = \delta(k,J_r) - \delta(k,I_r)$. So, *the algebraic number of particles* "k" *created through the collision* $I_r \Longrightarrow J_r$ is $\delta(k,J_r,I_r)$. With these notations the balance equation for the number density of molecules "k" has the following form :

$$\frac{\partial}{\partial t} N_k + \vec{u}_k.\vec{\nabla} N_k = \sum_{r=2}^{R} \sum_{I_r\in \mathcal{E}_r} \sum_{J_r\in \mathcal{E}_r} \delta(k,J_r,I_r)\, A_{I_r}^{J_r} N_{I_r} . \tag{3}$$

In the equation (3), we took into account the multiple collisions with r = 2,...,R particles. By using the assumption (II), and also the definition of $\delta(k,J_r,I_r)$, the right hand side of equation (3) can be written differently :

$$\frac{\partial}{\partial t} N_k + \vec{u}_k.\vec{\nabla} N_k = \sum_{r=2}^{R} \sum_{I_r\in \mathcal{E}_r} \sum_{J_r\in \mathcal{E}_r} \delta(k,I_r) \left(A_{J_r}^{I_r} N_{J_r} - A_{I_r}^{J_r} N_{I_r}\right) . \tag{4}$$

$$\frac{\partial}{\partial t} N_k + \vec{u}_k.\vec{\nabla} N_k = \sum_{r=2}^{R} \sum_{I_r\in \mathcal{E}_r} \sum_{J_r\in \mathcal{E}_r} \delta(k,I_r)\, A_{J_r}^{I_r} \left(N_{J_r} - N_{I_r}\right) . \tag{5}$$

The set of equations (3) with $1 \leqslant k \leqslant p$ can be written in a short form :

$$\frac{\partial}{\partial t} \mathbb{N} + \mathcal{A}\, \mathbb{N} = \sum_{r=2}^{R} \mathcal{F}^r(\mathbb{N},\dots,\mathbb{N}) \equiv C(\mathbb{N}) . \tag{6}$$

The r-collision operator $\mathcal{F}^r$ is an r-linear symmetrical operator from $\mathbb{R}^p\times\cdots\times\mathbb{R}^p$ in $\mathbb{R}^p$ defined by :

$$\mathcal{F}^r_k(\mathbb{U}^1,\cdots,\mathbb{U}^r) = \left(\frac{1}{r!}\right)^2 \sum_{i_1=1}^{p} \cdots \sum_{i_r=1}^{p} \sum_{j_1=1}^{p} \cdots \sum_{j_r=1}^{p} \delta(k,I_r)\, A_{J_r}^{I_r} \left(U^1_{j_1}\cdots U^r_{j_r} - U^1_{i_1}\cdots U^r_{i_r}\right) . \tag{7}$$

[Remark : The numbers $(i_1,\dots,i_r)$ of the r-set I_r are not arranged, so that $\Sigma_{I_r\in\mathcal{E}_r} = (1/r!)^2\, \Sigma^p_{i_1=1}\cdots\Sigma^p_{i_r=1}$]. For each p-component $\Phi = (\varphi_1,\dots,\varphi_p)$ of $\mathbb{R}^p$, we have :

$$\langle \Phi, \mathcal{F}^r(\mathbb{U}^1,\cdots,\mathbb{U}^r) \rangle = \left(\frac{1}{r!}\right)^2 \sum_{i_1=1}^{p} \cdots \sum_{i_r=1}^{p} \sum_{j_1=1}^{p} \cdots \sum_{j_r=1}^{p} \sum_{k=1}^{p} \delta(k,J_r,I_r)\ \varphi_k\ A_{J_r}^{I_r}\ U_{i_1}^1 \cdots U_{i_r}^r \ . \tag{8}$$

According to this last formula, all properties of the discrete Boltzmann equations with only binary collisions are proved to be general. At once we give the definition of the *summational invariants* : they are attached to the conservation properties through the collisions and are defined as the p-component vector Φ satisfying the following conditions :

$$A_{I_r}^{J_r} \left[\left[\varphi_{j_1} + \cdots + \varphi_{j_r}\right] - \left[\varphi_{i_1} + \cdots + \varphi_{i_r}\right] \right] = 0 \quad \forall\ I_r, J_r \in \mathcal{E}_r, \quad \forall\ r = 2,\ldots,R.$$

As a matter of fact, we have $\Sigma_{k=1}^{p} \delta(k,I_r)\ \varphi_k = \varphi_{i_1} + \cdots + \varphi_{i_r}$. So, using other notations, the summational invariants are also defined by :

$$A_{I_r}^{J_r} \sum_{k=1}^{p} \delta(k,J_r,I_r)\ \varphi_k = 0 \ , \qquad \forall\ I_r, J_r \in \mathcal{E}_r, \quad \forall\ r = 1,\ldots,R \ . \tag{9}$$

These summational invariants generate a linear subspace (of $\mathbb{R}^p$) denoted by $\mathbb{F}$. In the Reference [13], we also introduced the summational invariants of order r attached to the conservation properties through the r-collisions. A summational invariant of $\mathbb{F}$ is a summational invariant of order r for *each* r equal to 2,...,R. We denote by q the dimension of $\mathbb{F}$. In particular, Φ is an invariant if φ_i is equal to m, $m\vec{u}_i$, or $(1/2)\ m|\vec{u}_i|^2$. In contrast to the classical kinetic theory for monoatomic gases, the geometric character of the set of the given velocities may allow other summational invariants.

Before going on, the following theorem will be proved :

THEOREM 1 : The following properties are equivalent :

a) $\Phi \in \mathbb{F}$,

b) $\langle \Phi, \mathcal{F}^r(\mathbb{U}^1,\ldots,\mathbb{U}^r) \rangle = 0 \ , \quad \forall\ \mathbb{U}^1,\ldots,\mathbb{U}^r \in \mathbb{R}^p, \ \forall\ r = 2,\ldots,R,$

c) $\langle \Phi, C(\mathbb{N}) \rangle = 0 \ , \quad \forall\ \mathbb{N} \in \mathbb{R}^p.$

The implication b) $\Rightarrow$ c) is evident. Due to the definition (9) of the summationnal invariants and to the formula (8), the implication a) $\Rightarrow$ b) is also obvious. Finally, we must only prove c) $\Rightarrow$ a) . With the property c) and by using (8) we have :

$$< \Phi,\ C(\mathbb{N}) > = \sum_{r=2}^{R} \left(\frac{1}{r!}\right)^2 \sum_{I_r \in \mathcal{E}_r} \sum_{J_r \in \mathcal{E}_r} \sum_{k=1}^{p} \delta(k,J_r,I_r)\ \varphi_k\ A_{J_r}^{I_r}\ N_{I_r} = 0$$

for each $\mathbb{N}$ of $\mathbb{R}^p$. Consequently, that polynomial with the variables N_i, $i=1,\ldots,p$ is identical to zero. The fact that sequence $(i_1,\ldots,i_r)$ in I_r is not arranged shows that :

$$\sum_{J_r \in \mathcal{E}_r} \sum_{k=1}^{p} \delta(k,J_r,I_r)\ \varphi_k\ A_{J_r}^{I_r} = 0\ , \qquad \forall\ I_r \in \mathcal{E}_r .$$

For each I_r we multiply this equality by $\Sigma_{\ell=1}^{p} \delta(\ell,I_r)\ \varphi_\ell$, and we add up :

$$\sum_{I_r \in \mathcal{E}_r} \sum_{J_r \in \mathcal{E}_r} A_{J_r}^{I_r} \left(\sum_{k=1}^{p} \delta(k,J_r,I_r)\ \varphi_k\right) \left(\sum_{\ell=1}^{p} \delta(\ell,I_r)\ \varphi_\ell\right) = 0\ .$$

By using the relation $\delta(k,J_r,I_r) = \delta(k,J_r) - \delta(k,I_r)$, we can write :

$$\sum_{I_r \in \mathcal{E}_r} \sum_{J_r \in \mathcal{E}_r} A_{J_r}^{I_r} \left(\left(\sum_{k=1}^{p} \delta(k,J_r)\ \varphi_k\right) - \left(\sum_{k=1}^{p} \delta(k,I_r)\ \varphi_k\right)\right) \left(\sum_{\ell}^{p} \delta(\ell,I_r)\ \varphi_\ell\right) = 0$$

According to the assumptions (I) or (II) on the transition probabilities, I_r and J_r can be exchanged in the brackets :

$$\sum_{I_r \in \mathcal{E}_r} \sum_{J_r \in \mathcal{E}_r} A_{J_r}^{I_r} \left(\left(\sum_{k=1}^{p} \delta(k,I_r)\ \varphi_k\right) - \left(\sum_{k=1}^{p} \delta(k,J_r)\ \varphi_k\right)\right) \left(\sum_{\ell}^{p} \delta(\ell,J_r)\ \varphi_\ell\right) = 0$$

By adding the two last expressions, we get :

$$\sum_{I_r \in \mathcal{E}_r} \sum_{J_r \in \mathcal{E}_r} A_{J_r}^{I_r} \left(\left(\sum_{k=1}^{p} \delta(k,J_r)\ \varphi_k\right) - \left(\sum_{k=1}^{p} \delta(k,I_r)\ \varphi_k\right)\right)^2 = 0\ .$$

Since we have a sum of positive terms, we can conclude that :

$$A_{J_r}^{I_r} \left(\sum_{k=1}^{p} \delta(k,J_r,I_r)\ \varphi_k\right) = 0\ .$$

So, Φ has been proved to be a summational invariant. We note that if the hypothesis (I) on the transition probabilities is adopted, the prove of the theorem 1 becomes simplified.

3. MICROSCOPIC AND MACROSCOPIC DESCRIPTIONS OF THE DISCRETE GAS

Orthonormal bases in $\mathbb{F}$ and in $\mathbb{R}^p$ are introduced :

$$\mathbb{V}^1, \mathbb{V}^2, \ldots, \mathbb{V}^q \qquad \text{in } \mathbb{F} \quad , \tag{10a}$$

$$\mathbb{V}^1, \mathbb{V}^2, \ldots, \mathbb{V}^q, \mathbb{W}^{q+1}, \ldots, \mathbb{W}^p \ \text{in } \mathbb{R}^p \quad . \tag{10b}$$

We thus can write :

$$\mathbb{N} = \Sigma_{\alpha=1}^{q} \ a_\alpha \ \mathbb{V}^\alpha + \Sigma_{\beta=q+1}^{p} \ b_\beta \ \mathbb{W}^\beta \quad , \tag{11a}$$

$$N_i = \Sigma_{\alpha=1}^{q} \ a_\alpha \ V_i^\alpha + \Sigma_{\beta=q+1}^{p} \ b_\beta \ W_i^\beta \quad . \tag{11b}$$

The i-components of $\mathbb{V}^\alpha$ and $\mathbb{W}^\beta$ are denoted by V_i^α and W_i^β. The scalar product in $\mathbb{R}^p$ is denoted $\langle \mathbb{U} , \mathbb{V} \rangle = \Sigma_{i=1}^{p} U_i V_i$. Let be Φ a summational invariant. Using property c) of the Theorem 1, we deduce from the kinetic equations a conservation law for the mean quantity $\langle \Phi , \mathbb{N} \rangle = \Sigma_{i=1}^{p} N_i \varphi_i$:

$$\frac{\partial}{\partial t} \langle \Phi, \mathbb{N} \rangle + \langle \Phi, \mathcal{A}\mathbb{N} \rangle = 0 \ . \tag{12}$$

Two descriptions of the gas can be introduced : a *microscopic description* corresponding to the knowledge of the densities N_i or, equivalently, to the knowledge of the quantities a_α and b_β, and a *macroscopic description* corresponding to the knowledge of the q quantities a_α alone. Among them, there are the number density n, the mean velocity $\vec{u}$, and the temperature T . We emphazise there are q macroscopic conservation laws for the quantities a_α :

$$\frac{\partial}{\partial t} a_\alpha + \langle \mathcal{A}\mathbb{N}, \mathbb{V}^\alpha \rangle = 0 \ . \qquad \alpha = 1, \ldots, q \quad . \tag{13}$$

Particularly by taking the physical summational invariants, we obtain the macroscopic balance laws for the mass, momentum and energy in a classical form :

$$\left\{ \begin{aligned} & \frac{\partial}{\partial t} \rho + \vec{\nabla}.(\rho\vec{u}) = 0 \qquad\qquad (\text{with } \rho = nm) \\ & \frac{\partial}{\partial t} (\rho\vec{u}) + \vec{\nabla}.(\rho\overrightarrow{uu}) + \vec{\nabla}.\ \mathbb{P} = 0 \\ & \frac{\partial}{\partial t} \left[\rho(e + \frac{u^2}{2})\right] + \vec{\nabla}.\left[\rho(e + \frac{u^2}{2})\vec{u} + \vec{u}.\ \mathbb{P} + \vec{q}\ \right] = 0 \ , \end{aligned} \right. \tag{14}$$

where :

$$\begin{cases} n = \sum_{i=1}^{p} N_i \ , \qquad n\vec{u} = \sum_{i=1}^{p} N_i \vec{u}_i \ , \qquad \rho e = knT = \frac{m}{D} \sum_{i=1}^{p} N_i (\vec{u}_i - \vec{u})^2 \ , \\ \mathbb{P} = m \sum_{i=1}^{p} N_i (\vec{u}_i - \vec{u})(\vec{u}_i - \vec{u}) \ , \qquad \vec{q} = \frac{1}{2} \sum_{i=1}^{p} N_i (\vec{u}_i - \vec{u})^2 (\vec{u}_i - \vec{u}) \ , \end{cases} \tag{15}$$

D being the dimension of the physical space.

H-Theorem, Maxwellian State.

The Boltzmann H-function is defined by : $H = \sum_{i=1}^{p} N_i \operatorname{Log} N_i$. We denote by $\operatorname{Log} \mathbb{N}$ the p-component vector $(\operatorname{Log} N_1, \ldots, \operatorname{Log} N_p)$. For a uniform gas, the computations of dH/dt yields the H-Theorem :

$$\frac{dH}{dt} = \langle \operatorname{Log} \mathbb{N} \ , \ C(\mathbb{N}) \rangle \leqslant 0 \ . \tag{17}$$

Proof : By using the equality $\sum_{k=1}^{p} \delta(k, I_r) \operatorname{Log} N_k = \operatorname{Log} N_{I_r}$ and the formula (8), we have

$$\frac{dH}{dt} = \sum_{r=2}^{R} \sum_{I_r \in \mathcal{E}_r} \sum_{J_r \in \mathcal{E}_r} A_{I_r}^{J_r} \left[\operatorname{Log} N_{J_r} - \operatorname{Log} N_{I_r}\right] N_{I_r} \ .$$

With the hypothesis (II) it can be proved that :

$$\frac{dH}{dt} = \sum_{r=2}^{R} \sum_{I_r \in \mathcal{E}_r} \sum_{J_r \in \mathcal{E}_r} A_{I_r}^{J_r} \left[N_{I_r} \left[\operatorname{Log} N_{J_r} - \operatorname{Log} N_{I_r}\right] + N_{I_r} - N_{J_r} \right] . \tag{18}$$

It is important to note that each term of the sum of the right hand side is negative or zero [because $x(\operatorname{Log} y - \operatorname{Log} x) + x - y \leqslant 0$]. So, the H-Theorem is proved. The equality to zero is obtained if each term is equal to zero, that is if $A_{I_r}^{J_r}$ or $N_{I_r} - N_{J_r}$ is equal to zero.

THEOREM 2 : The following properties are equivalent :

a) $\operatorname{Log} \mathbb{N} \in \mathbb{F}$,

b) $\langle \operatorname{Log} \mathbb{N} \ , \ C(\mathbb{N}) \rangle = 0$,

c) $C(\mathbb{N}) = 0$,

d) $A_{I_r}^{J_r} \left[N_{J_r} - N_{I_r}\right] = 0$, $\quad \forall J_r, I_r \in \mathcal{E}_r$, $\quad \forall r = 1, \ldots, R$.

The implication c) ⇒ b) is evident. By using the definition of the summationnal invariants, we have

$$\mathrm{Log}\,\mathbb{N} \in \mathbb{F} \iff A_{I_r}^{J_r}\left[\mathrm{Log}\,N_{J_r} - \mathrm{Log}\,N_{I_r}\right] = 0\ ,\quad \forall\, J_r, I_r \in \mathcal{E}_r,\quad \forall\, r = 2,\dots,R.$$

$$\iff A_{I_r}^{J_r}\left[N_{J_r} - N_{I_r}\right] = 0\ ,\quad \forall\, J_r,\ I_r \in \mathcal{E}_r,\quad \forall\, r = 2,\dots,R\ ;$$

so, the implication a) $\iff$ d) is also proved. Now, we suppose that the property b) is satisfied; it is equivalent to say that the right hand side in Eq. (18) is equal to zero. Therefore, each term of the sum of this righ hand side is separatly equal to zero. In other words we obtain the property d) , and we have proved b) $\Rightarrow$ c) . Finally the implication d) $\Rightarrow$ c) is obvious by recalling the right hand side of Eq. (5).

As usual, the *Maxwellian state* is defined by dH/dt = 0 or Log $\mathbb{N} \in \mathbb{F}$.As in the binary collision theory, we introduce the bases (10) in $\mathbb{F}$ and in $\mathbb{R}^p$, and the decomposition (11) for $\mathbb{N}$. We also have the conservation laws (13). As in the binary collision theory, the densities N_i are function of the macroscopic variables only [11] in a Maxwellian state. The conservation laws for the quantities a_α form a hyperbolic system of equations. These equations are the *Euler equations associated with the model.*

If discrete gases are to be compared with real gases, two difficulties arise. The first one involves the number q of the summational invariants. For many models this number is too large : in addition to the physical invariants, there exists "**spurious**" invariants without physical interpretation. The second one is related to the anisotropic character of the models : in general the pressure tensor $\mathbb{P}$ and the flux vector $\vec{\mathbf{q}}$ are not isotropic. Moreover, in a model gas at rest ($\vec{u} = 0$), $\mathbb{P}$ is not spherical and $\vec{\mathbf{q}}$ not different of zero.

4. SYMMETRY PROPERTIES OF THE MODELS

The presence of multiple collisions reduces the dimension of the space of invariants. In favorable cases ([4],[5]) non physical invariants are removed and there only remain the usual conservation equations (14) of mass, momentum and energy. To compare the hydrodynamical equations derived from the discrete microscopic description with the equations of the classical fluid dynamics, as it was said before, there are some difficulties due to the lack of isotropy of the models. So, we use a particular class of models that generalize the models used in lattice gases ([8],[9]).

Let be $\mathcal{U}$ the given set of velocities $\vec{\mathbf{u}}_i$, i = 1,...,p, and let be $\mathcal{U}_\ell$ the sub-set of the velocities $\vec{\mathbf{u}}_i$ of magnitude c_ℓ with ℓ = 1,..., L. The velocities of $\mathcal{U}_\ell$ are $\vec{u}_i^{\,\ell}$, with $i = 1,\ldots,p(\ell)$ $(\Sigma_{\ell=1}^{L} p(\ell) = p)$. About the velocities $\vec{u}_i^{\,\ell}$ of $\mathcal{U}_\ell$, the following hypotheses are assumed :

H1) $\vec{u}_i^{\,\ell} \in \mathcal{U}_\ell \Rightarrow -\vec{u}_i^{\,\ell} \in \mathcal{U}_\ell$.

H2) $\sum_{i=1}^{p(\ell)} \vec{u}_i^{\,\ell}\vec{u}_i^{\,\ell}$ is a spherical tensor and $\mathbb{K}^\ell = \sum_{i=1}^{p(\ell)} \vec{u}_i^{\,\ell}\vec{u}_i^{\,\ell}\vec{u}_i^{\,\ell}\vec{u}_i^{\,\ell}$ is an isotropical tensor of order 4.

H3) The summational invariants are only the physical ones (mass, momentum and energy).

Due to assumption H2 we can write :

$$\frac{1}{p(\ell)} \sum_{i=1}^{p(\ell)} \vec{u}_i^{\,\ell}\vec{u}_i^{\,\ell} = \frac{1}{D} c_\ell^2 \, \mathbb{I} \, , \qquad \frac{1}{p(\ell)} \sum_{i=1}^{p(\ell)} \vec{u}_i^{\,\ell}\vec{u}_i^{\,\ell}\vec{u}_i^{\,\ell}\vec{u}_i^{\,\ell} = \frac{1}{D(D+2)} c_\ell^4 \, \mathbb{K} \, , \tag{19}$$

with $\mathbb{I}$ and $\mathbb{K}$ such that $I_{\alpha\beta} = \delta_{\alpha\beta}$ and $K_{\alpha\beta\gamma\delta} = \delta_{\alpha\beta}\delta_{\gamma\delta} + \delta_{\alpha\gamma}\delta_{\beta\delta} + \delta_{\alpha\delta}\delta_{\beta\gamma}$.

The hypotheses H1 and H2 correspond to properties of isotropy. We defined as **"good models"** the models satisfying the hypotheses H1, H2, and H3.

5. EULER EQUATIONS ASSOCIATED WITH THE "GOOD MODELS"

First we determine the densities in a Maxwellian state. In such a state it is known that Log $\mathbb{N}$ is in $\mathbb{F}$, or in other words :

$$N_i = \exp\,[h + \vec{q}.\vec{u}_i + k\,(|\vec{u}_i|^2 - c^2)] \tag{20}$$

with $n = \Sigma_{i=1}^{p} N_i$, $n\vec{u} = \Sigma_{i=1}^{p} N_i \vec{u}_i$, $ne = (1/2)\,\Sigma_{i=1}^{p} N_i |\vec{u}_i|^2$, $c^2 = (1/p)\,\Sigma_{i=1}^{p} |\vec{u}_i|^2$.
Equivalently, we have :

$$N_i = \frac{n}{p} \exp\,[\tilde{h} + \vec{q}.\vec{u}_i + k\,(|\vec{u}_i|^2 - c^2)] \tag{21}$$

where $\exp h = (n/p) \exp \tilde{h}$. We can write :

$$\exp \tilde{h} = 1 \Bigg/ \left(\frac{1}{p} \sum_{i=1}^{p} \exp\,[\vec{q}.\vec{u}_i + k\,(|\vec{u}_i|^2 - c^2)] \right) \tag{22a}$$

$$\vec{u} = \left(\sum_{i=1}^{p} \vec{u}_i \exp [\vec{q}.\vec{u}_i + k |\vec{u}_i|^2] \right) / \left(\sum_{i=1}^{p} \exp [\vec{q}.\vec{u}_i + k |\vec{u}_i|^2] \right), \quad (22b)$$

$$e = \left(\sum_{i=1}^{p} \frac{1}{2} |\vec{u}_i|^2 \exp [\vec{q}.\vec{u}_i + k |\vec{u}_i|^2] \right) / \left(\sum_{i=1}^{p} \exp [\vec{q}.\vec{u}_i + k |\vec{u}_i|^2] \right). \quad (22c)$$

The quantities $\vec{q}$ and k do not depend on n . Moreover, the matching between $(\vec{q}, k)$ and $(\vec{u}, e)$ defined by (22b) and (22c) is a bijection [12].

Now we assume :

H4) The gas is close to a "homogeneous state", which mean that it is close to a state in which all the densities are equal ($N_i = n/p$).

This last hypothesis, also used in lattice gas theory ([8],[9]), is important to obtain explicit expressions by performing expansions about a well known state of the gas. Due to the hypotheses **H1** and **H4**, we have $\vec{u} = 0$ and $e = e_0 = (1/2)c^2$ when $\vec{q} = 0$ and $k = 0$. We want to express $\vec{q}$ and k as functions of $\vec{u}$ and $e - e_0 \equiv \bar{e}$. The quantities $\vec{u}/c$ and $(e - e_0)/c^2$ are assumed very small and we denote by δ their order of magnitude. We have the following expansions :

$$\vec{q} = \vec{q}^{\,0} + \mathbb{Q}^1 . \vec{u} + \mathbb{Q}^2 . \vec{u}\, \bar{e} + \mathbb{Q}^3 : \vec{u}\, \vec{u} + \vec{q}^{\,1}\, \bar{e} + \vec{q}^{\,2} (\bar{e})^2 + c\, O(\delta^2) \quad (23)$$

$$k = k^0 + k^1\, \bar{e} + \vec{K}^1 . \vec{u} + \vec{K}^2 . \vec{u}\, \bar{e} + \mathbb{K}^1 : \vec{u}\, \vec{u} + k^2 (\bar{e})^2 + c^2\, O(\delta^2) , \quad (24)$$

where the scalars k^0, k^1, k^2, the vectors $\vec{q}^{\,0}$, $\vec{q}^{\,1}$, $\vec{q}^{\,2}$, $\vec{K}^1$, $\vec{K}^2$, the tensors of order two $\mathbb{Q}^1$, $\mathbb{Q}^2$, $\mathbb{K}^1$, and the tensor of order three $\mathbb{Q}^3$ are to be determined. We put (23) and (24) in (22b) and (22c), then perform the expansions of the exponential terms and finally identify the two sides. Taking into account the hypotheses **H1** and **H2** , and after some algebraic calculus, it can be shown that :

$$k^0 = 0, \quad ,k^1 = \frac{2}{a_2^4 - a_1^4}, \quad k^2 = \frac{2[3a_1^2 a_2^4 - 2a_1^6 - a_3^6]}{[a_2^4 - a_1^4]^3} ,$$

$$\vec{q}^{\,0} = 0, \quad \vec{q}^{\,1} = 0, \quad \vec{q}^{\,2} = 0, \quad \vec{K}^1 = 0, \quad \vec{K}^2 = 0 ,$$

$$\mathbb{Q}^1 = \frac{D}{a_1^2} \mathbb{I}, \quad \mathbb{Q}^2 = - \frac{2D}{a_1^4} \mathbb{I}, \quad \mathbb{Q}^3 = 0, \quad \mathbb{K}^1 = - \frac{D}{2a_1^4} \mathbb{I} ,$$

with $(1/p) \sum_{i=1}^{p} |\vec{u}_i|^{2k} = a_k^{2k}$ $(a_1 \equiv c)$. We remark that if the moduli of the velocities $\vec{u}_i$ are different, $a_2^4 - a_1^4$ are strictly positive.

In conclusion :

$$\vec{q} = \frac{D}{c^2}\vec{u} - \frac{2D}{c^4}\bar{e}\,\vec{u} + c^{-1}O(\delta^3)$$

$$k = \frac{2}{a_2^4 - a_1^4}\bar{e} - \frac{D}{2c^4}|\vec{u}|^2 + \frac{2[3a_1^2 a_2^4 - 2a_1^6 - a_3^6]}{[a_2^4 - a_1^4]^3}(\bar{e})^2 + c^{-2}O(\delta^3),$$

and also

$$\exp\tilde{h} = 1 - \frac{D}{2c^2}|\vec{u}|^2 - \frac{2}{a_2^4 - a_1^4}(\bar{e})^2 + O(\delta^3).$$

After some algebraic calculus, we get the final result for the Maxwellian densities N_i :

$$N_i = \frac{n}{p}\Bigg\{1 + \frac{D}{c^2}\vec{u}.\vec{u}_i + \frac{2[|\vec{u}_i|^2 - c^2]}{a_2^4 - a_1^4}\bar{e} + \frac{D^2}{2c^4}\left(\vec{u}_i\vec{u}_i - \frac{|\vec{u}_i|^2}{D}\mathbb{I}\right) : \vec{u}\,\vec{u}$$

$$+ \frac{2D}{c^4}\,\frac{c^2|\vec{u}_i|^2 - a_2^4}{a_2^4 - a_1^4}[\vec{u}.\vec{u}_i]\,\bar{e}$$

$$+ \frac{2\left([a_2^4 - a_1^4]\,|\vec{u}_i|^4 + [a_1^2 a_2^4 - a_3^6]\,|\vec{u}_i|^2 + [a_1^2 a_3^6 - a_2^8]\right)}{[a_2^4 - a_1^4]^3}(\bar{e})^2 + O(\delta^3)\Bigg\}.$$

With these expressions for the densities, we can give explicit expressions for the pressure tensor $\mathbb{P}$ and the heat flux vector $\vec{q}$:

$$\mathbb{P} = m\sum_{i=1}^{p} N_i(\vec{u}_i - \vec{u})(\vec{u}_i - \vec{u})$$

i.e.
$$\mathbb{P} = \rho\left[\frac{2e}{D} - \frac{1}{D+2}\frac{a_2^4}{a_1^4}|\vec{u}|^2\right]\mathbb{I} + \rho\frac{D}{D+2}\frac{a_2^4}{a_1^4}\vec{u}\,\vec{u} - \rho\,\vec{u}\,\vec{u}, \qquad (25)$$

$$\vec{q} = \frac{1}{2}m\sum_{i=1}^{p} N_i\,|(\vec{u}_i - \vec{u})|^2(\vec{u}_i - \vec{u})$$

i.e.
$$\vec{q} = \frac{1}{2}\rho\,c^2\left\{\left[\frac{2a_2^8 - a_2^4 a_1^4 - a_1^2 a_3^6}{a_1^4[a_2^4 - a_1^4]}\right] + \frac{2}{c^2}\left[\frac{a_3^6 a_1^2 - a_2^8}{a_1^2\left[a_1^4 a_2^4 - a_1^8\right]} - \frac{D+2}{D}\right]e\right\}\vec{u}. \qquad (26)$$

The Euler equations are the conservation laws (14) with the pressure tensor $\mathbb{P}$ and the heat flux vector $\vec{q}$ previously determined (at the order $O(\delta^3)$). Now it is possible to give the momentum and energy equations of the Euler equations in two equivalent forms. The first one is

$$\frac{\partial}{\partial t}(\rho\vec{u}) + \vec{\nabla}.(\rho\vec{u}\vec{u}) = -\vec{\nabla}.\ \mathbb{P} \tag{27}$$

$$\frac{\partial}{\partial t}(\rho e) + \vec{\nabla}.(\rho e\vec{u}) = -\vec{\nabla}.\ (\mathbb{P}.\vec{u} + \vec{q}) \tag{28}$$

with $\mathbb{P}$ and $\vec{q}$ given by the relations (25) and (26). The second form is where (ρ_0 is a reference density) :

$$\frac{\partial}{\partial t}(\rho\vec{u}) + \frac{a_2^4}{a_1^4}\frac{D}{D+2}\vec{\nabla}.(\rho\vec{u}\vec{u}) = -\vec{\nabla}\left[\frac{2}{D}\rho e - \frac{a_2^4}{a_1^4}\frac{1}{D+2}\rho|\vec{u}|^2\right] + \rho_0 c\ O(\delta^3)\ , \tag{29}$$

$$\frac{\partial}{\partial t}(\rho e) + \left[\frac{a_3^6 a_1^2 - a_2^8}{a_1^4[a_2^4 - a_1^4]} - \frac{2}{D}\right]\vec{\nabla}.(\rho e\vec{u}) = -\vec{\nabla}.[\frac{2}{D}\rho e\vec{u}]$$
$$+ \frac{a_1^2 a_3^6 + a_1^4 a_2^4 - 2a_2^8}{2a_1^4[a_2^4 - a_1^4]}\vec{\nabla}.(\rho\vec{u}) + \rho_0 c^2 O(\delta^3)\ . \tag{30}$$

The equations (27) and (28) correspond to the usual presentation in kinetic theory. $\mathbb{P}$ is not spherical; the hydrostatic pressure is defined as $\pi \equiv (1/D)\ \mathrm{Tr}\ \mathbb{P} = (2/D)\ \rho[e - |\vec{u}|^2/2]$. In a same way, $\vec{q}$ is not equal to zero. The latter ones, (29) and (30), have the form adopted in the lattice gas theory [9]. They are very close to the momentum and energy equations of the real fluid dynamics. However, the coefficient in front of $\vec{\nabla}.(\rho\vec{u}\vec{u})$ in (29) is $[a_2^4\ /\ a_1^4]\ (D/(D+2))$ and is different from 1. In lattice gas theory, in addition, this coefficient depends on ρ (its value is $D(p-2\rho)/[(D+2)(p-\rho)]$) if all the moduli are equal. The difference is due to the exclusion principle used in the lattice gas theory. In the energy equation, we remark that the coefficient of $\vec{\nabla}.(\rho e\vec{u})$ is not equal to 1. We have a constant coefficient depending on the model. Finally we make two other remarks :

1) It is possible to define a scalar pressure defined as :

$$\tilde{\pi} = \frac{2}{D}\rho e - \frac{a_2^4}{a_1^4}\frac{1}{D+2}\rho|\vec{u}|^2\ , \tag{31}$$

2) It is possible to introduce the heat flux vector $\vec{q}$ defined by :

$$\vec{q} = \left(\frac{2a_2^8 - a_1^2 a_3^6 - a_1^4 a_2^4}{2a_1^2 [a_2^4 - a_1^4]} + \frac{1}{D+2} \frac{a_2^4}{a_1^4} |\vec{u}|^2 \right) \rho\vec{u} \, . \tag{32}$$

The coefficient in the bracket is always positive. The Euler equations have the following "good" form :

$$\frac{\partial}{\partial t}(\rho\vec{u}) + \frac{a_2^4}{a_1^4} \frac{D}{D+2} \vec{\nabla}.(\rho\vec{u}\vec{u}) = - \vec{\nabla}\, \tilde{\pi} \, , \tag{33}$$

$$\frac{\partial}{\partial t}(\rho e) + \left[\frac{a_3^6 a_1^2 - a_2^8}{a_1^4 [a_2^4 - a_1^4]} - \frac{2}{D} \right] \vec{\nabla}.(\rho e\vec{u}) = - \vec{\nabla}.(\tilde{\pi}\, \vec{u}) - \vec{\nabla}.\vec{q} \, . \tag{34}$$

6. CONCLUSION

In conclusion, it should be emphasized that the discrete kinetic equations have a good structure. For a class of models with properties of isotropy and for which the summational invariants are reduced to the physical ones, we have a rather good agreement between the Euler equations associated with the model and the classical equations of the fluid dynamics. However, it is possible to differently organize the terms of the equations (27) and (28). Other forms for these Euler equations associated with the models have been proposed [3]. One question is : what is the best form for the comparison with the classical Euler equations ? To answer this question, it should be necessary to study the thermodynamics of the discrete gas and also some simple flow problems.

This work was supported by grants from DRET n° 89/1554.

REFERENCES

[1] C. BARDOS, F. GOLSE and D. LEVERMORE, "Fluid dynamic limits of discrete velocity kinetic equations", in *"Advances in Kinetic Theory and Continuum Mechanics"*, Ed. by R. Gatignol and Soubbaramayer, Springer-Verlag, 1991.

[2] J.M. BONY, "Existence globale et diffusion en théorie cinétique discrète", in *"Advances in Kinetic Theory and Continuum Mechanics"*, Ed. by R. Gatignol and Soubbaramayer, Springer-Verlag, 1991.

[3] P. CHAUVAT and R. GATIGNOL, "Transport coefficients in discrete kinetic theory and comparison with classical fluid dynamics", Proc. of 17th Int. Symp. on Rarefied Gas Dynamics, Aachen, 1990, (to appear).

[4] P. CHAUVAT, "Summational invariants in discrete kinetic theory with multiple collisions", Mechanics Research Communications, (to appear).

[5] P. CHAUVAT and R. GATIGNOL, "Coefficients de transport pour des modèles réguliers plans en théorie cinétique discrète", C. R. Acad. Sc., Paris, 309, Série II, (1989), p. 1457-1462.

[6] H.CORNILLE, "Exact positive (2+1)-dimensional solutions of discrete Boltzmann models", Proceedings of the 16th Int. Symp. on Rarefied Gas Dynamics, Pasedena, Vol. II, (1988), p. 131-154.

[7] F. COULOUVRAT and R.GATIGNOL, "Description hydrodynamique d'un gaz en théorie cinétique discrète : les modèles réguliers", C. R. Acad. Sc. Paris, 306, (1988), p. 393-398.

[8] U. FRISCH, B. HASSLACHER, and Y. POMEAU, "Lattice gas automata for Navier-Stokes equation", Physical Review Letters, 56, (1986), p. 1505 -1508.

[9] U. FRISCH, D. D'HUMIERES, B. HASSLACHER, P. LALLEMAND, Y. POMEAU and J.P. RIVET, "Lattice gas hydrodynamics in two and three dimensions", Complex Systems, 1, (1987), p. 649-707.

[10] R. GATIGNOL, "Théorie cinétique d'un gaz à répartition discrète de vitesses", Z. Flugwissenschaften, 18, (1970), p. 93-97.

[11] R. GATIGNOL, *"Théorie cinétique des gaz à répartition discrète de vitesses"*, Lectures Notes in Physics, 36, Springer-Verlag, 1975.

[12] R. GATIGNOL, "The hydrodynamical description for a discrete velocity model of gas", Complex Systems, 1, (1987), p. 709-725.

[13] R. GATIGNOL, "Constitutive laws for discrete velocity gases", Proc. 17th Int. Symp. on Rarefied Gas Dynamics, Aachen, 1990, (to appear).

[14] S. KAWASHIMA, and H. CABANNES, "Initial value problem in discrete kinetic theory", 16th Int. Symp. Rarefied Gas Dynamics, Pasedena, Vol. II, (1988), p. 148-154.

[15] S. KAWASHIMA,and N. BELLOMO, "On the Euler equation arriving in discrete kinetic theory", in *"Advances in Kinetic Theory and Continuum Mechanics"*, Ed. by R. Gatignol and Soubbaramayer, Springer-Verlag, 1991.

[16] T. PLAKOWSKY and R. ILLNER, "Discrete velocity models of the Boltzmann equation : a survey on the mathematical aspect of the theory", SIAM Review, 30, (1988), p. 213-255.

[17] Y. H. QIAN, D. D'HUMIERES, P. LALLEMAND, "A one dimension lattice Boltzmann equation with galilean invariance", in *"Advances in Kinetic Theory and Continuum Mechanics"*, Ed. by R. Gatignol and Soubbaramayer Springer-Verlag, 1991.

On the Semidiscrete Boltzmann Equation with Multiple Collisions

E. Longo and N. Bellomo

Dipartimento di Matematica, Politecnico di Torino,
Corso degli Abruzzi 24, I-10129 Torino, Italy

Abstract: This paper deals with the derivation, in the framework of the discrete kinetic theory, of an evolution equation for gas particles undergoing multiple collisions. In particular, we consider a gas of particles moving in all directions of the plane with only one velocity modulus and undergoing binary and triple collisions. An evolution equation is derived, the analysis of the Maxwellian state is studied in details, a formal H–theorem is derived and the formal derivation of the hydrodynamical equations is dealt with.

(Key words: Kinetic theory, Discrete Boltzman equation, H–theorem, Maxwellian state).

1. Introduction

One of the crucial topics in nonlinear kinetic theory is the derivation of kinetic evolution equations for dense gases. It is well known that if nd^3 (where n is the number density and d is the radius of the gas particles) is small with respect to the unity then one can use the classical Boltzmann equation [1,2]. On the other hand, as such a term becomes of the order of unity, Boltzmann's model can be reasonably replaced by Enskog–type models [3,4], which take into account the overall dimensions of the spheres, still neglecting the probability of triple collisions with respect to the one of binary collisions.

For higher densities triple and multiple collisions should also be taken into account as the probability of their event may become of an order non negligible with respect one of the probability of binary collisions.

Unfortunately, as well discussed in the papers by Cohen [4] and Sengers [5], the original Boltzmann arguments which is valid in the case of simple binary collisions cannot be straightforwardly extended to the case of multiple collisions. This is essentially due to the difficulty of a correct formulation of a three (or more) body collisions. As a matter of fact only formal expressions of the Boltzmann equation with both binary and triple collisions are available in the literature.

This difficulty can be partially overcome in the case of the semidiscrete Boltzmann equation, a mathematical model proposed by Cabannes [6], which defines the time–space evolution of the one particle number density distribution function of equal particles moving on a plane with only one velocity modulus.

This model, an integro–differential equation, was proposed for a simple monoatomic gas with binary collisions only [6]. Some generalizations of Cabannes' original model can be found in refs.[7, 8]. The initial and initial–boundary value problem has been studied by various authors [7, 8, 9, 10] under suitable assumptions on the initial and boundary conditions. Some related fluid dynamic applications are studied in [11, 12].

The analysis developed in this paper is essentially founded on the discrete kinetic theory with multiple collisions, a topic posed in Gatignol's lecture note [13] and developed in some recent papers [14, 15, 16, 17] which deal with various aspects of the discrete kinetic theory with binary and higher order collisions.

The most interesting aspects of the analysis developed in this paper are essentially two:

- It has been possible deriving, for the semidiscrete Boltzmann equation, an explicit expression of an evolution equation in the case of binary and triple collisions which is not possible to obtain for the full Boltzmann equation in some explicit form.
- The semidiscrete Boltzmann equation with binary and triple collisions is characterized by a unique definition of the maxwellian state in terms of the macroscopic observables, which is not the case of the semidiscrete Boltzmann equation with binary collisions only. This is due to the fact that the space of the collision invariants is characterized by the correct number of collision invariants corresponding to conservation of mass and momentum (the energy is implicitly conserved by the fact that the gas is characterized by only one velocity modulus).

Of course the model, which is characterized by a rather simple mathematical structure, shows some limitations (which are critically reviewed in the paper) essentially due to the limitation of a single velocity modulus. However because of the reasons which have been stated above and thanks to the fact that the model can describe some relevant moderately dense gas effects, the semidiscrete Boltzmann equation may be regarded, in the end, as a useful mathematical model in the kinetic theory of gases and can hopefully describe interesting fluid dynamic aspects.

The paper is in four sections. The first one is this introductory section. The second one proposes the formal structure of the model. The third one deals with the analysis of the Maxwellian state. The fourth one with the analysis of the hydrodynamical equations and finally a general discussion and some idea to construct specific models follow in the same Section.

2. The Semidiscrete Boltzmann Equation with Multiple Collisions

The semidiscrete Boltzmann equation is a mathematical model of the classical kinetic theory of gases which describes the time–space evolution of a system of particles with equal mass moving in all directions of the (x, y)–plane with velocity

$$\mathbf{v}(\theta) = c(\cos\theta\mathbf{i} + \sin\theta\mathbf{j}) \tag{2.1}$$

where c is a constant and $\mathbf{i}$, $\mathbf{j}$ are the unit vectors of the orthogonal axes x and y. $\mathbf{v}$ forms an angle θ with $\mathbf{i}$.

Such a model describes the evolution of the one particle number density

$$N = N(t, \mathbf{x}; \theta) \quad : \quad \Re_+ \times \Re^2 \times [0, 2\pi) \longmapsto \Re_+ \quad . \tag{2.2}$$

The evolution equation was derived, in the case of simple binary collisions only, by Cabannes [6] from the regular $2r$–velocity model due to Gatignol [13] simply letting r tend to infinity. Cabannes' model can be written as

$$(\frac{\partial}{\partial t} + \mathbf{v}(\theta) \cdot \nabla_{\mathbf{x}})N(t,\mathbf{x};\theta) =$$

$$= \frac{cS}{\pi} \int_{[0,2\pi)} [N(t,\mathbf{x};\varphi)N(t,\mathbf{x};\varphi+\pi) - N(t,\mathbf{x};\theta)N(t,\mathbf{x};\theta+\pi)]\, d\theta \quad . \tag{2.3}$$

A more general framework can be provided in order to derive a semidiscrete equation with binary and triple collisions in a fashion that eq.(2.3) can be regarded as a particular case of a more general class of evolution equations.

Keeping this in mind, some notations can now be introduced. Let then

$$\Theta_p = \{\theta_1, \dots, \theta_p\} \in I_\theta^p \quad , \quad p = 2,3 \quad , \quad \theta_i \in [0, 2\pi) \tag{2.4a}$$

$$\Phi_p = \{\varphi_1, \dots, \varphi_p\} \in I_\varphi^p \quad , \quad p = 2,3 \quad , \quad \varphi_i \in [0, 2\pi) \tag{2.4b}$$

be ,respectively, the sets of all pre–collisional and post–collisional, velocity directions corresponding to a collision with p particles. Moreover, let $a(\Theta_p, \Phi_p)$ be the transition probability density that a collision with a group of particles colliding with velocity directions Θ_p is re–emitted with velocity directions Φ_p The corresponding transition rates will be indicated with the notation $A(\Theta_p, \Phi_p)$.

Remark 2.1: The classical assumptions of indistinguishability of the particles and reversibility of the collisions yields

$$p = 2,3: \ a(\Theta_p, \Phi_p) = a(\Phi_p, \Theta_p) \tag{2.5}$$

for every permutation of $\Theta_p, \Phi_p \in I^p$. The same property holds for the terms A •

Remark 2.2: The terms a have to be normalized, as probability densities, with respect to the pre–collisional or post–collisional angles •

If one assumes that the terms A are given, then a formal expression of the semidiscrete Boltzmann equation can be written in the following fashion

$$(\frac{\partial}{\partial t} + \mathbf{v}(\theta) \cdot \nabla_{\mathbf{x}})N = J[N] = J_2[N] + J_3[N] \tag{2.6}$$

where

$$J_p[N](\theta_1 = \theta) = \int_\Omega A(\Theta_p, \Phi_p) \prod_{i=1}^{p} [N(t,\mathbf{x};\varphi_i) - N(t,\mathbf{x};\theta_i)] d\Omega \tag{2.7}$$

and $d\Omega = d\Phi_p d\Theta_p^\star$, where

$$d\Phi_p = d\varphi_1 \dots d\varphi_p \ \text{ for } \ p = 2,3 \tag{2.8}$$

$$p = 2 \ : \ d\Theta_p^\star = d\theta_2 \quad , \quad p = 3 \ : \ d\Theta_p^\star = d\theta_2 d\theta_3 \tag{2.9}$$

The model can then be specified by a detailed calculation of the terms A. An example of specific model will be presented in the last section. However, still remaining at a formal level, we can classify and compute the admissible collisions. With this in mind we can now consider binary and triple collisions separately.

Binary collisions: Consider now binary collisions of the type

$$\{\mathbf{v}(\theta_1), \mathbf{v}(\theta_2)\} \longleftrightarrow \{\mathbf{v}(\varphi_1), \mathbf{v}(\varphi_2)\} \quad .$$

It is simple to verify that a collision mechanics with conservation of momentum and energy yields

$$\theta_1 = \theta \ , \ \theta_2 = \theta + \pi \ , \ \varphi_1 = \varphi \ , \ \varphi_2 = \varphi + \pi \tag{2.10}$$

which are the only "non-trivial" collisions which give rise to changes in the fluxes.

Remark 3.2: The expression of the transition rates corresponding to this type of collisions is

$$A(\Theta_2, \Phi_2) = 2cSa(\Theta_2, \Phi_2) = 2cS\delta(\theta + pi - \theta_2)\frac{U(\varphi)}{2\pi}\delta(\varphi + \pi - \varphi_2) \tag{2.11}$$

where $\varphi, \theta \in [0, 2\pi)$, whereas δ denotes the Dirac delta function and U the uniform constant distribution •

Substituting (2.10,11) into (2.8-9) yields the expression of J_2 already defined by the right-hand side term of eq.(2.3).

Triple Collisions: Consider now triple collisions of the type

$$\{\mathbf{v}(\theta_1), \mathbf{v}(\theta_2), \mathbf{v}(\theta_3)\} \longleftrightarrow \{\mathbf{v}(\varphi_1), \mathbf{v}(\varphi_2), \mathbf{v}(\varphi_3)\}$$

characterized by a collision mechanics preserving momentum and energy. Technical calculations, based upon the aforementioned conservation equations, yield

$$\begin{aligned} &\theta_1, \theta_2, \theta_3 \in [0, 2\pi) \\ &\varphi_1 \in D_\varphi = [\varphi_m, \varphi_M](\Theta_3) \\ &\varphi_2 = \varphi_2(\Theta_3, \varphi_1) = \mu(\Theta_3, \varphi_1) + \gamma(\Theta_3, \varphi_1) \\ &\varphi_3 = \varphi_3(\Theta_3, \varphi_1) = \mu(\Theta_3, \varphi_1) - \gamma(\Theta_3, \varphi_1) \end{aligned} \tag{2.12}$$

where if

$$d = d(\Theta_3) = \Big[\Big(\sum_{i=1}^{3} \cos\theta_i\Big)^2 + \Big(\sum_{i=1}^{3} \sin\theta_i\Big)^2\Big]^{1/2} \quad , \quad d \in [0, 3] \tag{2.13}$$

and

$$\cos\theta_s = \frac{1}{d}\sum_{i=1}^{3} \cos\theta_i \quad , \quad \sin\theta_s = \frac{1}{d}\sum_{i=1}^{3} \sin\theta_i \tag{2.14}$$

then

$$\cos\mu = \frac{d\cos\theta_s - \cos\varphi_i}{2\cos\gamma}$$

$$\sin\mu = \frac{d\sin\theta_s - \sin\varphi_i}{2\cos\gamma} \tag{2.15}$$

$$\cos\gamma = \frac{1}{2}[d^2 + 1 - 2d\cos(\theta_s - \varphi_1)]^{1/2}$$

and where the domain D_φ of the variable φ_1 depends upon the pre–collisional angles as follows

$$\begin{aligned} d \in [0,1]: \quad & \varphi_m = 0 \quad , \quad \varphi_M = 2\pi \\ d \in (1,3]: \quad & \varphi_m = \theta_s - \arccos\left|\frac{d-3}{2d}\right| \quad , \quad \varphi_M = \theta_s + \arccos\left|\frac{d-3}{2d}\right| \end{aligned} \tag{2.16}$$

Remark 2.4: Once Θ_p and $\varphi_1 \in D_\varphi$ are fixed, then φ_2 and φ_3 are univocally determined by eqs.(2.12) •

It is useful to rescale eq.(2.7) in order to define a quantitative estimate of the order of magnitude of the two collision terms. Then, following [14], we introduce the two parameters

$$\epsilon = \frac{1}{\rho^2 \ell_c N_c} \quad , \quad \eta = \rho^3 N_c \tag{2.17}$$

where ρ is the radius of the gas particles and ℓ_c and N_c are, respectively, a reference length and number density. Therefore if t , x , $\mathbf{v}$ and N are normalized with respect to t_c , ℓ_c , ℓ_c/t_c and N_c, respectively, (where t_c is a reference time) and the cross section is normalized with respect to ρ^2, then the dimensionless equation is obtained as follows

$$(\frac{\partial}{\partial t} + \mathbf{v}(\theta)\cdot\nabla_{\mathbf{x}})N = \frac{1}{\epsilon}\{J_2[N] + \eta J_3[N]\} \quad . \tag{2.18}$$

Remark 2.5: ϵ is proportional to the Knudsen number referred to ℓ_c i. e. $\epsilon \propto Kn$, and has physical meaning if of a smaller order with respect to one. This shows that the term involving triple collisions is, for moderately dense gases, less relevant that the term involving binary collisions. Nevertheless it is, as we shall see, of a relevant importance in defining good properties of the model •

3. Collision Invariants and Maxwellian State

As known in the kinetic theory of gases [16], a function of the velocity $\psi = \psi(\mathbf{v})$ is defined collision invariant if its mean variation in the unit volume and time due to the collisions is equal to zero.

Referring to the mathematical model proposed in the second section, such a function can be regarded as a function of the angle θ

$$\psi = \psi(\mathbf{v}(\theta)) = \psi(\theta) \quad , \quad \theta \in [0, 2\pi) \tag{3.1}$$

The mean variation of ψ due to the binary and triple collision, respectively, is then

$$\int_{[0,2\pi)} \psi(\theta) J_2[N](\theta)\, d\theta \tag{3.2}$$

$$\int_{[0,2\pi)} \psi(\theta) J_3[N](\theta)\, d\theta \; . \tag{3.3}$$

Therefore, the following definition can be provided:

Definition 3.1: A function $\psi = \psi(\theta)$ of the admissible velocities (2.1) is a collision invariant if

$$\forall N(\theta) \geq 0 \;,\; p = 2,3 \;:\; \int_{[0,2\pi)} \psi(\theta) J_p[N](\theta)\, d\theta = 0 \quad . \tag{3.4}$$

This condition defines a linear subspace of the space of the 2π–periodic functions with respect to θ. Such a space is called Collision invariant space and will be indicated by $\mathcal{M}$, in what follows •

The behaviour of a collision invariant $\psi \in \mathcal{M}$ referred to the binary and triple collisions is defined by the following

LEMMA 1: *The following conditions hold:*

i) *The equality*

$$\int_{[0,2\pi)} \psi(\theta) J_2[N](\theta)\, d\theta = 0 \;\; , \;\; \forall N \geq 0$$

is verified iff the following condition

$$\psi(\theta) + \psi(\theta + \pi) = \psi(\varphi) + \psi(\theta + \pi) \; ; \;\; \theta, \varphi \in [0, 2\pi) \tag{3.5}$$

holds for every binary collision.

ii) *The equality*

$$\int_{[0,2\pi)} \psi(\theta) J_3[N](\theta)\, d\theta = 0 \;\; , \;\; \forall N \geq 0$$

is verified iff the following condition

$$\psi(\theta_1) + \psi(\theta_2) + \psi(\theta_3) = \psi(\varphi_1) + \psi(\varphi_2) + \psi(\varphi_3) \tag{3.6}$$

holds for every admissible triple collision.

In other words $\psi \in \mathcal{M}$ if and only if conditions (3.5, 3.6) hold.

Proof: The proof of condition i) is classical in the literature [9]. In fact it is an immediate extension to the semidiscrete Boltzmann equation of what is already known for the discrete Boltzmann equation. Condition ii) is proven exactly in the same fashion of i). That is using the definition (3.1) of collision invariant joined to the reversibility and indistinguishability properties stated in Remark 1.2.[]

The space L^p of the 2π–periodic functions (with respect to the variable θ) is defined by the basis

$$\{1, \cos(\theta), \sin(\theta), \cos(2\theta), \sin(2\theta), \ldots\} \quad . \tag{3.7}$$

Consequently the space $\mathcal{M} \in L^p$ needs to be referred to the basis (3.7). Keeping this in mind, the following can be proven

THEOREM 1 *The space $\mathcal{M}$ is the linear subspace of L^p spanned by the basis*

$$\{1, \cos\theta, \sin\theta, 0, \dots\} \quad . \tag{3.8}$$

In other words, $\psi = \psi(\theta)$ is a collision invariant iff it can be expressed in the form

$$\psi(\theta) = a_0 + a_1 \cos\theta + a_2 \sin\theta \tag{3.9}$$

where a_0 , a_1 , a_2 do not depend upon θ.

Proof: The functions 1 , $\cos\theta$, $\sin\theta$ all satisfy conditions (3.5-6). Then $\psi(\theta)$, in the form defined in (3.9), is a collision invariant. We have now to prove that only those functions defined by (3.9) are collision invariants, i. e.

$$\forall n \in \mathcal{N} \quad , \quad n \geq 2 \; : \quad \cos(n\theta) \notin \mathcal{M} \quad , \quad \sin(n\theta) \notin \mathcal{M} \quad . \tag{3.10}$$

Keeping this in mind, let us preliminarly state some remarks which particularize Lemma 1 in the case of the functions $\cos(n\theta)$ and $\sin(n\theta)$. In particular, one has

$$\forall n \in \mathcal{N} \; : \quad \cos(n\theta) \in \mathcal{M} \iff \sin(n\theta) \in \mathcal{M} \quad .$$

Moreover one has $\cos(n\theta) \in M$ (namely $\sin(n\theta) \in M$) if and only if both the following conditions hold

i) $\forall\theta \in [0, 2\pi) \; : \; \cos(n\theta) + \cos[n(\theta + \pi)] = \sin(n\theta) + \sin[n(\theta + \pi)] = 0 \quad ;$

ii) The collision

$$\{\mathbf{v}(n\theta_1), \mathbf{v}(n\theta_2), \mathbf{v}(n\theta_3)\} \longleftrightarrow \{\mathbf{v}(n\varphi_1), \mathbf{v}(n\varphi_2), \mathbf{v}(n\varphi_3)\}$$

is admissible $\forall\theta_1, \theta_2, \theta_3, \varphi_1, \varphi_2, \varphi_3 \in [0, 2\pi)$ such that also the collision

$$\{\mathbf{v}(\theta_1), \mathbf{v}(\theta_2), \mathbf{v}(\theta_3)\} \longleftrightarrow \{\mathbf{v}(\varphi_1), \mathbf{v}(\varphi_2), \mathbf{v}(\varphi_3)\}$$

is an admissible one in the terms defined in eqs.(2.12).

Note that condition ii) implies, in particular, the following $\forall\theta \in [0, 2\pi)$:

$$\cos(n\theta) + \cos[n(\theta + \frac{2}{3}\pi)] + \cos[n(\theta + \frac{4}{3}\pi)] = \sin(n\theta) + \sin[n(\theta + \frac{2}{3}\pi)] + \sin[n(\theta + \frac{4}{3}\pi)] = 0 \quad . \tag{3.11}$$

It is consequently immediate verifying the condition stated in (3.10) when $n = 2k$ and when $n = 3k$ where $k \in \mathcal{N}$. In fact if

$$n = 2k \; , \quad \forall\theta \in [0, 2\pi) \; : \quad \cos(2k\theta) = \cos[2k(\theta + \pi)]$$

therefore condition i) is violated. On the other hand if $n = 3k$ then

$$\cos(3k\theta) = \cos[3k(\theta + \frac{2}{3}\pi)] = \cos[3k(\theta + \frac{4}{3}\pi)]$$

therefore condition (3.11) is violated. At this end the only case left out is the case

$$n = (6k \pm 1) \quad , \quad \forall k \in \mathcal{N}$$

namely when $n > 1$ is odd and is not a multiple of 3.

Consider then the collision with pre-collisional velocities

$$\{\mathbf{v}(\theta_1 = \theta), \mathbf{v}(\theta_2 = \theta + \frac{\pi}{3}), \mathbf{v}(\theta_3 = \theta + \frac{2}{3}\pi)\}$$

and post-collisonal velocities

$$\{\mathbf{v}(\varphi_1 = \theta_1 + w_n), \mathbf{v}(\varphi_2), \mathbf{v}(\varphi_3)\}$$

where $\theta \in [0, 2\pi)$, φ_2 and φ_3 are the angles consistent with the collision dynamics defined in eqs.(2.12) and where

$$w_n = \frac{\pi}{2n} \quad \text{when } n = 6k - 1 \ , \ k \in \mathcal{N}$$

$$w_n = -\frac{\pi}{2n} \quad \text{when } n = 6k + 1 \ , \ k \in \mathcal{N} \quad .$$

According to Eqs. (2.12-16), this collision is admissible $\forall \theta \in [0, 2\pi)$ and therefore the angles φ_2 and φ_3 are univocally identified by the value of n and θ. On the other hand, this collision violates condition ii). In fact applying the change of variable

$$i = 1, 3 \ : \ \theta_i \to n\theta_i \ , \ \varphi_i \to n\varphi_i$$

condition ii) yelds the non-consistent inequality:

$$|\sin[n(\varphi_2 - \theta)] + \sin[n(\varphi_3 - \theta)]| = 1 + \sqrt{3} > 2 \quad .$$

Then condition (3.10) holds in this case, too, and the Theorem is proved as all admissible collisions have been taken into account.[]

Remark 3.1: The moments of the number density $N(t, \mathbf{x}; \theta)$ with respect to the basis defined in Theorem 1 correspond to the macroscopic hydrodynamic variables: number density n and mass velocity U. In fact

$$\int_{[0,2\pi)} N(\theta)\, d\theta = n \ , \ \int_{[0,2\pi)} N(\theta) \cos\theta \, d\theta = \frac{nU_1}{c} \ , \ \int_{[0,2\pi)} N(\theta) \sin\theta \, d\theta = \frac{nU_2}{c}$$

where U_1 and U_2 are the component of $\mathbf{U}$ along the two coordinate axes. Note that in some previous calculations and sometime in what follow θ replaces, for practical reasons, θ_1 •

We can now deal with the analysis of the Maxwellian state referred to the mathematical model described by eqs.(2.19). After the analysis which has been developed above the following definition can be proposed

Definition 3.2: The number densities $N = N(t, \mathbf{x}, \theta)$ which are strictly positive $N > 0$ and are such that

$$\log N \in \mathcal{M} \Rightarrow N(t, \mathbf{x}, \theta) = \exp\{u_o(t, \mathbf{x}) + u_1(t, \mathbf{x}) \cos\theta + u_2(t, \mathbf{x}) \sin\theta\} \qquad (3.12)$$

characterize the Maxwellian equilibrium state of the system •

The quantities $\mathbf{u} = \{u_0, u_1, u_2\}$ will be called, in what follows, the parameters of the Maxwellian state and the densities which satisfy the condition of Definition 3.2 will be written as $N = N(\mathbf{u}(t, \mathbf{x}), \theta)$.

A complete characterization of the Maxwellian state according to Definition 3.2 is then given by the following

THEOREM 2: *The following conditions are equivalent*

i) $\forall N > 0$ *and* $p = 2, 3$ *the following inequality holds*

$$\int_{[0,2\pi)} \log N(\theta) J_p[N](\theta) d\theta \leq 0 \tag{3.13}$$

where the equality sign holds iff $J_p[N] = 0$.

ii) Conditions $\log N \in \mathcal{M}$ *holds only for all and only those* $N > 0$ *such that*

$$J[N] = J_2[N] + J_3[N] = 0 \tag{3.14}$$

iii) $\forall N > 0$, *the spatially homogeneous Boltzmann H-function*

$$H(t) = \int_{[0,2\pi)} N(t, \theta) \log N(t, \theta) d\theta \tag{3.15}$$

is monotone decreasing $dH/dt \leq 0$, *and such that the equality sign* $dH/dt = 0$ *holds if and only if* $\log N \in \mathcal{M}$.

COROLLARY TO THEOREM 2: *The Maxwellian densities defined in (3.11) are univocally defined in terms of the fluid-dynamic parameters* n , U_1 *and* U_2. *In particular*

$$N = \frac{n}{2\pi} \frac{1}{I_o(u)} \exp\{\frac{u}{U}(\cos\theta U_1 + \sin\theta U_2)\} \tag{3.16}$$

where $u = (u_1^2 + u_2^2)^{1/2}$, $U = (U_1^2 + U_2^2)^{1/2}$ and $u = (I_1/I_o)^{-1}(\frac{U}{c})$, I_o and I_1 *are the first kind modified Bessel functions of zero and first order, respectively, and* $(I_1/I_o)^{-1}$ *denotes the inverse function of* I_1/I_o.

Proof of Theorem 2: The following equality follows from (2.10-2.16)

$$\int_{[0,2\pi)} \log N(\theta)\; J_p[N](\theta)\, d\theta = \frac{1}{2p} \int \frac{1 - B(\Theta_p)}{B(\Phi_p)} \log[\frac{B(\Theta_p)}{B(\Phi_p)}] B(\Theta_p) A(\Theta_p, \Phi_p)\, d\Theta_p d\Phi_p$$

where

$$B(\Theta_p) = \prod_{i=1}^{p} N(\theta_i) > 0 \quad , \quad B(\Phi_p) = \prod_{i=1}^{p} N(\varphi_i) > 0 \quad .$$

Therefore inequality (3.12) follows immediately from the well known inequality

$$\forall \alpha > 0 \; : \; (1 - \alpha) \log \alpha < 0$$

where $(1-\alpha)\log\alpha = 0$ iff $\alpha = 1$. Therefore

$$\alpha = \frac{B(\Theta_p)}{B(\Phi_p)} = 1 \iff J_p[N] = 0 \ \ \forall\Theta_p, \Phi_p \ \text{ such that } \ A(\Theta_p, \Phi_p) > 0 \quad .$$

Condition i) is then proven.

Condition ii) follows from (3.5-3.6), which imply that the condition $\log(N) \in M$ is equivalent to $J_2[N] = J_3[N] = 0$. In addition, inequality (3.13) implies

$$J[N] = J_2[N] + J_3[N] = 0 \Rightarrow J_2[N] = J_3[N] = 0 \quad .$$

Condition ii) is then proven.

The third condition iii) follows immediately from the kinetic equations and from the preceeding conditions i) and ii). Theorem 2 is then finally proven •

Proof of the Corollary to Theorem 2: The Fourier expansion of the Maxwellian densities (3.12) is

$$N(\theta) = \exp(u_o)\{I_o(u) + 2\sum_{k=1}^{\infty} I_k(u)\cos[k(\theta-\gamma)]\} \tag{3.17}$$

where $u_1 = u\cos\gamma$, $u_2 = u\sin\gamma$, $u \geq 0$ and I_k are the modified Bessel functions of order k. Referring (3.16) to the hydrodynamical observables yields

$$\exp(u_o) = \frac{n}{2\pi}\frac{1}{I_o(u)} \quad , \quad u_1 = \frac{uU_1}{U} \quad , \quad u_2 = \frac{uU_2}{U} \tag{3.18}$$

where

$$\frac{U}{c} = \frac{I_1}{I_o}(u) \iff u = \left(\frac{I_1}{I_o}\right)^{-1}\left(\frac{U}{c}\right) \tag{3.19}$$

the Corollary is then proven.[]

4. Conservation and Hydrodynamical Equations

The kinetic equations (2.6) and condition (3.4) yield the following conservation equations

$$\begin{aligned} \frac{\partial n}{\partial t} + \frac{\partial(nU_1)}{\partial x} + \frac{\partial(nU_2)}{\partial y} = 0 \\ \frac{\partial nU_1}{\partial t} + \frac{\partial}{\partial x}(p_{11} + nU_1^2) + \frac{\partial}{\partial y}(p_{12} + nU_1U_2) = 0 \\ \frac{\partial nU_2}{\partial t} + \frac{\partial}{\partial x}(p_{12} + nU_1U_2) + \frac{\partial}{\partial y}(p_{22} + nU_2^2) = 0 \end{aligned} \tag{4.1}$$

where n , U_1 and U_2 have been defined in Remark 3.1 and the term p_{ij}, with $i,j = 1,2$, is the stress tensor whose components are

$$p_{11} = c^2 \int_{[0,2\pi)} N(\theta)\cos^2\theta \, d\theta - nU_1^2$$

$$p_{12} = p_{21} = c^2 \int_{[0,2\pi)} N(\theta) \sin\theta \, \cos\theta \, d\theta - nU_1U_2 \tag{4.2}$$

$$p_{22} = c^2 \int_{[0,2\pi)} N(\theta) \sin^2\theta \, d\theta - nU_2^2 \quad .$$

It follows that the pressure p is given by

$$p = \frac{\mathrm{Tr}(p_{ij})}{2} = \frac{n}{2}(c^2 - U^2) \tag{4.3}$$

Remark 4.1: The pressure tensor can be written in terms of the viscous tensor as follows

$$p_{ij} = p\delta_{ij} - z_{ij} \tag{4.3'}$$

where z_{ij} is the viscosity tensor with components

$$z_{11} = -z_{22} = \frac{n}{2}(U_1^2 - U_2^2) - \frac{c^2}{2} \int_{[0,2\pi)} N(\theta) \cos 2\theta \, d\theta$$

$$z_{12} = z_{21} = nU_1U_2 - \frac{c^2}{2} \int_{[0,2\pi)} N(\theta) \sin 2\theta \, d\theta \quad . \tag{4.4}$$

In order to derive the hydrodynamical evolution equations, following the Chapman-Enskog theory [13], we look for solutions in the form

$$N = N^o + \epsilon N^1 \quad , \quad N^1 \in \mathcal{M}^\perp \tag{4.5}$$

where ϵ is the "small" dimensionless parameter defined in eq.(2.17) and $\mathcal{M}^\perp$ is the orthogonal complement to the space $\mathcal{M}$ of the collision invariants. Corresponding to the two terms of the expansion (4.5), we shall have

$$z_{ij} = z_{ij}^o + \epsilon z_{ij}^1 \quad . \tag{4.6}$$

Substituting (4.5) into (2.17) and equating the terms with the same power of ϵ, one has, for the terms corresponding to ϵ^o,

$$J[N^o] = 0 \tag{4.7}$$

therefore, in virtue of Theorem 2, N^o is a Maxwellian, $\log N^o \in \mathcal{M}$.

Remark 4.2: The viscous tensor z_{ij}^o corresponding to a maxwellian state is not equal to zero and is a function of the mass velocity. In fact technical calculations yield

$$\begin{aligned} z_{11}^o &= -z_{22}^o = \frac{n}{2}(U_1^2 - U_2^2) E\left(\frac{u}{c}\right) \\ z_{12}^o &= z_{21}^o = nU_1U_2 E\left(\frac{u}{c}\right) \end{aligned} \tag{4.8}$$

where

$$E\left(\frac{U}{c}\right) = 1 - \frac{c^2}{U^2}\frac{I_2}{I_o}\left(\left(\frac{I_1}{I_o}\right)^{-1}\left(\frac{U}{c}\right)\right) \quad . \tag{4.9}$$

Remark 4.3: The result of Remark 4.2 shows that, with reference to the viscous tensor, the physical behaviour of the fluid defined by the semidiscrete Boltzmann equation with triple collisions differs from the one of the Boltzmann equation with binary collisions. In fact it shows the non-newtonian effect of a residual "small" viscosity which is function of the mass velocity and tends monotonically to zero as the mass velocity tends to infinity.

We can now deduce the expression of the term N^1 defined in eq.(4.5). In fact equating the terms with the ϵ^1 power, recalling that N^o is a Maxwellian and setting $N^1 = \sqrt{N^o}\,F$ yields

$$\sqrt{N^o}\Big[\frac{\partial}{dt}\log N^o + c\cos\theta\frac{\partial}{dx}\log N^o + c\sin\theta\frac{\partial}{dy}\log N^o\Big] = L_{N^o}(F) \tag{4.10}$$

where the operator $L_{N^o}(F)(\theta)$ is defined $\forall\theta \in [0, 2\pi)$, as follows

$$\begin{aligned}
L_{N^O}(F) = &\int_{[0,2\pi)} \{\sqrt{N^o(\theta)N^o(\theta+\pi)}F(\theta+\pi) + N^o(\theta+\pi)F(\theta) - \\
&- \frac{1}{\sqrt{N^o(\theta)}}[N^o(\varphi)F(\varphi+\pi)\sqrt{N^o(\varphi+\pi)} + [N^o(\varphi+\pi)F(\varphi)\sqrt{N^o(\varphi)}]\}d\varphi \\
&+\eta\int_{[0,2\pi)^5}\{A^*(\Theta_p,\Phi_p)\{N^o(\theta_3)F(\theta_2)\sqrt{N^o(\theta)N^o(\theta_2)} + \\
&+N^o(\theta_2)F(\theta_3)\sqrt{N^o(\theta)N^o(\theta_3)} + N^o(\theta_2)N^o(\theta_3)F(\theta) - \\
&- \frac{1}{\sqrt{N^o(\theta)}}[F(\varphi_3)N^o(\varphi_1)N^o(\varphi_2)\sqrt{N^o(\varphi_3)} + F(\varphi_2)N^o(\varphi_1)N^o(\varphi_3)\sqrt{N^o(\varphi_2)} + \\
&+F(\varphi_1)N^o(\varphi_2)N^o(\varphi_3)\sqrt{N^o(\varphi_1)}]\}\,d\theta_2 d\theta_3 d\varphi_1 d\varphi_2 d\varphi_3
\end{aligned} \tag{4.11}$$

where the "small" parameter η has been defined in (2.17) and $A^* = \frac{A}{\eta}$.

Equation (4.9) holds for every $N^o > 0$. On the other hand since it has been verified that N^o is a Maxwellian it is possible to characterize some properties of the linear operator L_{N^o} in a fashion analogous to the one already known in the literature for the discrete Boltzmann equation with binary collision only [18].

Theorem 3: *Let $N^o > 0$ be a Maxwellian. Then the linear operator L_{N^o} defined in (4.11) is symmetric, positive semidefinite and such that*

$$L_{N^o}(F) = 0 \iff F \in \sqrt{N^o}\mathcal{M} \quad , \tag{4.12}$$

where $\mathcal{M}$ is the space of the collision invariants.[]

We here omit the proof of the theorem since it is of a merely technical type.

Eq. (4.12) implies that the linear equation

$$L_{N^o}(F) = b \tag{4.13}$$

defines a one-to-one mapping $F \mapsto b$ such that $F, b \in \mathcal{M}^\perp/\sqrt{N^o}$.

In addition considering that we have assumed $N^1 = \sqrt{N^o}F \in \mathcal{M}$, i.e. $F \in \mathcal{M}^\perp/\sqrt{N^o}$, then eq.(4.10) defines univocally F if also its left hand side term belongs to $\mathcal{M}^\perp/\sqrt{N^o}$. This is true taking into account the Corollary to Theorem 2.

The result stated above essentially consists in stating the invertibility of the operator L_{N^o}, or in other words, the existence of the inverse operator $L_{N^o}^{-1}$, so that after technical calculations one finally is able to define N^1

$$-\frac{1}{2}\sqrt{N^o(\theta)}L_{N^o}^{-1}[\sqrt{N^o(\theta)}\ \cos(2\theta)c(\frac{\partial u_1}{\partial x} - \frac{\partial u_2}{\partial y}) + \sqrt{N^o(\theta)}\ \sin(2\theta)c(\frac{\partial u_2}{\partial x} + \frac{\partial u_1}{\partial y})] \tag{4.14}$$

where u_1 and u_2 have been defined in (3.17-3.18).

The expression of N^1 defined in (4.14) enables us to write the terms z_{ij}^1 of the viscous tensor using the definition (4.3)

$$\begin{aligned} z_{11}^1 &= -z_{22}^1 = -\frac{c^2}{2}\int_{[0,2\pi)} N^1(\theta)\cos 2\theta\, d\theta \\ z_{12}^1 &= z_{21}^1 = -\frac{c^2}{2}\int_{[0,2\pi)} N^1(\theta)\sin 2\theta\, d\theta \quad . \end{aligned} \tag{4.15}$$

We have then reached the final result of this section. In fact, the analysis leads to the conclusion that eqs.(4.1) define the Euler and Navier Stokes equations if the calculation of the pressur tensor is related, respectively, to eqs.(4.4) for the Euler equations and (4.15) for the Navier Stokes equation.

It is now possible to derive some conclusions from the analysis developed throughout this paper which proposed a mathematical model of the semidiscrete Boltzmann equation with binary and triple collisions and provides the derivation of the hydrodynamical equations.

On the other hand the analysis developed in this paper stays at an abstract level, as the model has not been specialized in detail. However, the fact that the collision mechanics is univocally determined, in terms of input–output of the velocities in the collision, simplifies the procedure to obtain specific models.

In particular one can apply the same assumptions of [17] (also used in [14]) in order to characterize the transition rates in the triple collision scheme. This assumption states that a triple collision occurs when a third particle enters into the action volume of two colliding particles. In this case the transition rate is given as follows

$$A(\Theta_3, \Phi_3) = \frac{2}{3\sqrt{\pi}}S^{5/2}[G(\Theta_3) + G(\Phi_3)]a(\Theta_3, \Phi_3) \tag{5.1}$$

where

$$\begin{aligned} G(\Theta_3) &= [|\mathbf{v}(\theta_1) - \mathbf{v}(\theta_2)| + |\mathbf{v}(\theta_1) - \mathbf{v}(\theta_3)| + |\mathbf{v}(\theta_2) - \mathbf{v}(\theta_3)|] \\ G(\Phi_3) &= [|\mathbf{v}(\varphi_1) - \mathbf{v}(\varphi_2)| + |\mathbf{v}(\varphi_1) - \mathbf{v}(\varphi_3)| + |\mathbf{v}(\varphi_2) - \mathbf{v}(\varphi_3)|] \end{aligned} \tag{5.2}$$

and

$$a(\Theta_3,\Phi_3) = [\frac{U(\theta_2)}{2\pi}][\frac{U(\theta_3)}{2\pi}][\frac{U(\varphi_1)}{\text{Meas}(D_\varphi)}]\delta(\mu+\gamma-\varphi_2)\delta(\mu-\gamma-\varphi_3) \tag{5.3}$$

where D_φ, γ, μ are given in (2.12).

This modelling is only a preliminary attempt which can be improven on the basis of more general theories [4,5].

Acknowledgments: This paper has been partially supported by the Minister for the Scientific Research and by the National Council for the Research, CNR–GNFM (Project MMMI). Part of this work has been developed by the second Author during his visit at the "Laboratoire de Modelisation an Mecanique". The Authors are grateful to Proff. H. Cabannes, R. Gatignol and S. Kawashima for helpful discussions.

References

[1] Bellomo N., Palczewski A. and Toscani G., "Mathematical Topics in Nonlinear Kinetic Theory", World Sci., London, New Jersey, Singapore, (1988).
[2] Resibois P. and de Leener M., "Classical Kinetic Theory of Fluids", Wiley, London, New York, (1977).
[3] Bellomo N. and Lachowicz M., *Int. J. M. Phys. B*, **1**, (1987), 1193–1206.
[4] Cohen E. , *Acta Phys. Austr.*, **X**, (1973), 157–176.
[5] Sengers J., *Acta Phys. Austr.*, **X**, (1973), 177–208.
[6] Cabannes H., in "Mathematical Methods in the Kinetic Theory of Gases", Neunzert H. and D. Pack Eds., Lang, Frankfurt, (1979), 25–44.
[7] Monaco R. and Toscani G., *Meccanica*, **22**, (1987), 179–184.
[8] Toscani G., *Meccanica*, **20**, (1985), 249–252.
[9] Bellomo N., Illner R. and Toscani G.,*Comp. Rend. Acad. Sci. , Paris, I*, **299**, (1984), 835–839.
[10] Bellomo N. and de Socio L., *J. Math. Anal. Appl.*, **128**, (1987), 112–124.
[11] Longo E., Monaco R. and Platkowski T., *J. Mecan. Theor. Appl.*, **7**, (1987), 233–243.
[12] Piechor K., in "Discrete Kinetic Theory, Lattice Gas Dynamics and Foundation of Hydrodynamics", Monaco R. Ed., World Sci., London, New Jersey, Singapore, (1989), 238–247.
[13] Gatignol R., "Theorie Cinetique d'un Gaz a Repartition Discrete des Vitesses", Springer Lect. Notes Phys. n.**36**, (1975).
[14] Bellomo N. and Kawashima S., *J. Math. Phys.*, **31**, (1990), 245–253.
[15] Gatignol R. and Coulouvrat F., *Comp. Rend. Acad. Sci. Paris,I*, **306**, (1988), 169–174.
[16] Coulouvrat F. and Gatignol R., *Ibidem, I*, **306**, (1988), 393–398.
[17] Longo E. and Monaco R., In "Rarefied Gas Dynamics", *Progress in Astr. and Aeron.*, n.**118**, AIAA, Washington, (1989), 118–130.
[18] Kawashima S. and Shizuta Y., *J. Mecan. Theor. Appl.*, **7**, (1988), 169–174.

A Discrete Velocity Model for Gases with Chemical Reactions of Dissociation and Recombination

R. Monaco and M. Pandolfi Bianchi

Dipartimento di Matematica, Politecnico di Torino,
Corso degli Abruzzi 24, I-10129 Torino, Italy

In the last years, discrete velocity models in kinetic theory of gases have had a great development. A rather large bibliography on this subject can be found in the review paper [1]. This sucess is mainly due to the fact that discrete models can be used for the mathematical of rather complex physical phenomenologies such as multiple collisions between gas-particles or collisions between different gas-species (see, for instance, refs. [2,3]).

In the present paper the problem of modeling a gas dynamics where, besides mechanical collisions, chemical reactions occur is dealt with. In particular, our attention will be directed towards chemical reactions with molecules dissociation and recombination. Examples of such reactions are

$$I_2 + I_2 \rightarrow 4I$$
$$I_2 + I_2 \rightarrow I_2 + 2I$$
$$I_2 + I \rightarrow 3I$$
$$I + I \rightarrow I_2$$

which occur for iodine and other alogen gases [4].

In order to deduce the mathematical modelling of a gas with these chemical reactions a suitable discrete velocity model must be chosen. In fact it is necessary to consider a velocity discretization capable to assure momentum conservation for all the chemical reactions considered above. This is the case of the regular plane six velocity model [1] which will be here adopted.

The kinetic equations, which are derived, present a mathematical structure rather different from the one of discrete models with mechanical collisions only. In fact, besides the classical nonlinear terms due to mechanical collisions, nonlinear source and sink terms are also included, in a fashion that globally mass conservation is preserved.

The contents of the present paper can be summarized as follows : in the next section we derive the model; afterwards we discuss thermodynamical equilibrium in presence of chemical reactions of the afore-mentioned king.

THE MATHEMATICAL MODEL

Within the discrete velocities models in kinetic theory of gases [1], we propose here a mathematical model which takes into account chemical dissociation and recombination reactions of diatomic molecules. At this end consider a diatomic gas A_2 in presence of its monoatomic particles A and, conversely, atoms A can recombine themselves into molecules A_2 . According to this phenomenology the collisional scheme can be stated by the following hypotheses :

1. The collisions occur in absence of external force fields between two gas particles (binary collisions).

2. The pure mechanical collisions (without chemical reactions occur according to the scheme

$$\begin{array}{ll} a) & (A,A) \longleftrightarrow (A,A) \\ b) & (A_2,A_2) \longleftrightarrow (A_2,A_2) \\ c) & (A,A_2) \longleftrightarrow (A,A_2) \end{array}$$

with conservation of momentum and kinetic energy.

3. The collision with chemical dissociation are

$$\begin{array}{ll} d) & A_2 + A_2 \rightarrow A + A + A + A \\ e) & A_2 + A_2 \rightarrow A_2 + A + A \\ f) & A + A_2 \rightarrow A + A + A \end{array}$$

with conservation of momentum only.

4. The collisions with chemical recombination are

$$g) \quad A + A \rightarrow A_2$$

again with conservation of momentum only.

In order to provide different rates for the pure mechanical collisions and for the ones which produce chemical reactions, we will make the further hypothesis :

5. Collisions (a) and (g), which have the same encounters, will provide products with rates $\alpha \leqslant 1$ and $(1-\alpha)$, respectively. In the same fashion collisions (b) and (d,e) will have rates β and $(1-\beta)$ respectively; further on (c) and (f) rates γ and $(1-\gamma)$. Moreover α, β, γ are known functions of the temperature.

The proposed collisional scheme will be now applied to a plane six discrete velocity model [1]. In such a model the selected velocities in the plane are the following

- for atoms A of mass m :

$$i=1,\dots,6 : \quad \underset{\sim}{v}_i = c\, \underset{\sim}{e}_i = c\, [\cos(i-1)\pi/3\ \underset{\sim}{i} + \sin(i-1)\pi/3\ \underset{\sim}{j}] \tag{1}$$

- for molecules A_2 of mass 2m :

$$i=1,\dots,6 : \quad \underset{\sim}{w}_i = \underset{\sim}{v}_i/2 = c\, \underset{\sim}{e}_i/2 \tag{2}$$

To each set of velocities we join the corresponding number densities (number of particles which in unit time and volume have velocities $\underset{\sim}{v}_i$ and $\underset{\sim}{w}_i$, respectively :

$$N_i = N_i\ (t,\underset{\sim}{x}) \quad , \quad M_i = M_i\ (t,\underset{\sim}{x}) \tag{3}$$

<u>Remark 1</u> : In the diatomic molecules the effects due to vibrational and rotational degrees of freedom are neglected.

<u>Remark 2</u> : The discretization (1) and (2) takes into account two non-independent velocity moduli c and c/2 in order to allow the "interactive" collisions (c) with momentum conservation.

<u>Remark 3</u> : It is worth nothing that in the collisions with chemical reactions (d-g) a part of mechanical kinetic energy is transformed into chemical link energy and viceversa. Referring, for instance, to reaction (g) we note that before collision the kinetic energy of the two A-particles is equal to mc^2 while after collision the kinetic energy of the recombined molecule is equal to $mc^2/4$. Thus we can say that the difference $3mc^2/4$ is the energy of the chemical link of the molecule A_2 . It is easy to see that the same balance is also verified by reactions (d,e,f) . Therefore the constant parameter c can be identified by means of the chemical link energy of the considered diatomic gas-molecules.

Remark 4 : The knowledge of the number densities N_i and M_i at any position and time allows the recover of the macroscopic observables :

Numerical density

$$n_1(t,\underset{\sim}{x}) = \Sigma_i N_i \quad , \quad n_2(t,\underset{\sim}{x}) = \Sigma_i M_i \tag{4}$$

$$n(t,\underset{\sim}{x}) = n_1 + n_2$$

Mass density

$$\rho(t,\underset{\sim}{x}) = m \Sigma_i N_i + 2m \Sigma_i M_i \ , \tag{5}$$

Mean velocity

$$\underset{\sim}{u}(t,\underset{\sim}{x}) = \Sigma_i (\underset{\sim}{v}_i N_i + \underset{\sim}{w}_i M_i)/n \tag{6}$$

Temperature

$$T(t,\underset{\sim}{x}) = m[\Sigma_i (\underset{\sim}{v}_i - \underset{\sim}{u}_1)^2 N_i + \Sigma_i (\underset{\sim}{w}_i - \underset{\sim}{u}_2)^2 M_i]/2k_B n \tag{7}$$

where $\underset{\sim}{u}_1(t,\underset{\sim}{x}) = \Sigma_i \underset{\sim}{v}_i N_i/n_1$, $\underset{\sim}{u}_2(t,\underset{\sim}{x}) = \Sigma_i \underset{\sim}{w}_i M_i/n_2$, k_B being the Boltzmann constant.

Heat flux

$$\underset{\sim}{q}(t,\underset{\sim}{x}) = 3m \Sigma_i [N_i(\underset{\sim}{v}_i - \underset{\sim}{u}_1)^2(\underset{\sim}{v}_i - \underset{\sim}{u}_1) + M_i(\underset{\sim}{w}_i - \underset{\sim}{u}_2)^2(\underset{\sim}{w}_i - \underset{\sim}{u}_2)]/4 \tag{8}$$

The actual derivation of the kinetic equations can be realized following a procedure analogous to the one used in the discrete theory of inert gases [1]. Therefore the total derivatives with respect to time of N_i and M_i is set equal to the balance between the so-called "loss" and "gain" terms due either to mechanical collisions either to chemical reactions. The evolution equations will then assume the form

$$DN_i/Dt = (\partial/\partial t + \underset{\sim}{v}_i \cdot \underset{\sim}{\nabla}_x)N_i = J_i^{(1)}(\underset{\sim}{N},\underset{\sim}{M}) + S_i^{(1)}(\underset{\sim}{N},\underset{\sim}{M}) - D_i^{(1)}(\underset{\sim}{N},\underset{\sim}{M}) \tag{9a}$$

$$DM_i/Dt = (\partial/\partial t + \underset{\sim}{w}_i \cdot \underset{\sim}{\nabla}_x)M_i = J_i^{(2)}(\underset{\sim}{N},\underset{\sim}{M}) + S_i^{(2)}(\underset{\sim}{N},\underset{\sim}{M}) - D_i^{(2)}(\underset{\sim}{N},\underset{\sim}{M}) \tag{9b}$$

$$\underset{\sim}{N} = \{N_1,\ldots,N_6\} \quad , \quad \underset{\sim}{M} = \{M_1,\ldots,M_6\}$$

In Eqs. (9) the terms $J_i^{(1),(2)}$ are the collisional operators related to mechanical interactions. The terms $S_i^{(1),(2)}$ and $D_i^{(1),(2)}$ due to chemical reactions will be called "sources" and "sinks". A detailled calculation of the above terms will be now performed for each type of collisions (a-g).

The calculation of operators $J_i^{(1),(2)}$ is well known in literature [2] and is performed assuming that these terms are proportional to :

- the collisional cross-sectional areas;
- the relative velocity of particles before collision;
- the probability density of each admissible collision;
- both number densities of the colliding particles.

The scheme of all admissible mechanical collisions can be then given by

a. $(A,A) \longleftrightarrow (A,A)$

The admissible collisions are of the type

$$i=1,\ldots,6 : \qquad (N_i, N_{i+3}) \longleftrightarrow \begin{cases} (N_{i+1}, N_{i+4}) \\ (N_{i+2}, N_{i+5}) \end{cases}$$

with relative velocity of the colliding particles equal to 2c , probability density equal to 1/3 and rate α.

b. $(A_2, A_2) \longleftrightarrow (A_2, A_2)$

Similarly to case (a) the admissible collisions are

$$i=1,\ldots,6 : \qquad (M_i, M_{i+3}) \longleftrightarrow \begin{cases} (M_{i+1}, M_{i+4}) \\ (M_{i+2}, M_{i+5}) \end{cases}$$

with relative velocity c , probability density 1/3 and rate β.

c. $(A,A_2) \longleftrightarrow (A,A_2)$

We distinguish between

c1. "head-on" collisions

$$i=1,\ldots,6 : \qquad (N_i, M_{i+3}) \longleftrightarrow \begin{cases} (N_{i+1}, M_{i+4}) \\ (N_{i+4}, M_{i+1}) \\ (N_{i+2}, M_{i+5}) \\ (N_{i+5}, M_{i+2}) \\ (N_{i+3}, M_i) \end{cases}$$

with relative velocity $3c/2$, probability density $1/6$ and rate Υ.

c2. collisions "at angle"

$$i=1,\ldots,6 ,\ j\neq i, i+3 : \qquad (N_i, M_j) \longleftrightarrow (N_j, M_i)$$

with relative velocity

$$R_{ij} = [5/4 - \cos(i-j)\pi/3]^{1/2} c \quad , \tag{10}$$

probability density $1/2$ and rate Υ.

We can now give the actual expressions of the terms $J_i^{(1),(2)}$ to be casted in Eqs. (9) :

$$J_i^{(1)}(\underset{\sim}{N},\underset{\sim}{M}) = \frac{2}{3}\alpha c\sigma_{11}[N_{i+1}N_{i+4} + N_{i+2}N_{i+5} - 2N_iN_{i+3}] \tag{11a}$$

$$+ \frac{1}{4}\Upsilon c\sigma_{12}\sum_{j=1}^{5}[N_{i+j}M_{i+j+3} - N_iM_{i+3}] + \frac{1}{2}\Upsilon c\sigma_{12}\sum_{\substack{j=1\\ j\neq i,i+3}}^{6} R_{ij}[M_iN_j - N_iM_j]$$

$$J_i^{(2)}(\underset{\sim}{N},\underset{\sim}{M}) = \frac{1}{3}\beta c\sigma_{22}[M_{i+1}M_{i+4} + M_{i+2}M_{i+5} - 2M_iM_{i+3}] \tag{11b}$$

$$+ \frac{1}{4}\Upsilon c\sigma_{12}\sum_{j=1}^{5}[M_{i+j}N_{i+j+3} - M_iN_{i+3}] + \frac{1}{2}\Upsilon c\sigma_{12}\sum_{\substack{j=1\\ j\neq i,i+3}}^{6} R_{ij}[N_iM_j - M_iN_j]$$

where σ_{11}, σ_{22}, σ_{12} denote the collisional cross-sectional areas between atoms, molecules and atoms/molecules, respectively.

d. $A_2 + A_2 \rightarrow A + A + A + A$

Let us distinguish between

d1. "head-on" collisions

$$i=1,\dots,6: \qquad (M_i, M_{i+3}) \rightarrow \begin{cases} (N_i, N_i, N_{i+3}, N_{i+3}) \\ (N_i, N_{i+1}, N_{i+3}, N_{i+4}) \\ (N_i, N_{i+2}, N_{i+3}, N_{i+5}) \\ (N_{i+1}, N_{i+1}, N_{i+4}, N_{i+4}) \\ (N_{i+1}, N_{i+2}, N_{i+4}, N_{i+5}) \\ (N_{i+2}, N_{i+2}, N_{i+5}, N_{i+5}) \end{cases}$$

with relative velocity c, and rate $(1-\beta)$.

d2. collisions "at angle"

$$i=1,\dots,6, \; j \neq i, i+3: \qquad (M_i, M_j) \rightarrow \begin{cases} (N_i, N_i, N_{i+3}, N_j) \\ (N_i, N_{i+1}, N_{i+4}, N_j) \\ (N_i, N_{i+2}, N_{i+5}, N_j) \end{cases}$$

with relative velocity $1/\sqrt{2}\, R_{ij}$ and rate $(1-\beta)$.

e. $A_2 + A_2 \rightarrow A_2 + A + A$

These collisions are of the type

$$i=1,\dots,6, \; j=1,\dots,3: \qquad (M_i, M_{i+2}) \rightarrow (M_{i+1}, N_i, N_{j+3})$$

with relative velocity $\sqrt{3}/2\, c$ and rate $(1-\beta)$.

f. $A + A_2 \rightarrow A + A + A$

Such collisions are

$$i=1,\dots,6, \; j \neq i: \qquad (N_i, M_j) \rightarrow (N_{i-1}, N_{i+1}, N_{j+1})$$

with relative velocity R_{ij} and rate $(1-\gamma)$.

g. $A + A \rightarrow A_2$

These collisions are of the type

$$i=1,\ldots,6: \quad \begin{cases} (N_i, N_{i+2}) \rightarrow M_{i+1} \\ (N_i, N_{i+4}) \rightarrow M_{i+5} \end{cases}$$

with relative velocity $\sqrt{3}\,c$ and rate $(1-\alpha)$.

Remark 5 : Note that the encounters of the type (M_i, M_{i+2}) occur both in collisions (d2) (for $j=i+2$) and (e). Consequently each encounter will have probability 1/2 so that the outcomes (d2) and (e) are equally partitioned.

After this detailled analysis we can calculate each source term S_i and each sink term D_i.

"Head-on" collisions (d1) produce sink terms $D_i^{(2)1}$ can be inserted in the equations related to the diatomic particles with number densities M_i, namely in Eqs. (9b). The terms $D_i^{(2)1}$ can be expressed by :

$$i = 1,\ldots,6: \qquad D_i^{(2)1} = (1-\beta)c\sigma_{22}M_i M_{i+3} \tag{12}$$

Collisions "at angle" (d2) provide sink terms of molecules $D_i^{(2)2}$ in Eqs. (9b), namely

$$\begin{aligned} i = 1,\ldots,6: \quad D_i^{(2)2} = \frac{1}{\sqrt{2}}(1-\beta)\sigma_{22}[&R_{i,i+1}M_i M_{i+1} + R_{i,i+4}M_i M_{i+4} \\ &+ R_{i,i+5}M_i M_{i+5} + \frac{1}{2}R_{i,i+2}M_i M_{i+2}] \end{aligned} \tag{13}$$

Collisions (e) provide sink terms of molecules $D_i^{(2)3}$ in Eqs. (9b), namely :

$$i = 1,\ldots,6: \qquad D_i^{(2)3} = \frac{\sqrt{3}}{4}(1-\beta)c\sigma_{22}M_i M_{i+2} \tag{14}$$

Interactive collisions (f) provide sink terms of molecules $D_i^{(2)4}$ and sink terms of atoms $D_i^{(1)4}$ to be inserted in Eqs. (9b) and Eqs. (9a), respectively. Thus we have

$$i = 1,\dots,6: \qquad D_i^{(2)4} = (1-\gamma)\sigma_{12} \sum_{\substack{j=1 \\ j\neq i}}^{6} R_{ij} M_i N_j \tag{15}$$

$$i = 1,\dots,6: \qquad D_i^{(1)4} = (1-\gamma)\sigma_{12} \sum_{\substack{j=1 \\ j\neq i}}^{6} R_{ij} N_i M_j \tag{16}$$

Collisions "at angle" (g) provide sink terms of atoms $D_i^{(1)5}$ to be inserted in Eqs. (9a), given by

$$i = 1,\dots,6: \qquad D_i^{(1)5} = \sqrt{3}\,(1-\alpha)c\sigma_{11} N_i (N_{i+2} + N_{i+4}) \tag{17}$$

Accordingly in Eqs. (9a) the sink terms are given by

$$D_i^{(1)} = D_i^{(1)4} + D_i^{(1)5} \tag{18}$$

and in Eqs. (9b) by

$$D_i^{(2)} = D_i^{(2)1} + D_i^{(2)2} + D_i^{(2)3} + D_i^{(2)4} \tag{19}$$

In order to assure total mass conservation it is necessary that the sum of all the sink contributions must be equal to the sum of all the source terms. Thus collisional products arising from encounters (d-g) must be introduced as source terms in the equations of the speciae produced, that is atoms in Eqs. (9a), molecules in Eqs. (9b). According to the outcomes defined by the collisional scheme (d-g), such products of atoms and molecules will be then equally distributed in the appropriate equations as source terms.

Thus $\Sigma_i\, D_i^{(2)1}$ will be equally distributed in the set of Eqs. (9a), giving

$$i = 1,\dots,6: \qquad S_i^{(1)1} = \frac{1}{3}(1-\beta)c\sigma_{22} \sum_{k=1}^{3} M_k M_{k+3} \tag{20}$$

In the same fashion $\Sigma_i\, D_i^{(2)2}$ will be distributed again in Eqs. (9a), i.e.

$$i = 1,\dots,6: \quad S_i^{(1)2} = \frac{1}{6\sqrt{2}}(1-\beta)\sigma_{22}\sum_{k=1}^{6}[R_{k,k+1}M_kM_{k+1} + R_{k,k+4}M_kM_{k+4} + R_{k,k+5}M_kM_{k+5} + \frac{1}{2}R_{k,k+2}M_kM_{k+2}] \tag{21}$$

Moreover $\Sigma_i\, D_i^{(2)3}$ must be distributed both in Eqs. (9a) and (9b) in differents rates, since encounters (M_i, M_{i+2}) give rise to products (M_{i+1}, N_j, N_{j+3}), $j = 1,\dots,3$. Thus 1/3 of the above sum will be counted in Eqs. (9b) while 2/3 in Eqs. (9a). Then

$$S_i^{(1)3} = \frac{\sqrt{3}}{36}(1-\beta)c\sigma_{22}\sum_{k=1}^{6} M_kM_{k+2} \tag{22}$$

$$S_i^{(2)3} = \frac{\sqrt{3}}{72}(1-\beta)c\sigma_{22}\sum_{k=1}^{6} M_kM_{k+2} \tag{23}$$

Moreover both contributions $\Sigma_i\, D_i^{(1)4}$ and $\Sigma_i\, D_i^{(2)4}$ will partitioned in Eqs. (9a), i.e.

$$i = 1,\dots,6: \quad S_i^{(1)4} = \frac{1}{3}(1-\gamma)c\sigma_{12}\sum_{k=1}^{6}\sum_{\substack{j=1\\ j\neq k}}^{6} R_{kj}M_kN_j \tag{24}$$

Finally $\Sigma_i\, D_i^{(1)5}$ will be distributed in Eqs. (9b), giving

$$i = 1,\dots,6: \quad S_i^{(2)5} = \frac{\sqrt{3}}{6}(1-\alpha)c\sigma_{11}\sum_{k=1}^{6} N_k(N_{k+2} + N_{k+4}) \tag{25}$$

In conclusion the source terms in Eqs. (9a) are given by

$$S_i^{(1)} = S_i^{(1)1} + S_i^{(1)2} + S_i^{(1)3} + S_i^{(1)4} \tag{26}$$

while in Eqs. (9b) the source terms are

$$S_i^{(2)} = S_i^{(2)3} + S_i^{(2)5} \tag{27}$$

The kinetic equations can be now written in a final form by inserting in Eqs. (9a,9b) the mechanical collisional terms (11a-11b) and the contributions due to chemical reactions (12-27). Thus

$$[\partial/\partial t + c\underset{\sim}{e}_i \cdot \underset{\sim}{\nabla}_x] N_i = \frac{2}{3}\alpha c\sigma_{11}[N_{i+1}N_{i+4} + N_{i+2}N_{i+5} - 2N_iN_{i+3}]$$

$$+ \frac{1}{4}\gamma c\sigma_{12} \sum_{j=1}^{5} [N_{i+j}M_{i+j+3} - N_iM_{i+3}] + \frac{1}{2}\gamma\sigma_{12} \sum_{\substack{j=1 \\ j\neq i,i+3}}^{6} R_{ij}[M_iN_j - N_iM_j]$$

$$+ \frac{1}{3}(1-\beta)c\sigma_{22} \sum_{k=1}^{3} M_kM_{k+3} + \frac{\sqrt{3}}{36}(1-\beta)c\sigma_{22} \sum_{k=1}^{6} M_kM_{k+2}$$

$$+ \frac{1}{6\sqrt{2}}(1-\beta)\sigma_{22} \sum_{k=1}^{6} [R_{k,k+1}M_kM_{k+1} + R_{k,k+4}M_kM_{k+4} + R_{k,k+5}M_kM_{k+5}$$

$$+ \frac{1}{2}R_{k,k+2}M_kM_{k+2}] + \frac{1}{3}(1-\gamma)\sigma_{12} \sum_{k=1}^{6}\sum_{\substack{j=1 \\ j\neq k}}^{6} R_{kj}M_kN_j$$

$$- (1-\gamma)\sigma_{12} \sum_{\substack{j=1 \\ j\neq i}}^{6} R_{ij}N_iM_j - \sqrt{3}(1-\alpha)c\sigma_{11}N_i(N_{i+2} + N_{i+4}) \quad (28a)$$

$$[\partial/\partial t + \frac{1}{2}c\underset{\sim}{e}_i \cdot \underset{\sim}{\nabla}_x] M_i = \frac{1}{3}\beta c\sigma_{22}[M_{i+1}M_{i+4} + M_{i+2}M_{i+5} - 2M_iM_{i+3}]$$

$$+ \frac{1}{4}\gamma c\sigma_{12} \sum_{j=1}^{5} [M_{i+j}N_{i+j+3} - M_iN_{i+3}] + \frac{1}{2}\gamma\sigma_{12} \sum_{\substack{j=1 \\ j\neq i,i+3}}^{6} R_{ij}[N_iM_j - M_iN_j]$$

$$+ \frac{\sqrt{3}}{72}(1-\beta)c\sigma_{22} \sum_{k=1}^{6} M_kM_{k+2} + \frac{\sqrt{3}}{6}(1-\alpha)c\sigma_{11} \sum_{k=1}^{6} N_k(N_{k+2} + N_{k+4})$$

$$- \frac{1}{\sqrt{2}}(1-\beta)\sigma_{22} [R_{i,i+1}M_iM_{i+1} + R_{i,i+4}M_iM_{i+4} + R_{i,i+5}M_iM_{i+5}$$

$$+ \frac{1}{2}R_{i,i+2}M_iM_{i+2}] - (1-\beta)c\sigma_{22} M_iM_{i+3} - \frac{\sqrt{3}}{4}(1-\beta)c\sigma_{22}M_iM_{i+2}$$

$$- (1-\gamma)\sigma_{12} \sum_{\substack{j=1 \\ j\neq i}}^{6} R_{ij}M_iN_j \quad (28b)$$

THERMODYNAMICAL EQUILIBRIUM

It is well known [2] that the plane six velocity model for inert gases admits a Maxwellian equilibrium state with zero mean velocity when

$$i = 1,\dots,6 : \qquad N_i = N_0 \; , \quad M_i = M_0 \; , \qquad N_0 , \, M_0 \in \mathbb{R}^+ \tag{29}$$

The pure mechanical collision operators $J_i^{(1),(2)}$ vanish when conditions (29) are inserted in their expressions (11a,11b).

On the other hand, thermodynamical equilibrium in presence of chemical reactions implies that in the whole gas the number of dissociating particles equals the number of recombination reactions. Therefore in order to obtain such a chemical equilibrium it is necessary to find the relationship between N_0 and M_0.

At this end casting N_0 and M_0 in the kinetic equations (28a,28b) leads to

$$DN_i/Dt = - DM_i/Dt = - K_1(T)\, N_0^2 + K_2(T)\, N_0 M_0 + K_3(T)\, M_0^2 \tag{30}$$

where

$$K_1(T) = 2\sqrt{3}\,[1-\alpha(T)]\, c\sigma_{11}$$

$$K_2(T) = \frac{1}{2}\,[3 + 2\sqrt{3} + 2\sqrt{7}]\,[1-\gamma(T)]\, c\sigma_{12}$$

$$K_3(T) = [1 + \frac{3}{4}\sqrt{\frac{7}{2}} + (\frac{\sqrt{2}}{2} + \frac{1}{6})\sqrt{3}]\,[1-\beta(T)]\, c\sigma_{22}$$

are known functions of temperature T, thanks to hypothesis (5).

Thus, setting Eq. (30) equal to zero, the equlibrium rate $Q(T) = M_0/N_0$ is obtained by

$$Q(T) = [(K_2 + 4K_1K_3)^{1/2} - K_2] \,/\, (2K_3) \tag{31}$$

Q(T) represents the rate between the numerical densities (see expression (4)) of molecules and atoms in the thermodynamical and chemical equilibrium states at temperature T. Such rate will be uniquely determined once the expressions of α, β and γ as functions of temperature are privided.

Acknowledgements

This paper has been partially supported by the Project MMMI of GNFM (CNR) and by the Ministery of University and Scientific Research and Technology The authors wish to thank Prof. Maurizio Pandolfi for very interesting and useful discussions which gave rise to this paper.

REFERENCES

[1] T. PLASKOWSKI and R. ILLNER, SIAM Review, 30, 1988, p. 213.

[2] E. LONGO and R. MONACO, Transp. Theory Stat. Phys., 17, 1988, p. 423.

[3] E. LONGO and R. MONACO, in *Rarefied Gas Dynamics : Theoretical and Computational Techniques,* Ed. E. P. Muntz, AIAA Publ., 118,; Washington, 1989, p. 118.

[4] J. N. BRADLEY, *Shock Waves in Chemistry and Physics,* John Wiley Inc., New York, 1962.

Part III

Applied Fluid Mechanics

Frozen and Equilibrium Speeds of Sound in Non-equilibrium Flows

M. Pandolfi and R. Marsilio

Dipartimento di Ingegneria Aeronautica e Spaziale,
Politecnico di Torino, I-10129 Torino, Italy

1 Introduction

Efficient techniques have been developed, in the last decade, to achieve numerical solutions for the system of conservation laws that describe compressible inviscid flows. These techniques are founded on **upwind** formulations, such as the flux-vector or the flux-difference splitting. The high quality of the numerical results is related to the respect of the domains of dependence of propagating signals, a physical feature that is retained and preserved in the numerical procedure. The speed of sound plays a preminent role in describing the propagation of signals.

The speed of sound represents the velocity of an infinitesimal front of a perturbation that propagates through an upstream uniform region and leaves behind a perturbed, but still uniform, region. The process is assumed to be reversible and particles that go across the perturbation front do not vary the entropy level. Such a propagation does not occur in a reacting **non-equilibrium** medium, where the flow is not uniform behind the head of the perturbation and entropy is generated because of the chemical relaxation. Therefore, the speed of sound can not be defined in non-equilibrium media.

However, in the two limiting cases of reacting flows which are represented by the **frozen** and the **equilibrium** flows, the region behind the front of the perturbation is uniform and there is no generation of entropy. In these cases the frozen (a_f) and the equilibrium (a_e) speeds of sound can be defined correctly.

In the following we recall the definition of the frozen and equilibrium speeds of sound and the compatibility equations which governe the propagation of signals for the frozen, equilibrium and non-equilibrium flows. An **upwind** conservative procedure for predicting numerically non-equilibrium flows is briefly shown. Then we describe the physics of the propagation, through a reacting medium, of a perturbation generated by an impulsively accelerated piston. Also, we remind the results of a classical approximate analytical solution. Finally, we present results of the numerical simulation of the above physical problem and we comment on them, with reference to the significance of the speed of sound in the upwind numerical procedure.

2 Speeds of Sound

Consider a perturbation that propagates through a reacting medium. Let the medium be in equilibrium and at rest. The front of the perturbation is characterized by small jumps in pressure (dp) and density ($d\rho$). For a reversible process, as it occurs in frozen or equilibrium flows, the velocity of propagation of the perturbation front is given by:

$$a^2 = (\frac{\partial p}{\partial \rho})_{S=const}$$

This is the definition of the speed of sound.

In the case of frozen flow, the enthalpy h is only depending on the pressure p and density ρ:

$$h = h(p, \rho) \tag{1}$$

and the concentrations Y_i of the species are constant. The frozen speed of sound is defined as:

$$a_f^2 = \frac{h_\rho}{1/\rho - h_p} \tag{2}$$

For the equilibrium flow, the concentrations Y_i^* are only depending on the local values of pressure and density:

$$Y_i = Y_i^*(p, \rho) \tag{3}$$

Now, the enthalpy is also depending on the concentrations, namely the equilibrium values Y_i^*, just defined. Therefore, we have:

$$h = h(p, \rho, Y_i^*(p, \rho)) \tag{4}$$

and the equilibrium speed of sound is given by:

$$a_e^2 = \frac{h_\rho + \sum_{i=1}^5 h_{Y_i} {Y_i^*}_\rho}{1/\rho - h_p - \sum_{i=1}^5 h_{Y_i} {Y_i^*}_p} \tag{5}$$

It can be shown that always is $a_f > a_e$.

3 Compatibility Equations

By arranging properly the governing equations written in the quasi-linear form, we obtain the compatibility equations. Here we focus the attention on the **acoustic** waves (1,3), where the speed of sound plays a fundamental role.

For the frozen flow, the signals are defined as:

$$d(R_f)_{1,3} = dp \mp \rho\, a_f\, du \tag{6}$$

They propagate on characteristic lines with slopes:

$$(\lambda_f)_{1,3} = u \mp a_f \tag{7}$$

according to the compatibility equations:

$$\frac{\partial (R_f)_{1,3}}{\partial t} + (\lambda_f)_{1,3} \frac{\partial (R_f)_{1,3}}{\partial x} = 0 \tag{8}$$

On the other hand, for the equilibrium flow, the signals are given by:

$$d(R_e)_{1,3} = dp \mp \rho\, a_e\, du \tag{9}$$

and the slope of the characteristic lines are provided by:

$$(\lambda_e)_{1,3} = u \mp a_e \tag{10}$$

The compatibility equations for equilibrium flows are:

$$\frac{\partial (R_e)_{1,3}}{\partial t} + (\lambda_e)_{1,3} \frac{\partial (R_e)_{1,3}}{\partial x} = 0 \tag{11}$$

For non-equilibrium flows, it can be shown that signals and characteristic are based upon the frozen speed of sound

$$dR_{1,3} = d(R_f)_{1,3} = dp \mp \rho\, a_f\, du \tag{12}$$

They propagate on characteristic lines with slopes:

$$\lambda_{1,3} = (\lambda_f)_{1,3} = u \mp a_f \tag{13}$$

The compatibility equations for non-equilibrium flows are written as:

$$\frac{\partial R_{1,3}}{\partial t} + \lambda_{1,3} \frac{\partial R_{1,3}}{\partial x} = a_f^2\, \psi \tag{14}$$

These equations are not homogeneous, as in the previous cases. Therefore, the signals are not anymore **Invariants**, since they change, along the propagation, owing to the source term. The latter (ψ) incorporates the rates of production of the species and, in particular, the constants of equilibrium. In conclusion, the propagation of waves in non-equilibrium flows (Eqs. 12, 13 and 14) includes features of the frozen (a_f) and equilibrium (ψ) flows.

4 The Numerical Method

Let us briefly review the numerical method we used to get the prediction of the propagation of perturbations through reacting media.

The governing equations are written as conservation laws. The continuity, momentum and energy equations describe the fluid dynamics. The finite rate equations provide the production of the species, according to the rates of the chemical reactions that are included in the source terms on the right hand side terms. The discretization is based upon the finite volumes approach. The averaged density, momentum, total energy, and partial density of each species, over a given volume, are obtained from the integral of appropriate fluxes over the faces of the volume.

The above fluxes are evaluated on the basis of an **upwind** formulation, the **flux difference splitting** (FDS). A Riemann problem is defined at each face that separates two neighboring volumes by a proper interpretation of the initial data at volumes located on the two sides of the face. The collapse of the discontinuity of the Riemann problem provides the values of the fluxes to be inserted in the integration scheme. Such a collapse is described by the governing equations where, however, the source terms are set equal to zero. Therefore, the resulting solution of the Riemann problem corresponds to the frozen flow model. Let us point out that any kind of upwind formulation is characterized by this point. The diagonalization of the matrix $\partial f/\partial w$ (derivative of the flux with respect to the conservative variable) in the **flux vector splitting** formulation accounts for the homogeneous form of the governing equations. Equivalent procedures are followed in the quasi linear formulation as the **lambda** or the **split coefficient matrix** method.

Since the solution of the Riemann problem is based on the quasi linear form of the frozen flow model, the frozen speed of sound a_f appears as the basic variable that contributes to the determination of the splitting of the difference of the fluxes. The non-equilibrium effects appear only when the source terms are integrated over the volumes. It is evident that the respect of the domains of dependence is founded only on the frozen speed of sound, even in conditions very close to the equilibrium, when we would expect the equilibrium speed of sound be the more significant variable. This point turns out to be very important in the following discussion of numerical results.

5 Propagation of a Perturbation

5.1 Physical Description

Let us discuss a simple one-dimensional problem. A constant area duct is filled with a gas mixture at rest and in high enthalpy equilibrium conditions, hereafter denoted by ∞. A piston is impulsively accelerated up to a constant speed V_P along the duct. Let the velocity of the piston be very slow compared with any speed of sound. So, the perturbation generated by the piston moves ahead of it, as an acoustic wave. Because of the high temperature regime, the perturbation triggers chemical reactions and related variations of the concentrations.

Consider an extremely small time t_S, after the starting of the piston at the time $t_0 = 0$. Since no appreciable variations of concentrations can occur within the time t_S, the flow can be considered as frozen. Note that the rates of reactions are far to be zero, but the time t_S allowed to generate new concentrations is too small. The front of the perturbation, ahead of the piston, moves according to the inert gas model, that is at a velocity equal to the frozen speed of sound. The flow is uniform all over the perturbed region, between the piston and the front. The velocity of the gas is equal to V_P and the pressure is provided by the compatibility equation (Eq. 8) along the backwards running characteristic:

$$p_S = p_\infty + a_f \, \rho_\infty \, V_P$$

It is convenient to normalize, at any time t, the abscissa along the duct with respect to a reference length proportional to the time, for instance $a_f \cdot t$. Therefore, at times about t_S, the piston and the front of the perturbation are located respectively at V_P/a_f and 1.0.

Let us now look at the situation at a very large time t_L. Of course, times are defined small or large with respect to a typical chemical relaxation time t_{CH}. The finite rate of the reactions, induced by the perturbation, have enough time available to move the concentrations from the initial equilibrium levels ($Y^*_{i\infty}$) up to new equilibrium values (Y^*_i). The perturbed region looks uniform, from the piston up to the front, except the layer of chemical relaxation just behind it. However, note that the thickness of this layer is constant in time, approximately equal to the speed of the front multiplied by the relaxation time t_{CH}. Since the width of the perturbed region grows linearly with time, the relative extension of this layer vanishes. At such very large times, the overall picture is dominated by the equilibrium effects. The front of the perturbation travels at the equilibrium speed of sound a_e. The perturbed region is confined by the piston, at V_P/a_f, and the normalized location of the front at a_e/a_f, less than 1.0. The velocity of the gas is still equal to V_P and the pressure can be evaluated from the compatibility equation of the equilibrium flow (Eq. 11):

$$p_L = p_\infty + a_e \, \rho_\infty \, V_P$$

The pressure p_L, typical of the equilibrium flow, is lower than the level p_S, characteristic of the frozen flow, because $a_e < a_f$.

At times between t_S and t_L, which correspond to the extreme cases of respectively frozen and equilibrium configurations, the non-equilibrium transition occurs. We expect the perturbation travelling at a speed somewhat between a_f and a_e. Also, the concentrations of the species move towards the new equilibrium values (Y^*_i), being these values reached earlier by the gas near the piston. The pressure on the piston in the non-equilibrium regime can be evaluated from the compatibility equation on the backwards running characteristic (Eq. 14):

$$p_{NE} = p_\infty + a_f\, \rho_\infty\, V_P + \int a_f^2\, \psi\, dt$$

The source term ψ in the integral is strictly related to the rates of production of the species. Therefore the integral contributes where the chemical relaxation is more active, that is behind the head of the perturbation. At earlier times, moderately larger than t_S, the integral is small. Therefore the non-equilibrium pressure p_{NE} is close to the frozen level p_S. As the time proceeds, the integral grows and the non-equilibrium pressure moves towards the equilibrium level p_L. Later, the chemical activity tends to be confined in the relaxation layer behind the head of the perturbation. At very large times, as t_L, the term $(a_f\, \rho_\infty\, V_P + \int a_f^2\, \psi\, dt)$ tends up to $(a_e\, \rho_\infty\, V_P)$, so that the non-equilibrium pressure tends to the equilibrium level $(p_{NE} \rightarrow p_L)$.

5.2 Analytical Results

The previous physical description of the problem is poor in providing information on the non-equilibrium transition from the frozen configuration to the equilibrium one. A much more detailed description has been proposed in the past, based on analytical investigations (see [1] and [2]). The transition should occur as it follows.

Initially, the front of the perturbation travels at the frozen speed of sound and the perturbed region is uniform, according to the frozen flow configuration. Then, the front of the perturbation tends to be dispersed as soon as the non-equilibrium effects begin to develop. The front of the perturbation travels always at the speed a_f, but its strenght decays proportionally to $exp(-t/t_{CH})$. The decay is faster as much as the ratio a_f/a_e is larger than the unity. Meanwhile, the uniform frozen region is distorted and the dispersed perturbation tends to show a new front that develops about the location $x = a_e \cdot t$, behind the original front which, in turn, is still travelling at a_f. At larger times, the initial front disappears and the perturbation becomes headed by the front travelling at a_e. However such a new front is not shaped as a sharp step, like the frozen front of the initial perturbation. On the contrary, it is diffused and S-shaped. Its thickness grows with time proportionally to $\sqrt{t \cdot t_{CH}}$ so that it becomes flatter as the time increases.

It is interesting to look at the distribution along the abscissa, normalized with respect to $a_f \cdot t$, of the velocity over the perturbated region. The initial step distribution of the frozen regime is bounded by the piston, at V_P/a_f, and 1.0. During the non-equilibrium transition, the intensity of the front decays as $exp(-t/t_{CH})$ and is always located at

1.0. As this head of the perturbation vanishes, the new diffusive front develops about the location $a_e/a_f < 1.0$. Note that the physical thickness of the diffusive front increases in time as $\sqrt{t \cdot t_{CH}}$, but its normalized thickness decreases proportionally to $1/\sqrt{t}$. Therefore, in the normalized coordinate and at very large times, the new front is located at a_e/a_f and looks sharp, as the initial frozen one, since the non-equilibrium effects are confined in a layer which vanishes as $1/\sqrt{t}$.

The picture of the non-equilibrium transition looks now described more clearly than in the previous paragraph. Nevertheless, such a description has been obtained on the basis of several assumption and it would be worthwhile to compare it with the one provide by the numerical method.

5.3 Numerical Results

The computations we have performed are based on the following data. The undisturbed reacting medium is air, at $p_\infty = 20(N/m^2)$ and $T_\infty = 3500(°K)$. According to the chemical model we have assumed [3], the equilibrium concentrations show that the oxygen is fully dissociated, the nitrogen only weakly dissociated and a very small amount of nitric oxide exists. The speed of sound are respectively, the frozen $a_f = 1315(m/s)$ and the equilibrium one $a_e = 1204(m/s)$. The piston is accelerated to the velocity $V_P = 24(m/s)$, a very low value, consistent with the assumption of a small perturbation.

The results we present refer to the distribution of the gas velocity, along the normalized abscissa, at different times. Therefore the piston is located at $V_P/a_f = 0.018$, the equilibrium acoustic perturbation at $a_e/a_f = 0.916$ and the frozen one at 1.0. The computational domain extends from the piston at 0.018 up to an undisturbed boundary located at 1.2.

We have started the computation at extremely small times, so that the chemical effects have no possibility to develop. At the number of steps of integration $K = 1000$, the structure of the frozen configuration is well established. The velocity profile is shown in Fig.1. The front of the perturbation is predicted correctly at the location of 1.0 and the perturbed region looks perfectly uniform. At $K = 2000$, the fluid dynamics behind the head of the perturbation is affected by non-equilibrium effects, which begin to show up. At $k = 3000$, the trend is more significant. The velocity of the gas decreases strongly, closely behind the front and weakly back near the piston. At $K = 5000$, the picture changes. The initial frozen front tends to disappear, the velocity behind it goes to zero faster than before, but, more to the left, it starts to recover up to the boundary value V_P. By carrying out the computation further in time, the distribution of the velocity shows the formation of the new front, about the equilibrium acoustic location of 0.916, as reported in Fig.2, at $K = 10000$. We note that the new distribution of the velocity shows a rather smooth S-shaped transition, well centered at a_e/a_f. Such a distribution looks very similar to the corresponding analytical result. By proceeding in the computation , we expect the smooth S-shaped transition become sharper, since its normalized thickness should decrease with time, as $1/\sqrt{t}$, according to the analytical prediction. In fact, such a trend is confirmed by the numerical results obtained at $K =$ 40000. However, by performing further integration steps, the transition does not become any sharper and the distribution at $K = 40000$ represents the final stable configuration reachable asymptotically in time. We remark that the front of the perturbation at such large times is expected to be located at a_e/a_f, just as the numerical results indicate, but with a much steeper transition, as the front of the frozen configuration presents in Fig.1, at $K = 1000$.

To understand the nature of the diffusive front in the non equilibrium flow, at $k = 40000$, where we expect a configuration close to the equilibrium one, we have developed a numerical code for predicting equilibrium flows. This code, hereafter denoted by *EQ1*, is obtained by taking the non-equilibrium flow code and replacing the evaluation of the concentrations, at any point and time, with the equilibrium concentrations (Y_i^*). Such values are obtained from the equilibrium constants consistent with the non-equilibrium model. We have carried out a computation with this *EQ1* code. The difference between these results and those for the non-equilibrium at $K = 40000$ is practically undiscernible. Therefore we can argue that the diffusion of the front for the non-equilibrium has not to be ascribed to the way the equations of the non-equilibrium (Eq. 14) tend to the equilibrium ones (Eq. 11), at very high Damkhöler number. The reason for the spreading of the front still has to be explained.

We remind that the **upwind** numerical procedure we have used requires the definition of the speed of sound in order to proceed to the splitting of the difference of the flux, that ensures the respect of the domains of dependence. For the non-equilibrium code, the speed of sound to be considered is the frozen one (see Eqs. 12, 13, 14). For the equilibrium flow, we have to introduce the equilibrium speed of sound (see Eqs. 9, 10, 11).

However, in the code *EQ1*, where the integration of the conservation laws is matched to the algebraic prediction of the equilibrium concentrations (Y_i^*), the splitting has been achieved with the Riemann solver of the non-equilibrium flow, that retains the frozen speed of sound. This is not consistent with equilibrium concepts and algorithms.

Therefore, we have generated a second equilibrium code, denoted *EQ2*, where the speed of sound introduced in the Riemann solver is the equilibrium one. The results obtained with this second and wholly consistent code (*EQ2*) presents now the sharpness of the front, as in the frozen flow. The results are shown in Fig.3, where the distribution of the velocity are reported for the frozen flow ($K = 1000$), for the non-equilibrium flow at $K = 40000$ and for the equilibrium flow predicted by the code *EQ2*.

In the case of a strong shock, the frozen speed of sound used in the non-equilibrium flow at high Damkhöler number does not bring to appreciable difference in the splitting. However, in the present problem, when the propagation travels at the equilibrium speed of sound, the splitting worked out with the frozen speed of sound violates the domains of dependence. Therefore, the FDS formulation looses its peculiar capability of capturing sharply and neatly perturbation front as weak shocks, a feature that is based on the respect of the domains of dependence.

We anticipate that similar problems are found also in the case of the expansion (not necessarly weak) generated by a piston moving in the opposite direction. At very large times, where the overall picture is dominated by the equilibrium conditions, most of the expansion fan is computed correctly by the non-equilibrium code. Nevertheless the gradient discontinuities, that travel on the two caracteristics which bound the fan, are predicted with a smeared transition of flow derivatives, just as diffusive is computed the front of the perturbation investigated here.

References

[1] Vincenti,W.G. and Kruger,C.H., "Introduction to Physical Gas Dynamics", John Wiley & Sons, New York,1965.

[2] Clarke,J.F. and McChesney,M. ,"The Dynamics of Real Gas", Butterworths,London,1964.

[3] Park,C., "On Convergence of Computation of Chemically Reacting Flows", AIAA Paper-85-0247, Jan. 1985.

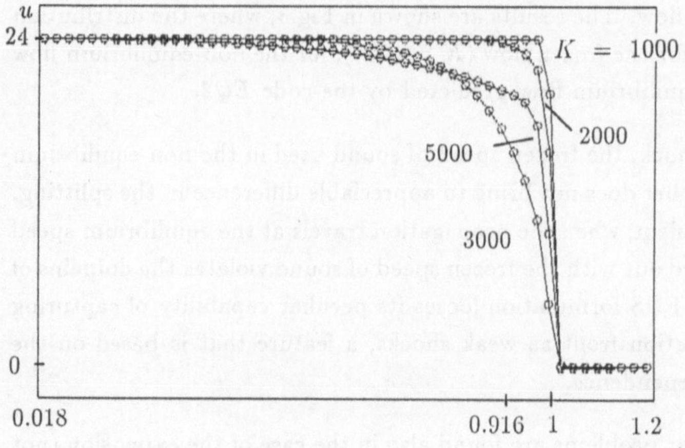

Fig. 1

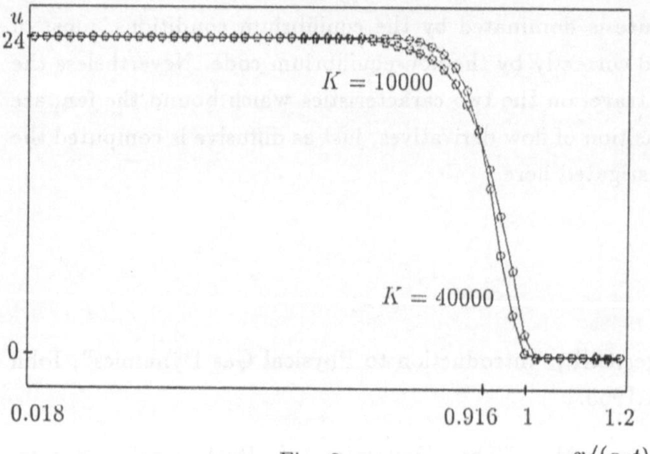

Fig. 2

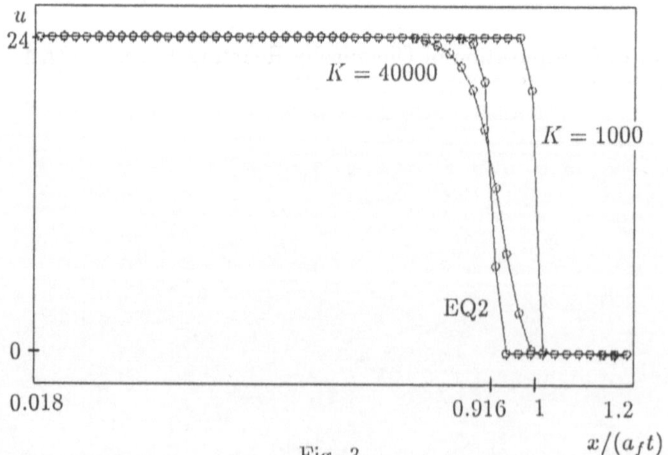

Fig. 3

Nonlinear Propagation of Acoustic and Internal Waves in a Stratified Fluid

J. Naze-Tjøtta and *S. Tjøtta*

Department of Mathematics, The University of Bergen, 5000-Bergen, Norway, and Applied Research Laboratory, The University of Texas at Austin, Austin, TX 78713, USA

INTRODUCTION

The present article is a theoretical study of the combined propagation of acoustic and internal waves of finite amplitude in a (horizontally, say) stratified fluid. The linear theory of small amplitude waves predicts that the two waves (i.e., the fluctuations in acoustic pressure and in vertical particle velocity or displacement) are in general coupled. The waves are decoupled whenever the typical wavenumbers are large compared with the inverse of the scale height for the stratification. Furthermore, a horizontally propagating acoustic wave with zero vertical particle displacement (or velocity), when this is possible, is always decoupled from the internal waves, according to linear theory. In a weakly nonlinear theory, however, the waves may be coupled. A time independent radiation pressure is formed within a standing, horizontally directed acoustic wave, which produces a stationary vertical displacement of the particles. Experimental evidence of this effect has been reported elsewhere [1].

In the following, we first present the basic equations of nonlinear acoustic-internal waves in a way that shows the structure of the coupling terms, and their origin. Model equations are then derived under the assumption of weak nonlinearity. The nonlinear terms combine to form source terms in the governing wave equations. The motion is described within the framework of a thermoviscous fluid, although generalizations accounting for various relaxation effects may readily be carried out. An example is given that shows how effects of nonlinear coupling can be observed at rather moderate acoustic pressure level. However, dramatic changes may occur as the pressure is increased beyond some critical level. Also associated with finite amplitude waves in a dissipative fluid, is generation of nonlinear vorticity and formation of steady flow (acoustic streaming). A solution of the fully nonlinear equations is often required to explain these phenomena.

BASIC EQUATIONS

The basic equations are the Navier-Stokes equation, the equation of continuity, the heat-exchange equation for a thermoviscous fluid, and various equations of state in thermodynamics:

$$\rho\frac{d\mathbf{v}}{dt}+\nabla p-\mathbf{g}\rho=\mathbf{F}\,, \tag{1}$$

$$\frac{d\rho}{dt}+\rho\nabla\cdot\mathbf{v}=0\,, \tag{2}$$

$$\rho\Theta\frac{ds}{dt}=\nabla\cdot(\rho c_v K\nabla\Theta)+\Delta\,, \tag{3}$$

$$p=p(\rho,\Theta)\,,\qquad s=s(\rho,\Theta)\,, \tag{4}$$

where $\mathbf{v}$ is the particle velocity, p,ρ,Θ,s, the pressure, density, temperature and specific entropy, respectively, t the time, and $d/dt=\partial/\partial t+\mathbf{v}\cdot\nabla$. The gravity force is $\mathbf{g}\rho$, where $\mathbf{g}=(0,0,-g)$ is directed downwards. The viscous force is $\mathbf{F}$ and the dissipation function Δ:

$$\mathbf{F}=(\kappa+\frac{\mu}{3})\nabla\nabla\cdot\mathbf{v}+\mu\nabla^2\mathbf{v}=(\kappa+\frac{4}{3}\mu)\nabla\nabla\cdot\mathbf{v}-\mu\nabla\times\nabla\times\mathbf{v}\,, \tag{5}$$

$$\Delta=\mathrm{tr}\Big[\nabla\mathbf{v}\cdot(\nabla\mathbf{v}+\overline{\nabla\mathbf{v}})+(\kappa-\frac{2}{3}\mu)\underset{\sim}{\mathbf{I}}(\nabla\cdot\mathbf{v})^2\Big]=\frac{\mu}{2}\sum_{k,j}\Big[\partial_k v_j+\partial_j v_k-\frac{2}{3}\nabla\cdot\mathbf{v}\delta_{jk}\Big]^2+\kappa(\nabla\cdot\mathbf{v})^2\,, \tag{6}$$

where κ and μ are the coefficient of bulk and shear viscosity, and K is the coefficient of thermal conductivity. For simplicity, constant viscosity coefficients κ,μ are assumed (variations can be accounted for, see [2]). The tensors $\underset{\sim}{\mathbf{I}}$, $\nabla\mathbf{v}$, and $\overline{\nabla\mathbf{v}}$ have elements $\delta_{jk},\partial_j v_k$, and $\partial_k v_j$, and $\mathrm{tr}\,\underset{\sim}{\mathbf{A}}=\Sigma_j A_{jj}$. From Eq. (4) follow the thermodynamic identities

$$dp=\frac{b}{\rho}d\rho+\eta b\,d\Theta\,,\quad dp=c^2d\rho+\frac{\eta b\Theta}{c_v}ds\,, \tag{7}$$

$$b=\frac{\rho c^2}{\gamma}\,,\qquad \eta^2 b=\frac{\gamma-1}{\Theta}\rho c_v\,, \tag{8}$$

where γ is the ratio c_p/c_v of the specific heat at constant pressure and volume, respectively, η is the coefficient of thermal expansion at constant pressure, b the isothermal bulk modulus, and c the isentropic sound speed.

Using Eq. (7) to eleminate s, we can write Eq. (3) on the form

$$\frac{dp}{dt}-c^2\frac{d\rho}{dt}=Q\,, \tag{9}$$

where

$$Q=\frac{\eta b\Theta}{\rho c_v}\Big[\nabla\cdot(\frac{\rho c_v K}{\Theta}\nabla\Theta)+\sigma\Big]\,,\quad \sigma=\rho c_v K\frac{|\nabla\Theta|^2}{\Theta^2}+\frac{\Delta}{\Theta}\geq 0\,. \tag{10}$$

Here σ is the sum of thermal and viscous dissipation divided by the temperature. Note that Δ and σ are non-negative. Let the subscript zero denote the ambient values in the unperturbed, homogeneous fluid at rest: $\mathbf{v}_0 = 0$, $\rho_0 = \rho_0(z)$, $p_0'(z) = -g\rho_0$ (the prime denotes the derivative with respect to z, the coordinate along $\mathbf{g}$). The velocity and the excess values $p - p_0, \rho - \rho_0, \Theta - \Theta_0$, etc., are referred to as the acoustic variables. We introduce the following nonlinear quantities

$$\mathbf{K} = -(\rho - \rho_0)\frac{\partial \mathbf{v}}{\partial t} - \rho \mathbf{v} \cdot \boldsymbol{\nabla} \mathbf{v}\,, \tag{11}$$

$$M = -\mathbf{v} \cdot [\boldsymbol{\nabla}(p - p_0) - c^2 \boldsymbol{\nabla}(\rho - \rho_0)] + (c^2 - c_0^2)\left(\frac{\partial \rho}{\partial t} + \rho_0' w\right). \tag{12}$$

Here w is the vertical component of the velocity. The governing equations can then be rewritten on the following form ($\perp$ refers to the horizontal component):

$$\rho_0 \frac{\partial \mathbf{v}_\perp}{\partial t} + \boldsymbol{\nabla}_\perp (p - p_0) = \mathbf{F}_\perp + \mathbf{K}_\perp\,, \tag{13}$$

which is the momentum equation in the horizontal plane, and

$$\frac{\partial p}{\partial t} + \rho_0 c_0^2 \boldsymbol{\nabla}_\perp \cdot \mathbf{v}_\perp + c_0^2 \left(\frac{\partial}{\partial z} + \frac{N^2}{g}\right)(\rho_0 w) = M + Q - c_0^2 \boldsymbol{\nabla} \cdot [(\rho - \rho_0)\mathbf{v}]\,, \tag{14}$$

which is obtained by combining Eqs. (2) and (9). Here N is the Brunt-Väisälä frequency, defined by

$$p_0' - c_0^2 \rho_0' = \frac{\rho_0 c_0^2}{g} N^2\,, \qquad N^2 = -g\left(\frac{\rho_0'}{\rho_0} + \frac{g}{c_0^2}\right). \tag{15}$$

Combining Eqs. (13) and (14), we obtain

$$\left[\nabla_\perp^2 - \frac{1}{c_0^2}\frac{\partial^2}{\partial t^2}\right](p - p_0) - \frac{\partial}{\partial t}\left[\frac{\partial}{\partial z} + \frac{N^2}{g}\right](\rho_0 w)$$
$$= \boldsymbol{\nabla}_\perp \cdot (\mathbf{F}_\perp + \mathbf{K}_\perp) - \frac{1}{c_0^2}\frac{\partial}{\partial t}(M + Q) + \frac{\partial}{\partial t}\boldsymbol{\nabla} \cdot [(\rho - \rho_0)\mathbf{v}]\,. \tag{16}$$

(Note a printing error in the coefficient for $\rho_0 w$ in Eq. (75) of [3]). Furthermore, we have the z-component of the momentum equation,

$$\rho_0 \frac{\partial w}{\partial t} + \frac{\partial}{\partial z}(p - p_0) + g(\rho - \rho_0) = F_z + K_z\,, \tag{17}$$

and from Eq. (9),

$$\frac{\partial}{\partial t}(p - p_0) - c_0^2 \frac{\partial}{\partial t}(\rho - \rho_0) + \frac{\rho_0 c_0^2 N^2}{g} w = M + Q\,. \tag{18}$$

We can here eliminate $\rho - \rho_0$ on the left-hand side by combining the two equations:

$$\left[\frac{\partial^2}{\partial t^2} + N^2\right](\rho_0 w) + \frac{\partial}{\partial t}\left[\frac{\partial}{\partial z} + \frac{g}{c_0^2}\right](p - p_0) = \frac{\partial}{\partial t}(F_z + K_z) + \frac{g}{c_0^2}(M + Q)\,. \tag{19}$$

In terms of the vertical displacement $\zeta = \int^t w\,dt$, we have also (discarding the integration constant)

$$\left[\frac{\partial^2}{\partial t^2} + N^2\right](\rho_0\zeta) + \left[\frac{\partial}{\partial z} + \frac{g}{c_0^2}\right](p - p_0) = F_z + K_z + \frac{g}{c_0^2}\int^t (M + Q)\,. \tag{20}$$

The two fundamental, coupled equations in $p - p_0$ and $\rho_0 w$, Eqs. (16) and (19), govern the combined propagation of sound and internal waves of finite amplitude in a horizontally stratified, dissipative fluid. We can use Eq. (16) to eliminate the pressure from Eq. (19) to obtain

$$\left[\nabla^2\frac{\partial^2}{\partial t^2} + N^2\nabla_\perp^2 - \frac{1}{c_0^2}\frac{\partial^4}{\partial t^4} + \left(\frac{N^2}{g} + \frac{g}{c_0^2}\right)\frac{\partial^3}{\partial z\partial t^2} + \left(\frac{N^2}{g}\right)'\frac{\partial^2}{\partial t^2}\right](\rho_0 w) =$$
$$-\left(\frac{\partial}{\partial z} + \frac{g}{c_0^2}\right)\frac{\partial}{\partial t}\boldsymbol{\nabla}_\perp\cdot(\mathbf{K}_\perp + \mathbf{F}_\perp) + \left(\nabla_\perp^2 - \frac{1}{c_0^2}\frac{\partial^2}{\partial t^2}\right)\frac{\partial}{\partial t}(K_z + F_z)$$
$$+\left(\frac{1}{c_0^2}\frac{\partial^3}{\partial z\partial t^2} + \frac{g}{c_0^2}\nabla_\perp^2\right)(M + Q) - \left(\frac{\partial}{\partial z} + \frac{g}{c_0^2}\right)\frac{\partial^2}{\partial t^2}\boldsymbol{\nabla}\cdot[(\rho - \rho_0)\mathbf{v}]\,. \tag{21}$$

Similarly, we may use Eq. (19) to eliminate the vertical mass flux, $\rho_0 w$, from the left-hand side of Eq. (16), to obtain

$$\left[\left(\nabla^2 - \frac{1}{c_0^2}\frac{\partial^2}{\partial t^2}\right)\frac{\partial^2}{\partial t^2} + N^2\nabla_\perp^2 + \left(\frac{N^2}{g} + \frac{g}{c_0^2}\right)\frac{\partial^3}{\partial z\partial t^2} + \left(\frac{g}{c_0^2}\right)'\frac{\partial^2}{\partial t^2}\right](p - p_0) =$$
$$\left(\frac{\partial^2}{\partial t^2} + N^2\right)\boldsymbol{\nabla}_\perp\cdot(\mathbf{K}_\perp + \mathbf{F}_\perp) + \left(\frac{\partial}{\partial z} + \frac{N^2}{g}\right)\frac{\partial^2}{\partial t^2}(K_z + F_z)$$
$$+\left[\frac{1}{c_0^2}\left(g\frac{\partial}{\partial z} - \frac{1}{c_0^2}\frac{\partial^2}{\partial t^2}\right) + \left(\frac{g}{c_0^2}\right)'\right]\frac{\partial}{\partial t}(M + Q) + \left(\frac{\partial^2}{\partial t^2} + N^2\right)\frac{\partial}{\partial t}\boldsymbol{\nabla}\cdot[(\rho - \rho_0)\mathbf{v}]\,. \tag{22}$$

We have also the exact vorticity equation,

$$\nu\nabla^2\boldsymbol{\nabla}\times\mathbf{v} - \frac{\partial}{\partial t}\boldsymbol{\nabla}\times\mathbf{v} = -\boldsymbol{\nabla}\times(\mathbf{v}\times\boldsymbol{\nabla}\times\mathbf{v}) + \frac{\boldsymbol{\nabla}\rho}{\rho}\times\left(\frac{d\mathbf{v}}{dt} - \mathbf{g}\right) + \nu\frac{\rho - \rho_0}{\rho}\nabla^2\boldsymbol{\nabla}\times\mathbf{v}$$
$$= -\boldsymbol{\nabla}\times(\mathbf{v}\times\boldsymbol{\nabla}\times\mathbf{v}) + \frac{\boldsymbol{\nabla}\rho}{\rho}\times\left(-\frac{\boldsymbol{\nabla}p}{\rho} + \mathbf{F}\right) + \nu\frac{\rho - \rho_0}{\rho}\nabla^2\boldsymbol{\nabla}\times\mathbf{v}\,, \tag{23}$$

where $\nu = \mu/\rho_0$ is the kinematic viscosity coefficient. Sofar, all equations are exact, as long as the viscosity coefficients are assumed constant. These equations can be used to study the nonlinear propagation and generation of internal and acoustic waves. Further approximations have to be introduced, however, in order to make the equations tractable.

If we discard the right-hand side of Eqs. (21) and (22), we are left with the linearized equation for the combined waves in a non-dissipative fluid. The first two terms on the left-hand side of Eq. (21) correspond to the Boussinesq approximation. When c_0 and the scale height ρ_0'/ρ_0 are constant, both equations yield the dispersion relation

$$\frac{\omega^4}{c_0^2} - (k_x^2 + k_y^2 + k_z^2)\omega^2 + (k_x^2 + k_y^2)N^2 + ik_z\left(\frac{N^2}{g} + \frac{g}{c_0^2}\right)\omega^2 = 0\,. \tag{24}$$

The terms on the right-hand side in Eqs. (21) and (22) come from the equation of momentum (second line), equation of state (first term on third line), and equation of continuity (last term). If we discard the nonlinear terms, but keep the (linearized) terms in $\mathbf{F}$ and Q, Eqs. (16) and (19) provide the linearized equations for wave propagation in a dissipative fluid. Equation (23) then reduces to a diffusion type equation, which shows that in the linear approximation, vorticity can be generated only in presence of viscosity, and only at the boundaries, within a boundary layer of thickness $\delta_{ac} = (\nu/\omega)^{1/2}$ (Stokes layer). Vorticity may, however, be generated nonlinearly away from the boundary layer, as well as inside the boundary layer.

MODEL EQUATIONS

We now introduce the following approximation: We keep square order terms in the acoustic variables $p - p_0, \rho - \rho_0$, and $\mathbf{v}$, but neglect all terms of cubic or higher order, and in addition, we take into account dissipative effects only through linear terms. The nonlinear terms are thus the same as in the nondissipative case. We obtain

$$\mathbf{K}_\perp = -\boldsymbol{\nabla}_\perp\left[\mathcal{L} - \frac{1}{2}\rho_0 N^2\left(\int^t w\right)^2\right] - \frac{\partial}{\partial t}\left[\left(\int^t w\right)\boldsymbol{\nabla}_\perp\left(\frac{\rho_0'}{\rho_0}\int^t (p - p_0) - \frac{p_0'}{\rho_0}\int^t(\rho - \rho_0)\right)\right], \quad (25)$$

$$K_z = -\frac{1}{c_0^2}\frac{\partial w}{\partial t}\left[p - p_0 + \frac{\rho_0 c_0^2 N^2}{g}\int^t w\right] - \rho_0\frac{\partial}{\partial z}\frac{w^2}{2} + \boldsymbol{\nabla}_\perp\int^t (p - p_0)\cdot\boldsymbol{\nabla}_\perp w\,, \quad (26)$$

$$M = \frac{c^2 - c_0^2}{c_0^2}\left[\frac{\partial p}{\partial t} - g\rho_0 w\right] - w\frac{\partial}{\partial z}(p - p_0) - \frac{c_0^2}{g}w\left[\frac{\partial^2}{\partial t\partial z}(\rho_0 w) + \frac{\partial^2}{\partial z^2}(p - p_0)\right]$$
$$- \frac{c_0^2 N^2}{g}\left(\boldsymbol{\nabla}_\perp\int^t (p - p_0)\cdot\boldsymbol{\nabla}_\perp\int^t w\right), \quad (27)$$

$$-\frac{M}{c_0^2} + \boldsymbol{\nabla}\cdot[(\rho - \rho_0)\mathbf{v}] = -\frac{\partial}{\partial t}\left[\frac{\mathcal{L}}{c_0^2} + \frac{(p - p_0)^2}{\rho_0 c_0^4}\right] - \frac{c^2 - c_0^2}{c_0^4}\left[\frac{\partial p}{\partial t} - g\rho_0 w\right] - \frac{N^2}{g c_0^2}\left(\int^t w\right)\frac{\partial p}{\partial t}, \quad (28)$$

and for the dissipative terms,

$$F_z = (\kappa + \frac{4}{3}\mu)\nabla^2 w - \frac{N^2}{\rho_0 g}(\kappa + \frac{\mu}{3})\nabla_\perp^2\int^t (p - p_0 - g\rho_0\int^t w)\,, \quad (29)$$

$$\boldsymbol{\nabla}_\perp\cdot\mathbf{F}_\perp = -\frac{\kappa + \frac{4}{3}\mu}{\rho_0 c_0^2}\nabla_\perp^2\left(\frac{\partial p}{\partial t} - g\rho_0 w\right) + \mu\nabla_\perp^2\frac{\partial}{\partial z}\left[\frac{\rho_0'}{\rho_0^2}\int^t (p - p_0) + \frac{g}{\rho_0}\int^t(\rho - \rho_0)\right], \quad (30)$$

$$Q = \frac{\eta b\Theta_0}{\rho_0 c_v}\frac{\partial}{\partial z}\left[\frac{\rho_0 c_v K}{\Theta_0}\frac{\partial}{\partial z}\left(\frac{\gamma - 1}{\gamma\eta b}(p - p_0) - \frac{\rho_0 c_0^2}{\gamma\eta b}\frac{N^2}{g}\int^t w\right)\right]$$
$$+ K\nabla_\perp^2\left[\frac{\gamma - 1}{\gamma}(p - p_0) - \rho_0 c_0^2\frac{N^2}{\gamma g}\int^t w\right]. \quad (31)$$

Here $\mathcal{L}$ is equal to

$$\mathcal{L} = \frac{\rho_0\mathbf{v}^2}{2} - \frac{(p - p_0)^2}{2\rho_0 c_0^2} = \frac{\rho_0 w^2}{2} + \frac{1}{2\rho_0}|\boldsymbol{\nabla}_\perp\int^t (p - p_0)|^2 - \frac{(p - p_0)^2}{2\rho_0 c_0^2}, \quad (32)$$

which, in the present approximation, is the Lagrangian. In the first term on the right-hand side of Eq. (25), we also recognize the internal wave potential energy,

$$\frac{1}{2}\rho_0 N^2 \Big(\int^t w\Big)^2 .$$

At this point, we have calculated all terms on the right-hand side of Eqs. (16), (19), (21) and (22), consistently within our approximation scheme. All nonlinear terms are quadratic in the variables $p - p_0$ and w (provided c^2 is developed). In particular, insertion of the result into Eqs. (16) and (19) yields a system of two coupled nonlinear equations in $p - p_0$ and w.

TIME-AVERAGED EQUATIONS

If we assume periodicity with respect to time, we obtain from Eq. (13)

$$\boldsymbol{\nabla}_\perp < p - p_0 > = < \mathbf{K}_\perp + \mathbf{F}_\perp > , \tag{33}$$

where $<>$ denotes the time-average. Since $< \mathbf{K}_\perp > = -\boldsymbol{\nabla}_\perp < \mathcal{L} - \frac{1}{2}\rho_0 N^2 (\int^t w)^2 >$ according to the approximations in our model equations [see Eq. (25)], we obtain

$$\boldsymbol{\nabla}_\perp < p - p_0 + \mathcal{L} - \frac{1}{2}\rho_0 N^2 \Big(\int^t w\Big)^2 > = < \mathbf{F}_\perp > , \tag{34}$$

which shows that

$$\boldsymbol{\nabla}_\perp \times < \mathbf{F}_\perp > = 0 , \qquad \text{or} \qquad \mu \nabla^2 \boldsymbol{\nabla}_\perp \times \mathbf{v}_\perp = 0 . \tag{35}$$

Thus, $\boldsymbol{\nabla}_\perp \times \mathbf{v}_\perp = 0$ (or $(\boldsymbol{\nabla} \times \mathbf{v})_z = 0$) for an unbounded space. Further, if we put $\mathbf{v}_\perp = \boldsymbol{\nabla}_\perp \psi$, we have

$$\mathbf{F}_\perp = (\kappa + \frac{\mu}{3}) \boldsymbol{\nabla}_\perp \boldsymbol{\nabla} \cdot \mathbf{v} + \mu \boldsymbol{\nabla}_\perp \nabla^2 \psi , \tag{36}$$

which yields

$$< p - p_0 > = - < \mathcal{L} - \frac{1}{2}\rho_0 N^2 \Big(\int^t w\Big)^2 - (\kappa + \frac{\mu}{3}) \boldsymbol{\nabla} \cdot \mathbf{v} - \mu \nabla^2 \psi > . \tag{37}$$

This can be used to determine the time-average of the vertical displacement ζ (when evaluated through Eq. (20)):

$$\begin{aligned} \rho_0 N^2 < \zeta > = & \left(\frac{\partial}{\partial z} + \frac{g}{c_0^2} \right) < \mathcal{L} - \frac{1}{2}\rho_0 N^2 \Big(\int^t w\Big)^2 > + < K_z > + \frac{g}{c_0^2} < \int^t M > \\ & + < F_z > + \frac{g}{c_0^2} < \int^t Q > - \left(\frac{\partial}{\partial z} + \frac{g}{c_0^2} \right) < (\kappa + \frac{\mu}{3}) \boldsymbol{\nabla} \cdot \mathbf{v} + \mu \nabla^2 \psi > , \end{aligned} \tag{38}$$

where K_z and M are given by Eqs. (26) and (27). This is a very general result. The terms in the last line are due to dissipation. [The dominant dissipative term is $(\kappa + \frac{4}{3}\mu) \nabla^2 < w >$ when the approximations of Eqs. (39) to (41) below are used. This result will lead to Eq. (55) for the special case of a standing wave, Eq. (55).]

SIMPLIFIED MODEL EQUATIONS

The equations can be considerably simplified by introducing further approximations:
(1) If we discard effects of stratification in the dissipative terms, we have instead of Eqs. (29) to (31)

$$F_z = (\kappa + \frac{4}{3}\mu)\nabla^2 w\,, \tag{39}$$

$$\boldsymbol{\nabla}_\perp \cdot \mathbf{F}_\perp = -\frac{\kappa + \frac{4}{3}\mu}{\rho_0 c_0^2}\nabla_\perp^2 \frac{\partial p}{\partial t}\,, \tag{40}$$

$$Q = \frac{\gamma - 1}{\gamma} K \nabla^2 (p - p_0)\,. \tag{41}$$

(2) If we assume (which is often a good approximation) that

$$\frac{g}{kc_0^2} \ll \frac{N^2}{kg}\,, \tag{42}$$

where k is the characteristic wavenumber, we can discard all terms containing the factor g in Eqs. (25) to (28).
(3) Let H be the scale height, $H^{-1} = O(\rho_0'/\rho_0)$ for the case of an isothermal equilibrium, or $H^{-1} = O(\Theta_0'/\Theta_0)$ if stratification is caused by a gradient in temperature. We may consider the case where the characteristic wavelength is small compared with H,

$$\frac{N^2}{kg} = O(\frac{1}{kH}) \ll 1\,, \tag{43}$$

and keep terms of relative order $1/kH$ only in the linear terms (left-hand side) of Eqs. (16) and (19). We then obtain

$$\left[\nabla_\perp^2 - \frac{1}{c_0^2}\frac{\partial^2}{\partial t^2} + \frac{1}{c_0^2}\left(D\nabla_\perp^2 + \frac{\gamma - 1}{\gamma}K\frac{\partial^2}{\partial z^2}\right)\frac{\partial}{\partial t}\right](p - p_0) - \left[\frac{\partial}{\partial z} + \frac{N^2}{g}\right]\frac{\partial}{\partial t}(\rho_0 w) = - \left[\nabla_\perp^2 + \frac{1}{c_0^2}\frac{\partial^2}{\partial t^2}\right]\mathcal{L} - (1 + \frac{B}{2A})\frac{1}{\rho_0 c_0^4}\frac{\partial^2}{\partial t^2}(p - p_0)^2\,, \tag{44}$$

$$\left[\frac{\partial^2}{\partial t^2} + N^2\right](\rho_0 w) + \left[\frac{\partial}{\partial z} + \frac{g}{c_0^2}\right]\frac{\partial}{\partial t}(p - p_0) - (\kappa + \frac{4}{3}\mu)\nabla^2\frac{\partial w}{\partial t} = -\left(\frac{\partial^2}{\partial t \partial z}\mathcal{L}\right)_h\,. \tag{45}$$

Here D is the sound diffusivity, and $1 + B/2A$ is the coefficient of nonlinearity,

$$D = \frac{\kappa + \frac{4}{3}\mu}{\rho_0} + K\frac{\gamma - 1}{\gamma}\,, \qquad \frac{B}{2A} = \frac{\rho_0}{c_0^2}\left(\frac{\partial^2 p}{\partial \rho^2}\right)_{\rho=\rho_0}\,. \tag{46}$$

On the right-hand side of Eq. (45), the subscript h indicates that the homogeneous, nondissipative approximation applies, for the sake of consistency with the present approximation, Eq. (43). Apart from the coupling terms in w and the two-dimensional form of the Laplace operator. Eq. (44) is similar to the equation we obtained for the homogeneous fluid in [3] (Eqs. (36) or (37), but without the curl term). We have a similar equation for the vertical particle displacement ζ.

$$\left[\frac{\partial^2}{\partial t^2}+N^2\right](\rho_0\zeta)+\left[\frac{\partial}{\partial z}+\frac{g}{c_0^2}\right](p-p_0)-(\kappa+\frac{4}{3}\mu)\nabla^2\frac{\partial\zeta}{\partial t}=-\left(\frac{\partial}{\partial z}\mathcal{L}\right)_h . \qquad (47)$$

Thus, when $kH \gg 1$, Eqs. (44) and (45) can be used as the basic equations of nonlinear acoustics in a stratified fluid. The right-hand side of each equation is composed of product terms in the acoustic variables $p - p_0$ and w, whereas the left-hand side, although of second order, contains only linear terms.

ACOUSTIC STREAMING

We note that acoustic streaming is not accounted for in these model equations. Equation (45) predicts a zero mean value (over one period) for the vertical component of the velocity, $< w >= 0$, and from Eq. (13) it follows that

$$\rho_0\frac{\partial}{\partial t}\nabla_\perp\times\mathbf{v}_\perp-\mu\nabla^2\nabla_\perp\times\mathbf{v}_\perp=0\,, \qquad (48)$$

where terms of order $(kH)^{-1}$ have been neglected. This shows that no stationary vorticity component in the z-direction, essential in the generation of acoustic streaming, can exist. Equations (13) and (25) also show that $\nabla_\perp\times<\mathbf{v}_\perp>=0$ even when terms of order $(kH)^{-1}$ are included.

Let us decompose the velocity into its steady and unsteady part: $\mathbf{v} = \mathbf{v}_s + \mathbf{v}_u$, where $<\mathbf{v}_u>=0. <\mathbf{v}>=\mathbf{v}_s$. Equation (23) leads to the exact equation in the steady vorticity,

$$\nu\nabla^2\nabla\times\mathbf{v}_s+\nabla\times(\mathbf{v}_s\times\nabla\times\mathbf{v}_s)=<\frac{\nabla\rho}{\rho}\times(\frac{d\mathbf{v}}{dt}-\mathbf{g})>+\nu<\frac{\rho-\rho_0}{\rho}\nabla^2\nabla\times\mathbf{v}>-\nabla\times<\mathbf{v}_u\times\nabla\times\mathbf{v}_u>. \qquad (49)$$

Outside the Stokes layer, where $\nabla\times\mathbf{v}_u = 0$ in the linear approximation, the last two terms on the right-hand side can be neglected. Within the Stokes layer, however, they dominate as a source of vorticity.

A model equation that governs $\nabla\times\mathbf{v}_s$ to a good approximation is then obtained by inserting the linear solution, including the dissipative terms, into the source term on the right-hand side of Eq. (49). The second term on the left-hand side is of order the streaming Reynolds number, $R_s = V_s/k\nu$, compared to the first term, and it can accordingly be neglected when $R_s \ll 1$ [Here V_s is a characteristic streaming velocity]. Streaming within this approximation was discussed for a homogeneous fluid in [4,5]. If $R_s \geq 1$, this term has to be accounted for, and the streaming caused by the wave can be studied only within a fully nonlinear theory. For homogeneous fluids it has been observed in experiments how symmetric streaming patterns generated in the viscous Stokes layer near an oscillating cylinder at low streaming Reynolds number turn into an asymmetric jet-type streaming pattern when $R_s \gg 1$ [6,7]. This development may also be predicted by using the double boundary layer theory developed by Stuart [8]. To compute $\mathbf{v}_s$ we must know $\nabla \cdot \mathbf{v}_s$ in addition to $\nabla\times\mathbf{v}_s$, since in general $\nabla\cdot\mathbf{v}_s \neq 0$. However, in experiments on acoustic streaming, one measures (by tracing suspended particles) the Lagrangian velocity $\mathbf{v}_L$ specified in a reference system for which $\nabla\cdot <\rho_0\mathbf{v}_L>= 0$. When we evaluated $\mathbf{K}$, as in Eqs. (25), (26), we discarded the effects of viscosity and heat conduction in the nonlinear terms. To account for streaming, we have to include these effects in the nonlinear terms. The resulting momentum equation then yields a vorticity equation which is consistent with Eq. (49).

STANDING WAVE IN A STRATIFIED FLUID

Equation (45) was used in [1] to explain some experimental results in a horizontally directed standing wave. For the special case of a horizontally directed standing acoustic wave propagating in a stratified fluid, and in the linear, nondissipative approximation, Eqs. (44) and (45) reduce to

$$\nabla_\perp^2 p - \frac{1}{c_0^2}\frac{\partial^2 p}{\partial t^2} = 0 \quad . \qquad \frac{\partial p}{\partial z} + \frac{g}{c_0^2} p = 0\,, \tag{50}$$

with solution

$$w = 0\,, \qquad p = \widehat{p} e^{-gz/c_0^2} \sin kx \sin \omega t\,, \qquad \text{where} \qquad k = \omega/c_0\,, \tag{51}$$

if we assume for simplicity that c_0 and the scale height H are constant. The corresponding Lagrangian, $\mathcal{L}$, has a time independent part,

$$<\mathcal{L}>= \frac{\widehat{p}^2}{4\rho_0 c_0^2} e^{-2gz/c_0^2} \cos^2 kx\,. \tag{52}$$

It follows from Eqs. (13) and (25) that the second order pressure and density also have a time independent part

$$<p^{(2)}>= -<\mathcal{L} + f(z)>\,, \qquad <\rho^{(2)}>= -\frac{1}{g}\frac{\partial <p>}{\partial z}\,. \tag{53}$$

The integration constant $f(z)$ is determined by the condition that $<p^{(2)}>= 0$ at $x = 0$, in accordance with the chosen linear solution, Eq. (51), and we obtain

$$<p^{(2)}>= \frac{\widehat{p}^2}{2\rho_0 c_0^2} e^{-2gz/c_0^2} \sin^2 kx\,. \tag{54}$$

From Eq. (45) it follows that $< w^{(2)} >= 0$, i.e., there is no vertical streaming. However, integration of Eq. (18) with respect to t and use of Eq. (47) shows that there is a steady vertical displacement,

$$<\zeta^{(2)}>= \left(1 + \frac{B}{2A} + \frac{c_0^2}{g}\frac{\rho_0'}{\rho_0}\right)\frac{g}{N^2}\frac{\widehat{p}^2}{2\rho_0^2 c_0^4} e^{-2gz/c_0^2} \sin^2 kx\,. \tag{55}$$

If $\rho_0'/\rho_0 \gg g/c_0^2$, as is the case in the experiment reported in [1], we have

$$<\zeta^{(2)}>\sim -\frac{\widehat{p}^2}{2\rho_0^2 c_0^2 g} e^{-2gz/c_0^2} \sin^2 kx \,, \tag{56}$$

which leads to results in good agreement with the experimental observations.

SUMMARY

The fundamental equations of nonlinear acoustic-internal waves in a horizontally stratified, thermoviscous fluid have been presented in a way that shows the structure of the coupling terms, and their origin. The nonlinear terms have been combined to form source terms in the governing wave equation. Model equations have been derived under the assumption of weak nonlinearity. The equations can be used to study the propagation and nonlinear generation of acoustic and internal waves. An example has been presented that shows how a standing acoustic wave can generate an internal wave, in favorable agreement with experimental observations. The generation of nonlinear vorticity and formation of steady flow (acoustic streaming) has also been discussed briefly.

REFERENCES

[1] H. Hobæk, J. Naze Tjøtta, & S. Tjøtta, Nonlinear effects from standing sound waves in a stratified fluid, Frontiers of Nonlinear Acoustics – 12th ISNA, edited by M. F. Hamilton and D. T. Blackstock (Elsevier, London, 1990), pp. 159–164.

[2] S. Tjøtta, On some nonlinear effects in sound fields, with special emphasis on the generation of vorticity and the formation of streaming patterns, Archiv for Mathematik og Naturvidenskab, B. LIV.Nr. 1 & 2, Oslo (1959).

[3] J. Naze Tjøtta & S. Tjøtta, Nonlinear equations of acoustics, Frontiers of Nonlinear Acoustics – 12th ISNA, edited by M. F. Hamilton and D. T. Blackstock (Elsevier, London, 1990), pp. 80–97.

[4] J. Naze Tjøtta & S. Tjøtta, Sur le transport de masse produit par des oscillations en milieu compressible, dissipatif et inhomogène, C. R. Acad. Sci. Paris **277** série A, 61–64 (1974).

[5] J. Naze Tjøtta & S. Tjøtta, Sur le transport de masse produit par des oscillations en milieu stratifié, C. R. Acad. Sci. Paris **278** série B, 1107–1110 (1974).

[6] A. Bertelsen, A. Svardal, and S. Tjøtta, Nonlinear streaming effects associated with oscillating cylinders, J. Fluid Mechanics, **59**, 493–511 (1973).

[7] A. Bertelsen, An experimental investigation of high Reynolds number steady streaming generated by oscillating cylinders, J. Fluid Mechanics, **64**, 589–597 (1974).

[8] J. T. Stuart, Double boundary layers in oscillatory viscous flow, J. Fluid Mech., **74**, 673–687 (1966).

A Higher Order Panel Method for Nonlinear Gravity Wave Simulation

J. Broeze[1], *E.F.G. van Daalen*[2], *and P.J. Zandbergen*[3]

[1]Delft Hydraulics, P.O. Box 152, 8300 AD Emmeloord, The Netherlands
[2]Maritime Research Institute Netherlands,
P.O. Box 28, 6700 AA Wageningen, The Netherlands
[3]Twente University, Faculty of Applied Mathematics,
P.O. Box 217, 7500 AE Enschede, The Netherlands

Summary : We present an efficient higher order panel method for the numerical simulation of nonlinear gravity waves. The method is based on a Green's formulation for the velocity potential that is introduced under the assumptions of an ideal fluid and an irrotational flow. This panel method gives accurate results for both linear and highly nonlinear waves. Test results are shown for a highly nonlinear Stokes wave and an overturning sinusoidal wave.

1. INTRODUCTION

In the past decade boundary element methods have proven to be very suitable for nonlinear gravity waves simulations, where moving boundaries are involved. Two dimensional boundary element methods have been developed by Dold and Peregrine [5], Grilli, Skourup and Svendsen [6] and by Vinje and Brevig [15]. For most of these methods, some kind of smoothing is necessary in order to avoid numerical instabilities.

An efficient higher order panel method for three dimensional gravity waves simulation has been developed by Romate [11, 12]. His method gives stable and accurate results for linear and weakly nonlinear waves. The program written by Romate runs on a CRAY-XMP and, since 1989, on a NEC-SX2. The method is extremely fast, due to the intensive use of the vectorization utilities of these supercomputers.

Romate's investigations have been continued in order to obtain an efficient three dimensional panel method for highly nonlinear gravity waves and the wave interactions with floating moving bodies.

In this paper we shall give a brief description of the panel method as developed by Romate, and we shall discuss our investigations, which are a basis to some recent modifications in the model. Results obtained with an improved 2D-version of the program will be presented and discussed.

2. PROBLEM DEFINITION

Under the assumptions that the fluid is ideal (i.e. inviscid and incompressible) and that the fluid flow is irrotational, a potential ϕ for the fluid velocity $\mathbf{v}$ can be introduced, where ϕ satisfies Laplace's equation

$$\nabla^2 \phi = 0 \tag{1}$$

throughout the fluid domain Ω.

The time dependence comes in by the boundary conditions for the free surface. The dynamic boundary condition on the free surface S_f states that the pressure equals the atmospheric pressure p_0, which will be taken zero: $p = p_0 = 0$. Expressed in a Lagrangian notation, we obtain from Bernoulli's equation:

$$\frac{D\phi}{Dt} = \frac{1}{2}(\nabla\phi)^2 - gz \quad \text{on } S_f \ . \tag{2}$$

The kinematic boundary condition states that the normal velocity of the free surface must equal the normal fluid velocity. Following a fluid particle at S_f, this means that the particle remains at the free surface:

$$\frac{D\mathbf{x}_f}{Dt} = \mathbf{v} \quad \text{on } S_f \ , \tag{3}$$

where $\mathbf{x}_f$ denotes the position of the free surface particle.

On solid fixed boundaries (such as the bottom) the normal component of the fluid velocity must be zero (no-flux condition):

$$\frac{\partial\phi}{\partial n} = \mathbf{v}\cdot\mathbf{n} = 0 \ , \tag{4}$$

where $\mathbf{n}$ is the outward normal.

The boundary condition on the wetted part of the surface of a solid moving body reads:

$$\frac{\partial\phi}{\partial n} = V_n \ . \tag{5}$$

The local normal velocity V_n of the body is either given or must be determined as part of the solution. In case of a rigid body floating freely in the fluid, the motion is part of the solution, and extra equations of motion are needed to determine the trajectory and orientation of the body. These equations can be found in for instance Landweber [9].

In general, the fluid domain Ω will extend to infinity in the horizontal directions. For computational reasons however, it is necessary to truncate the fluid domain at some distance from the area of interest. The artificial boundaries thus introduced should simulate the behaviour of the excluded part of the fluid domain as well as possible. For the free surface problem, this last requirement means that surface waves approaching an artificial boundary should be transmitted at the boundary.

Romate has followed the ideas of Higdon [7, 8] to derive absorbing boundary conditions for linear free surface flows. Both first and second order absorbing boundary conditions have been implemented in the model. Investigations on the well-posedness of the problem with these boundary conditions and numerical test results have been presented in [1], [11], [14] and [3]. More recently, van Daalen, Broeze and van Groesen derived absorbing boundary conditions for general wave equations in a more fundamental way. By making use of variational principles and conservation laws, a nonlinear first order absorbing (energy-transmitting) boundary condition was obtained for the linear wave equation. This boundary condition provides smaller reflections, for both single and composed incident waves, than the first order absorbing boundary condition proposed by Higdon [7]. For a detailed description of this theory and its applications the reader is referred to [4] and [2].

3. METHOD OF SOLUTION

Due to the fact that the time dependence comes in only by the boundary conditions on the free surface, the problem can be split into a number of subproblems which can be solved one by one.

In the first place we have to solve the spatial problem together with the boundary conditions. Romate has chosen a higher order panel method, which is based on Green's second identity for the velocity potential. The problem as described in the previous section is reformulated in terms of integral equations and source and dipole distributions. Since the quantities ϕ and $\partial\phi/\partial n$ are of direct interest (we refer to the boundary conditions given in section 1), a Green's formulation (source ϕ, dipole $\partial\phi/\partial n$) is preferable to for instance a source-only formulation. This choice leads us to the following integral equations:

$$\frac{1}{2}\phi(\mathbf{x}) = \int_S \left[\frac{\partial\phi}{\partial n}G - \phi\frac{\partial G}{\partial n}\right] dS \ , \tag{6}$$

$$\frac{1}{2}\frac{\partial\phi}{\partial n_x}(\mathbf{x}) = \int_S \left[\frac{\partial\phi}{\partial n}\frac{\partial G}{\partial n_x} - \phi\frac{\partial^2 G}{\partial n_x \partial n}\right] dS \ , \tag{7}$$

where G denotes Green's function (depending on the field point $\mathbf{x}$ and the integration point ξ). The integration is over the surface S of the fluid domain. The panel method divides the surface into a finite number of panels. In the center of each panel a collocation point is chosen. Higher order approximations are used for the integral in (6).

If ϕ and $\partial\phi/\partial n$ are known in all collocation points on S, the tangential derivatives are calculated using numerical differentiation. Also ϕ and $\nabla\phi$ can be calculated anywhere in the computational domain Ω, as well as other quantities of interest, such as the pressure.

In the second place we have to solve for the time dependency of the problem, especially for the shape of the free surface. In this problem questions of numerical instability arise, especially if artificial boundaries (see section 1) play a role.

Romate has used the classical fourth order Runge-Kutta method for integrating the system in time. He decided to determine the influence coefficients once every time step for nonlinear simulations, without updating the influence coefficients for the intermediate stages ("frozen coefficients"), thus saving computer time. However, this in effect reduces the order of the time integration method and it may have a bad influence on the stability of the system.

The Runge-Kutta method has the disadvantage that four evaluations of the influence coefficients (the most expensive part of the method) per time step are required in order to maintain its high order of accuracy. A higher order Taylor method does not have this disadvantage, but it requires rather complex expressions for the higher order (material) time derivatives.

In order to avoid these disadvantages, we have decided to develop another fourth order time integration method, which combines the properties of the two previous methods. It uses the first and second order time derivatives, and one intermediate time level. It can be shown that this method has the same stability region as the fourth order Runge-Kutta method. This new method has been compared with the fourth order Runge-Kutta method and the third order Taylor method. Test results indicate that the new method is faster than the other two methods, with the same accuracy.

For each time step a new surface shape together with the distribution of the panels has to be determined with some kind of geometrical modelling. Romate has chosen bicubic splines for the description of the panels on the boundary S. From this panel description various geometrical quantities, such as the area and curvature, can be determined.

An essential part of the method is to make sure that no gaps occur in the surface, especially in the corners. Romate calculated the position of the corner points from the intersection of the extrapolated splines. Recently, a simple iterative algorithm has been developed for the calculation of the corner points positions. It is based on a series of corner point approximations from higher order extrapolations of the outer collocation points on the adjacent boundaries.

For each time step the influence coefficients have to be determined. For an extensive treatment of this part of the problem, the reader is referred to the work of Romate [11, 13]. Finally, Romate has implemented three methods for solving the resulting matrix problem:

- Gaussian elimination
- Successive Over-Relaxation (S.O.R.)
- Preconditioned Conjugate Gradients Squared method (P.C.G.S.)

Although developed for sparse systems, it appears that for large systems the PCGS-method is much faster than the other two methods, with the same order of accuracy as Gaussian elimination.

Although in practice the method follows the steps described above, we have found it beneficial to consider a variational formulation of the problem, which gives a compact statement about the well-posedness of the problem, and in a certain sense unites the items mentioned above.

The field equation (1) and the free surface boundary conditions (2) and (3) can easily be obtained from appropriate variations in the Lagrangian functional

$$L = -\rho \int_T \int_{\Omega(t)} \left\{ \phi_t + \frac{1}{2} (\nabla \phi)^2 + gz \right\} d\Omega \, dt \ . \tag{8}$$

It was concluded that for well-posedness of the problem, a Lagrangian description should be used not only for the free surface, but also for the inlet and outlet boundaries. As a result of this new approach we decided to follow the lateral boundaries in a Lagrangian way.

4. TEST RESULTS

In order to test the various items given above, a 2D-version of the program has been developed, which has been tested for two cases.

The first test case is a highly nonlinear wave, based on a Fourier approximation method (see Rienecker and Fenton [10]). The physical properties of the wave read:

still water depth	:	10.0 m
wave length	:	60.0 m
wave period (Eulerian)	:	6.55 s
phase velocity	:	9.16 m/s

The model has a length of one wave length, with 32 panels in horizontal direction and 10 in vertical direction and the chosen time step Δt is 0.05 s. The wave crest is initially at the center of the model. On the lateral boundaries the velocity in normal direction is prescribed. The results based on a Lagrangian description and the new time integration method as described above after 1 period are compared to the initial solution.

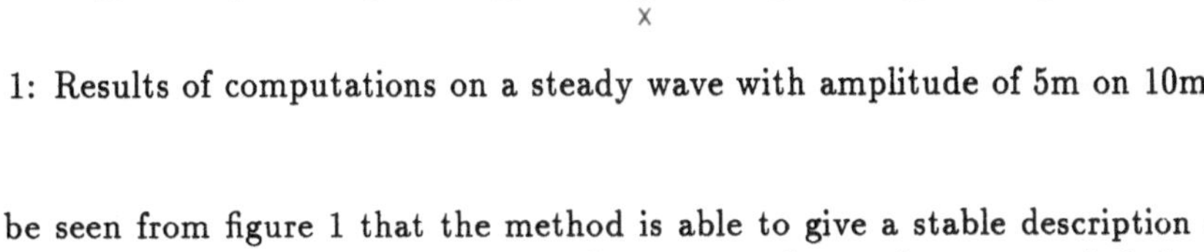

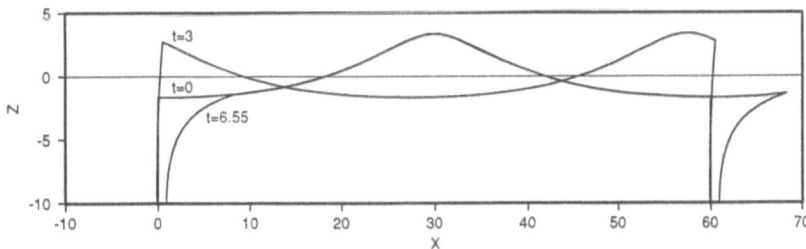

Figure 1: Results of computations on a steady wave with amplitude of 5m on 10m water depth.

It can be seen from figure 1 that the method is able to give a stable description of this wave. Although the panel distribution is rather coarse, the maximum error is below 1.5 % of the amplitude of the wave. However, due to the large Stokes drift near the free surface, after one period the lateral boundaries are strongly curved. Therefore, another treatment of the lateral boundaries is necessary, if one would like to evaluate this wave for several periods.

Another test was done on an overturning wave. It is our aim to follow this wave until it breaks (collapses). Like Grilli et al. [6], we start with a cosine solution for the potential of the linearized problem, giving a sinusoidal profile at the free surface. The physical properties of this wave read:

initial wave height	:	20.0 m
still water depth	:	60.0 m
wave length	:	110.0 m
wave period (Eulerian)	:	8.40 s

Grilli's calculations show that in this case wave breaking occurs after 5 seconds. Since we do not have exact boundary conditions for this test case, the model must be large enough so that there remains sufficient distance between the overturning wave crest and the boundaries. This is why we have chosen the model length equal to 1.25 λ, where λ denotes the wave length.

On the lateral boundaries the normal velocity is prescribed, based on the solution of the linearized problem. The number of panels is 55 at the free surface, 20 at the lateral boundaries and 40 at the bottom. The initial time step is 0.05 s.

Due to the concentration of panels near the wave crest (caused by the Lagrangian description of the fluid motion), numerical instabilities arise after 3.0s. Re-starting the computations with $\Delta t = 0.025$s enables us to evaluate the wave until $t = 4.5$s. No further evaluation of this wave can be simulated at this moment, due to the large curvature of the surface near the wave crest. We expect that this problem can be solved by including extra higher order terms for the panel curvature in the calculation of the influence coefficients. Figure 2 shows the shape of the surface each second from 1 to 4s and after 4.5s. The results show a good agreement with the results obtained by Grilli et al. [6].

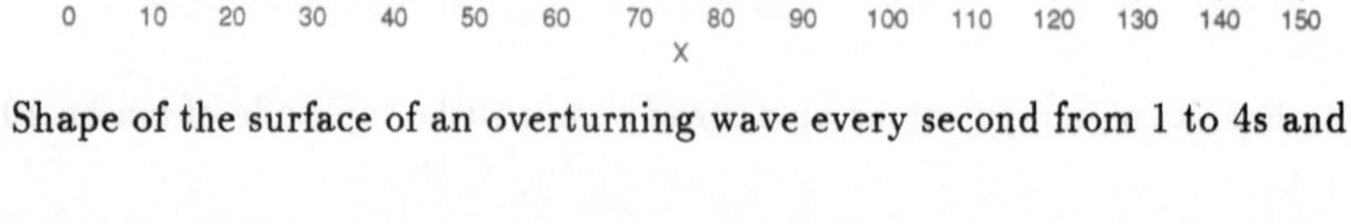

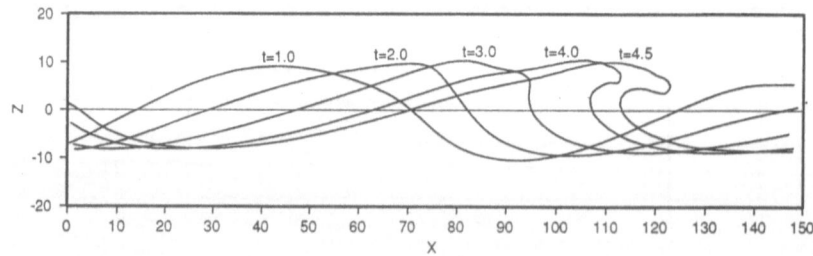

Figure 2: Shape of the surface of an overturning wave every second from 1 to 4s and after 4.5s.

5. CONCLUSIONS

We conclude from the tests done that this wave model gives good results, without the use of artificial smoothing. We believe that this model can be used very well for studying some theoretical aspects of overturning waves. In the near future, we expect the model to be capable of simulating the wave interactions with floating and fixed objects.

Since the 2D-version of the computer program has been derived from the original 3D-version, the methods presented here can be extended to the new 3D-version rather easily. We expect that we shall be able to present similar 3D-results in the near future.

ACKNOWLEDGEMENTS

These investigations were supported by the Netherlands Technology Foundation (STW).

References

[1] J. Broeze, *Absorbing Boundary Conditions for the Simulation of Surface Waves with the Boundary Element Method*, Delft Hydraulics, Report H475, Marknesse, 1988.

[2] J. Broeze and E.F.G. van Daalen, *Radiation Boundary Conditions for the Two Dimensional Wave Equation from a Variational Principle*, to appear in Mathematics of Computation.

[3] J. Broeze and J.E. Romate, *Absorbing Boundary Conditions for Free Surface Waves. Part 2: Numerical Results*, to appear in Journal of Computational Physics.

[4] E.F.G. van Daalen, J. Broeze and E.W.C. van Groesen, *Variational Methods and Conservation Laws in the Derivation of Radiation Boundary Conditions for Wave Equations*, to appear in Mathematics of Computation.

[5] J.W. Dold and D.H. Peregrine, *An Efficient Boundary-integral Method for Steep Unsteady Water Waves*, Numerical Methods for Fluid Dynamics II (Eds. K.W. Morton & M.J. Baines), Clarendon, Oxford, (1986), 671-679.

[6] S.T. Grilli, J. Skourup and I.A. Svendsen, *An Efficient Boundary Element Method for Nonlinear Water Waves*, Engineering Analysis with Boundary Elements **6** (1989), 97-107.

[7] R.L. Higdon, *Absorbing Boundary Conditions for Difference Approximations to the Multi-Dimensional Wave Equation*, Mathematics of Computation **47** (1986), 437-459.

[8] R.L. Higdon, *Numerical Absorbing Boundary Conditions for the Wave Equation*, Mathematics of Computation **49** (1987), 65-90.

[9] L. Landweber, *Motion of Immersed and Floating Bodies*, In: Handbook of Fluid Dynamics (ed. V.L. Streeter), McGraw-Hill, New York, 1961, 13.1-13.50.

[10] M.M. Rienecker and J.D. Fenton, *A Fourier Approximation Method for Steady Water Waves*, Journal of Fluid Mechanics **104** (1981), 119-137.

[11] J.E. Romate, *The Numerical Simulation of Nonlinear Gravity Waves in Three Dimensions Using a Higher Order Panel Method*, Ph.D. Thesis, University of Twente, Enschede, 1989.

[12] J.E. Romate and P.J. Zandbergen, *Boundary Integral Equation Formulations for Free Surface Flow Problems in Two and Three Dimensions*, Advanced Boundary Element Methods (Ed. Th.A. Cruse), Springer-Verlag, Berlin, 1988, 359-367.

[13] J.E. Romate, *Local Error Analysis in 3-D Panel Methods*, Journal of Engineering Mathematics **22** (1988), 123-142.

[14] J.E. Romate, *Absorbing Boundary Conditions for Free Surface Waves. Part 1: Analytic Equations*, to appear in Journal of Computational Physics.

[15] T. Vinje and P. Brevig, *Numerical Simulation of Breaking Waves*, Adv. Water Resources **4** (1981), 77-82.

Numerical Reliability of MHD Flow Calculations at High Hartmann Numbers

K.G. Roesner and W.U. Würfel

Institut für Mechanik, Technische Hochschule Darmstadt,
Hochschulstr. 1, W-6100 Darmstadt, Fed. Rep. of Germany

Abstract : In this paper, a new numerical investigation is performed for the two-dimensional MHD flow in a rectangular duct and an error analysis of the traditional calculation of solution is given.

Arbitrary values of the flow parameters : Hartmann number and wall conduction ratio can now be chosen, and the singular perturbation problem solved and analysed using the interval mathematics and verified inclusion methods (E-Methods). Furthermore, the error analysis of the traditional calculation applied to this MHD flow shows that there is a lack of reliability of the known numerical result of this physical problem, even for Hartmann numbers $M \leqslant 1000$. These results indicate that the reliability of numerical data should, at least, be proved for any calculation with the control of rounding errors using an accurate floating-point arithmetic and inclusion methods of interval mathematics.

TWO-DIMENSIONAL MHD FLOW IN RECTANGULAR DUCT.

The fully developed flow of an incompressible, electrically conducting liquid in an infinitely long, rectangular duct is investigated. The walls of the channel are assumed to be very thin but electrically conducting ($t_w \ll a$, t_w is the wall thickness, and a the half lenght of the side walls). The surrounding medium is assumed to be electrically insulating. A homogeneous magnetic induction **B** is applied parallel to the side walls of the duct (see Fig. 1).

Hunt [2] (1969) has shown, including the usual assumption of magnetohydrodynamics, that a time-independent flow can be described in the following typical and unique form :

$$V = (0,\ 0,\ V(x,y))\ ,$$

$$\Phi = \Phi(x,y)\ ,$$

$$\mathbf{p} = (0,\ 0,\ P_o' - P_o z)\ ,$$

$$-\ \nabla\mathbf{p} = (0,\ 0,\ P_o)\ ,$$

where V represents the velocity field, Φ the electrical potential, **p** the pressure, and ∇**p** the pressure gradient. The unknown velocity V and the electrical potential Φ are defined by the following equations :

$$\Delta\Phi(x,y) = -\ \frac{\partial}{\partial x}\, V(x,y)$$

$$\Delta V(x,y) - M^2\ V(x,y) = M^2\ \left(\frac{\partial}{\partial x}\,\Phi(x,y) - P_o\right)\ . \tag{1}$$

with the Hartmann number $M = a\ B_o \sqrt{\frac{\sigma}{\eta}}$, where a is the half length of the side walls, B_o the absolute value of the magnetic induction, η the viscosity of the fluid, and σ the electrical conductivity of the fluid.

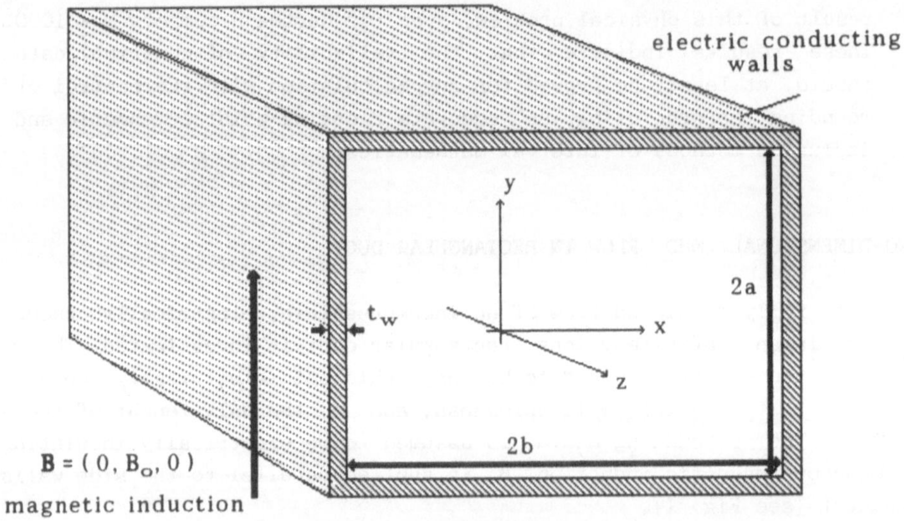

Fig. 1 : Infinitely long, rectangular duct and applied magnetic induction **B**

The boundary conditions are given by

$$V \;|_{wall} = 0$$

$$\frac{\partial \Phi}{\partial n}\;|_{wall} = \varphi \,\frac{\partial \Phi^2}{\partial s^2}\;|_{wall} \tag{2}$$

where φ is called the wall conduction ratio and defined by $\varphi = \frac{\sigma_w}{\sigma} t_w$, n describes the normal direction in points of the walls and s the tangential direction, σ_w is the electrical conductivity of the wall and t_w the wall thickness [7].

This system has a singular behavior for large Hartmann numbers M. Another problem arises because of the very unusual boundary conditions which have the same order of derivation as the governing equations. Therefore in contrast to a usual formulation of a Dirichlet problem or a system with Neumann conditions, the question of existence and uniqueness of the solution of problem (1),(2) is unsolved up to now.

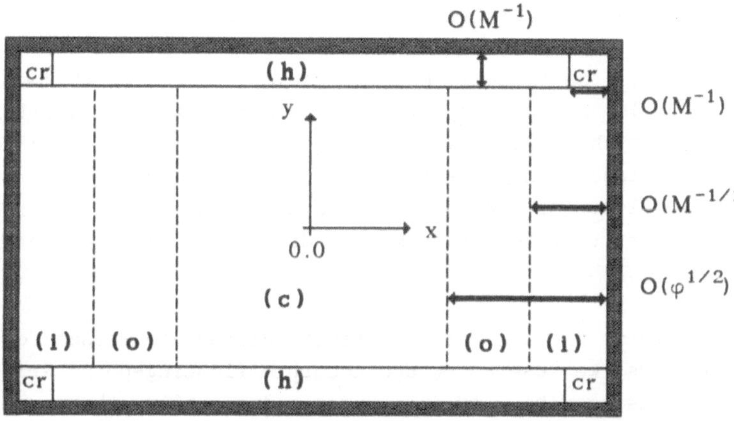

Fig. 2 : Subregions of the duct flow for large Hartmann numbers (c) core, (h) Hartmann layers, (i) inner side layers, (o) outer side layers, (cr) region of corner.

The fully developed MHD flow in a rectangular duct has been analysed by Walker ([7], 1981) with the asymptotic method for singular perturbation problems for an arbitrary wall conduction ratio φ. One important result of this study was that the interior of the duct can be divided into the following subregions : The core of the duct with nearly constant axial velocity, the Hartmann layers adjacent to the top and bottom, the inner and outer side layers adjacent to the side walls and the corner regions. The thickness of the boundary layers can be estimated depending of the Hartmann number and the wall conduction ratio (see Fig. 2).

It is well known that the velocity reaches its global maximum in the regions of the inner side layers, and becomes negative in the outer side layers. Therefore the velocity profile has a typical M-shape if plotted against the x-direction of the duct.

NUMERICAL METHOD AND FIRST NUMERICAL RESULTS.

Numerical results of the two-dimensional MHD flow in a rectangular duct are only known for Hartmann numbers $M < 1000$ (Sterl [6]). These results were calculated with both the method of finite differences and the method of finite elements.

The unusual boundary conditions of the electrical potential can only be discretized with finite differences. Using the method of finite elements, the boundary condition must be iterated by an additional step. Therefore we prefer in this study the finite difference method.

Additionally, the typical boundary layer structure of the flow field leads to variable and/or adaptive grid generations. But similar to other numerical methods the traditional computation of the difference method with variable grids leads only in the case of Hartmann numbers $M \leqslant 1000$ to consistent results. For $M > 1000$, the numerical data of different grids are inconsistent, therefore this method divergences for Hartmann numbers $M > 1000$.

To solve this problem with the finite difference method for arbitrary Hartmann numbers, we use a new and accurate floating-point arithmetic. In this way, it is possible to introduce an interval arithmetic on the computer, and therefore the rounding errors of the numerical calculations can be controlled (Kulisch [4]). The discretization of the partial differential equations (1), (2), are determined with high accuracy, and all coefficients of the discrete system are represented on the computer as intervals, which include the exact value of the coefficients. The discrete formulation of the problem is now a linear system of interval equations, and this system can be solved with high accuracy by so-called E-Methods, which guarantee the enclosure, existence and uniqueness of the solution (Kaucher [3]). The accurate floating-point arithmetic can be used by the programming-languages

"PASCAL-SC" or "FORTRAN-SC", the E-Method for solving the linear system of interval equations is based on the routine "ILIN" of the ACRITH-library, a software product of IBM.

The result of this accurate interval computation is a verified enclosure of the solution of the exact discrete system, and thus the reliability of the numerical data is proved.

The application of this E-Method of interval calculation of the method of finite differences to different variable grids leads to consistent numerical results of the two-dimensional MHD flow for arbitrary Hartmann numbers and any wall conduction ratio. As a simple example of this succesful interval method, Fig. 3 shows the velocity distribution in a square duct for the Hartmann number M = 10 000 .

The perfect agreement with the asymptotic results of Walker [7] (1981) and Hunt [1] (1964) are also shown as the " M-shaped" velocity profile in x-direction and the subregions at the walls, like the Hartmann layer or the inner and outer side layers.

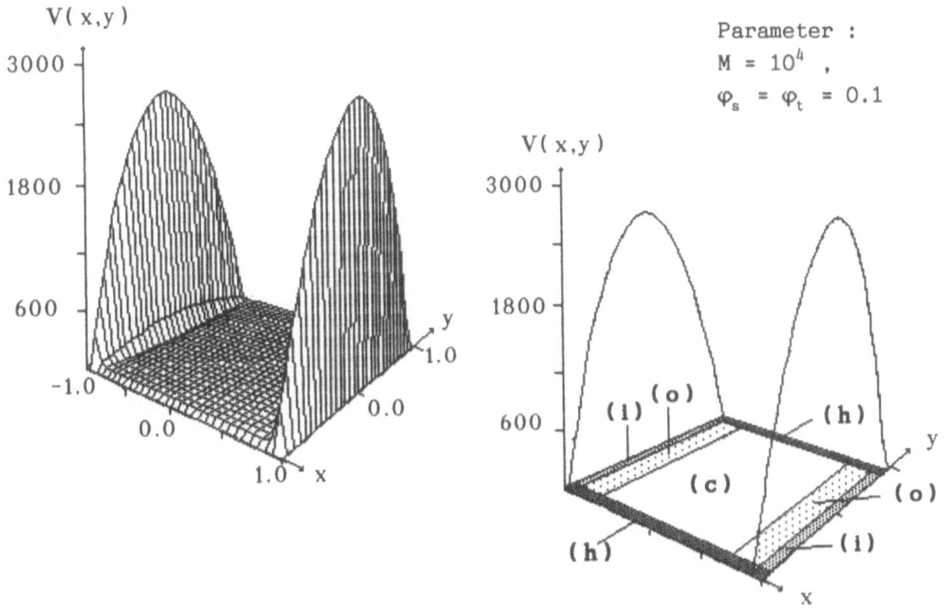

Fig. 3 : Velocity profile in a square duct

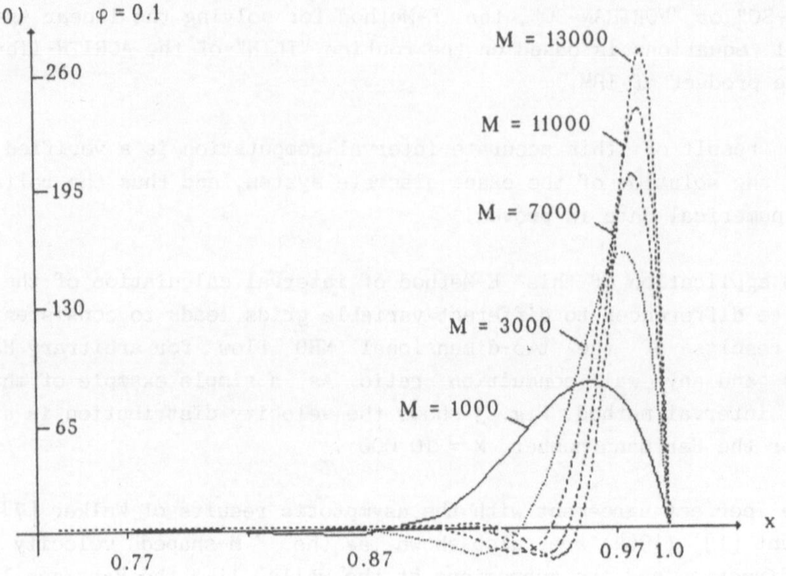

Figure 4 : Axial velocity distribution along the x-axis in a square duct in the vicinity of the right side wall (M : Hartmann number)

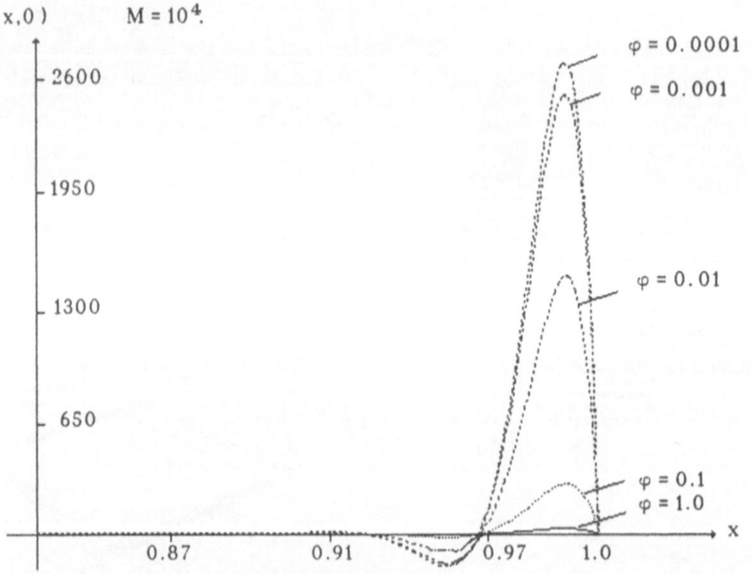

Figure 5 : Axial velocity distribution along the x-axis in a square duct close to the right side wall (φ : wall conduction ratios)

Now, we can also analyse the relationship between the velocity and the Hartmann number (see Fig. 4).

There are two aspects of the dependence :

1. An increase of the Hartmann number M results in a decrease of the thickness of the side layers.
2. The absolute values of the global extrema, maximum in the inner side layers and minimum in the outer side layers increase.

The influence of the wall conduction ratio is shown in Fig. 5. In this case, only the absolute values of the global maximun and minimun of velocity are depending of the wall conduction ratio. There is no influence to the boundary later structure of the wall conduction ratio φ.

These are some global results of the numerical analysis of the two-dimensional MHD flow at high Hartmann numbers.

ERROR ANALYSIS OF TRADITIONAL COMPUTATION.

Finally, we want to answer the question, why there is a lack of convergence of numerical calculations for Hartmann numbers $M > 1000$ in traditional computation, and under which circumstances it is necessary and useful to apply accurate and interval methods.

Therefore we analyse the two steps of the finite difference method :

(I) Transformation of the system of partial differential equations (1), (2), to a system of linear equations.

(II) Determination of the solution of the system of linear algebraic equations.

ad (I) :

If we compute the solution of the discrete system with an accurate floating-point arithmetic (Kulisch [4]) using the programming languages "PASCAL-SC" or "FORTRAN-SC" and interval methods (Kaucher [3]), the coefficients of the discrete system are calculated with 14 exact decimal digits using a double real representation independent of the Hartmann number and independent of the special grid generation. The bounds of the computed inclusions of the coefficients differ only in the 15th digit. Thus, we get a verified inclusion of the discrete problem with high accuracy. If we compare these calculations with the traditional computation, we get the important relationship between the Hartmann number and the reached accuracy of the coefficient calculation (Fig. 6). But, only for $M < 1000$, the traditional computed system disturbed by appearing rounding errors is of regular type, and the system is solvable. For $M \geq 1000$, the computed discrete system of linear equations gets singular as a result of the perturbation by rounding errors. The proof of this singularity of the disturbed system of linear equations can be analysed with the verified E-Method of solving systems of linear equations due to Rump [5] (procedure "DLIN" of the ACRITH-library). But, it is very interesting that the region of regularity of traditional computation agrees very well with the region of known numerical results for Hartmann numbers less than 1000 .

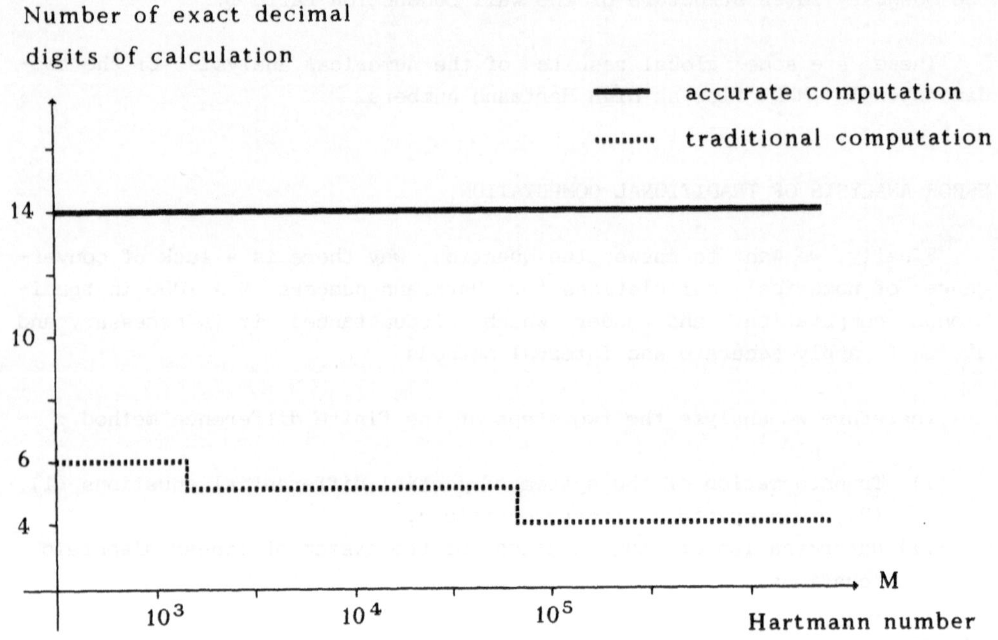

Fig. 6 : Accuracy of accurate interval analysis and traditional computation of a coefficient calculation of the discrete linear problem

ad (II) :

The first step of the error analysis shows that the representation of the exact discretized system is disturbed by rounding errors on the computer in the following way :

$$(A + F) \times (x + h) = b$$

$A \times x = b$ - exact discrete system
F - matrix of error rounding
h - error of the approximate solution .

The relative error of the approximate solution is estimated by the following inequality :

$$\frac{\|h\|}{\|x\|} \leqslant \frac{10^{-t} \times n \times \|A\| \, \|A^{-1}\|}{1 - 10^{-t} \times n \times \|A\| \, \|A^{-1}\|} ,$$

where t is the number of decimal digits of exact representation of the matrix A , and $\|\ldots\|$ an arbitrary compatible matrix/vector norm (Wilkinson [8]).

The condition number of A ($\mathrm{cond}(A) = \|A\| \, \|A^{-1}\|$) can be estimated depending of the Hartmann number M (Würfel [9]). This leads to the important relationship :

$$t \times \log(n) \times \log(2M) \leqslant k$$

t - number of exact decimal digits of the calculation
n - dimension of the discrete problem
M - Hartmann number
k - accuracy of the approximate solution of the discrete system

Using this inequality, it is now possible to find an optimal relationship between the number of necessary grid points ($= n$: the dimension of the discrete problem), the accuracy of the calculation ($= t$: number of exact digits), and the quality of the approximate solution depending of the interesting Hartmann number M .

Example : For $M = 1000$, we want to determine the approximate solution with an accuracy of the first two digits ($k = 2$). In the case of 1000 grid points ($n = 1000$), we need an accuracy of the discretized calculation with 12 exact decimal digits.

The first step of the error analysis has shown that it is possible to calculate some results with 6 exact digits. For the same discrete problem, we need 12 exact digits to get a reliability of the approximate solution in two digits. Therefore the numerical results of the MHD flow are doubtful even for Hartmann numbers $M \leqslant 1000$ on the basis of a traditional computation. Only an interval computation and error analysis with high accuracy combined with methods of the interval mathematics lead to reliable numerical results.

REFERENCES.

[1] J. C. R. HUNT. - "Magnetohydrodynamic flow in rectangular ducts", J. Fluid Mech., 21 (4), 1964, p. 577-590.

[2] J. C. R. HUNT. - "A uniqueness theorem for magnetohydrodynamic duct flow", Proc. Camb. Phil. Soc., 65, 1969, p. 319-327.

[3] E. W. KAUCHER and W. L. MIRANKER. - *"Self validating numerics for function space problems"*, Academic Press, New York, 1984.

[4] U. KULISCH and W. L. MIRANKER. - *"Computer arithmetic in theory and practice"*, Academic Press, New York, 1981.

[5] S. M. RUMP. - "Solving algebraic problems with high accuracy", in *"New approach to scientific computation"*, by U. Kulisch and W. L. Miranker (Ed.), Academic Press, New York, 1983.

[6] A. STERL. - "Numerische Simulation magnetohydrodynamischer Flüssig-Metall-Strömungen im rechteckigen Rohr bei grossen Hartmann-Zahlen", Dissertation an der Fakultät für Maschinenbau der Universität Karlsruhe, 1989.

[7] J. S. WALKER. - "Magnetohydrodynamic flows in rectangular ducts with thin walls, Part I : Constant area and variable area ducts with strong uniform magnetic fields", Journal de Mécanique, 20, 1981, p. 79-112.

[8] J. H. WILKINSON. - *"Rundungsfehler"*, Heidelberger Taschenbücher, 44, Springer-Verlag, 1969.

[9] W. U. WURFEL. - "Numerische Berechnung der zweidimensionalen MHD-Strömung für beliebig grosse Hartmannzahlen mit E- Methoden", Dissertation, TH Darmstadt, Fachbereich Mechanik, 1990.

Interaction Between an Oblique Shock and a Detached Shock Upstream of a Cylinder in Supersonic Flow

M. Holt and M.P. Loomis

Mechanical Engineering, University of California, Berkeley, CA 94720, USA

This paper describes an experimental study concerning the interaction, in supersonic flow, between an oblique shock, generated by a wedge, and a detached shock upstream of a cylinder. The experiments were performed in the Berkeley supersonic wind tunnel at a Mach number of 2.4. The experiments differ from those carried out earlier in the application of new interferometric techniques to measure the density. These experimental results were compared with established numerical results. The data obtained were used in order to understand the phenomena which appear, for example, when an oblique shock is generated at the entry to a duct on the aerospace plane. It is known that these interactions can cause significant increases in pressure and heat transfer rate.

The principal aim of our experiments is to develop interferometric techniques and their application in providing data about different types of interaction.

We have carried out two types of interferometry, the first is holographic while the second is called dark central ground. In the first type, holographic interferograms are made of two photos of the working section of the wind tunnel ; the first is taken when the tunnel is operating under test conditions, while the second photo is taken with tunnel at rest. The interferogram is then formed at the conclusion of the experiment by comparison of the two images. In the second type of interferometry, a new technique is used called dark central ground interferometry : a spatial filter is placed at the focal point of the system, after the laser beam has passed through the working section. Since the technique produces interferograms instantaneously rather than at the end of a reconstruction process, it appears to be better than holographic interferometry. The present work represents one of the first applications of this new technique in wind tunnel experiments.

BACKGROUND

Edney (1968) listed several examples of the interaction of a weak oblique shock with a detached shock upstream of a blunt body (cowl- engine inlet, fuselage-fin, missile-booster, tension shell). His experiments were made at a Mach number of 4.6 (hypersonic threshhold). He found unexpected maxima of pressure and heat transfer, much greater than the values found in flow without interaction. Edney studied the flow past a sphere disturbed by a weak shock.

Twenty years later, there was renewed interest in Edney's work, arising from investigation of the cowl-lip problem in the National Aerospace Plane. This is more serious than the earlier problem owing to an increase of Mach number in excess of value 8 . The more recent problem was investigated numerically by Klopfer and Yee (1988) and by Moon and Holt (1989).

DETAILS OF EXPERIMENT

Fig. 1 shows a general sketch of the National Aerospace Plane (NASP), including the entry to the propulsion system towards the rear of the lower wing surface. The flow near the cowl lip in one of the entry ducts of the system is shown in Fig. 2, with an oblique shock, generated by the ramp compression region, and a detached shock upstream of the cowl lip. The type of interaction between the oblique and detached shocks depends on their relative positions. Edney identified six possible types of interaction, and these are shown in Fig. 3 (presented by Keyes and Hains (1973)). Types III and IV are the most critical in connection with the cowl lip problem, but since Type IV requires a tunnel Mach number greater than 3 , the present investigation concentrates on Types III and II. In Type II (shown in Fig. 3), the interaction results in two triple shock intersection points, downstream of which two shear layers develop. In Type III interaction only one triple shock intersection point appears, but the shear layer downstream impinges on the cowl lip surface.

Two types of interferometry were used to measure the density field resulting from shock-shock interaction. The first is holographic, of dual plate type, in which holograms are used to store two images of the wind tunnel test section, one made at the test condition, the other under no flow condition. The second type, called dark central ground interferometry, was first proposed by Anderson and Milton (1989), and produces real time interferograms. At the focal point of the optical system generated by a pulsed laser beam, a glass plate coated with photo emulsion is introduced. The intensity of the laser beam is then increased as much as possible to burn a dark spot on the emulsion and so create a spatial filter, located exactly at the spot needed, simultaneously blocking out unwanted light. A full description of the two systems of interferometry is given in Loomis (1990).

An example of holographic interferometry is shown in Fig. 4, corresponding to a Type III interaction. The density contours are clearly defined, including those across the shear layer downstream of the shock interaction point. Fig. 5 shows a dark central ground interferogram for a Type II interaction with the double shear layer downstream of the shock intersection points easily observed.

CONCLUSIONS

1. The effectiveness of two systems of interferometry has been demonstrated in application to the study of two types of shock interaction.
2. The measured values of the density, in various interaction modes, agree well with corresponding values derived from numerical investigations.
3. In order to study interaction of Type IV, it is necessary to increase the test Mach number above a value of 5 . To achieve this, the whole interferometric system should be transferred to a hypersonic wind tunnel.

REFERENCES

[1] R. C. ANDERSON and J. E. MILTON, Proc. Int. Cong. on Instrumentation in Aerospace simulation facilities, Göttingen, RFA, (1989), p. 394-399.

[2] B. EDNEY, FFA Report 115, Aero Res. Institute of Sweden, 1968.

[3] J. N. KEYES and F. D. HAINS, NASA TN D-7139, 1973.

[4] Mark P. LOOMIS, Ph. D. Thesis, University of California, Berkeley, 1990.

[5] M. P. LOOMIS and M. HOLT, "Interferometric study of supersonic flow fields with shock-shock interactions", 5th International Symposium on Application of laser techniques to fluid mechanics, Lisbon, Portugal, July 1990.

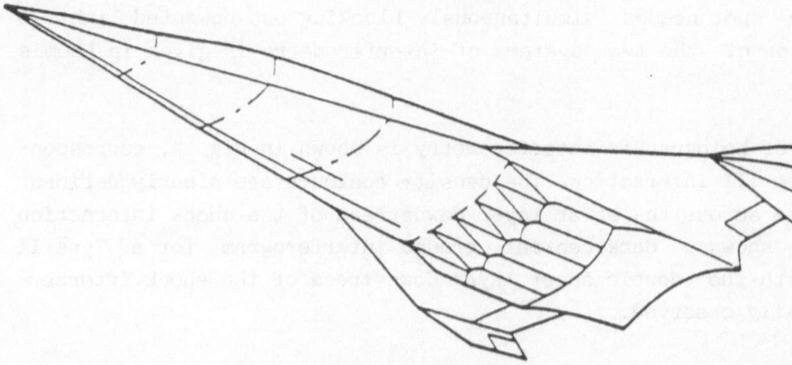

Figure 1 : NASP Schematic

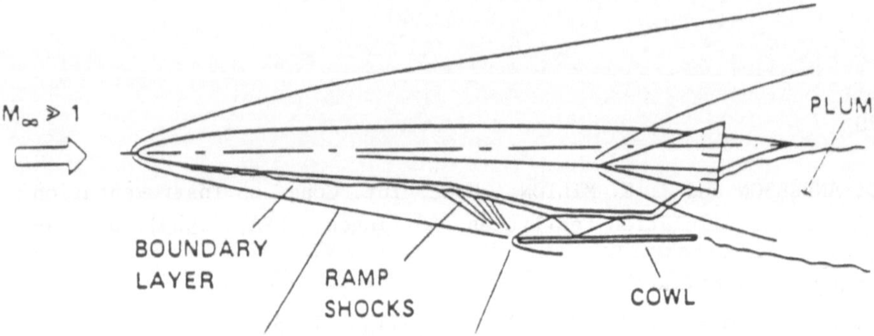

Figure 2 : NASP Flow Field-Side View

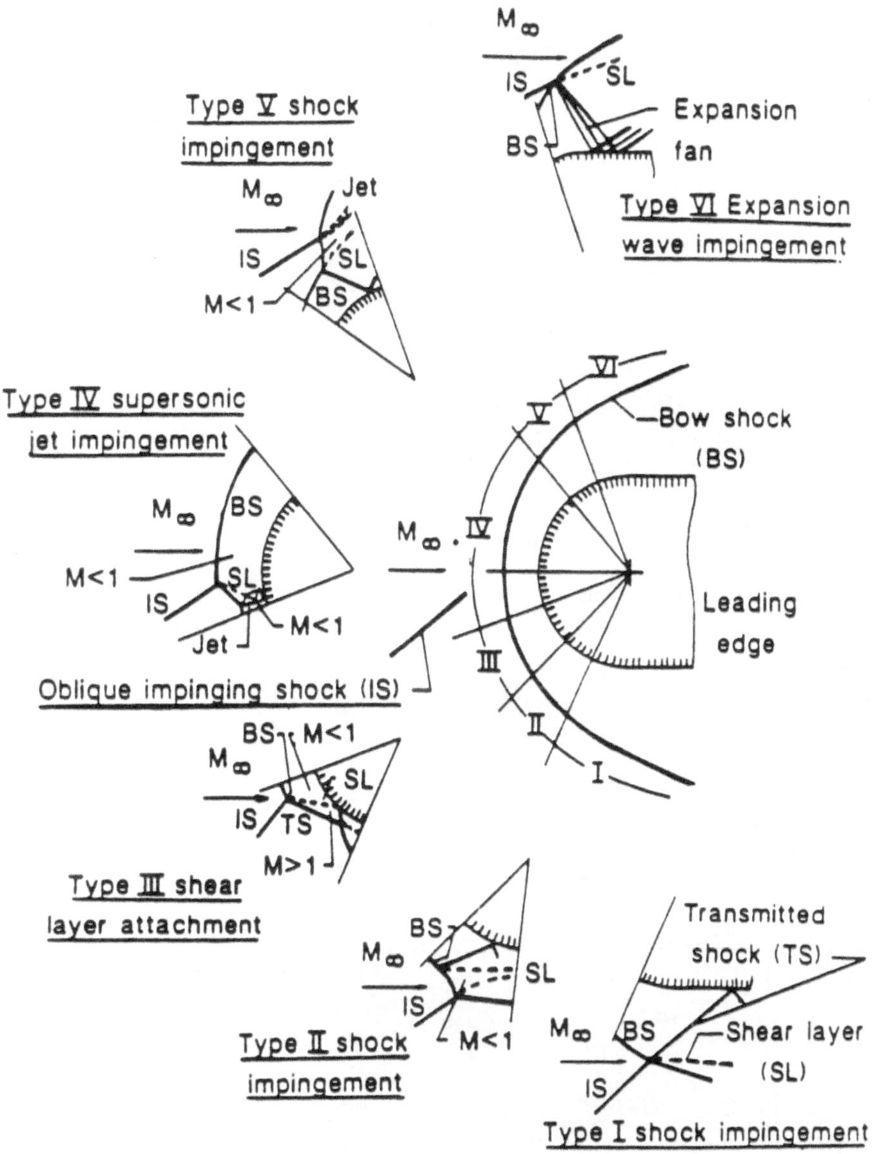

Figure 3 : Edney's Six Types of Interactions (Keyes and Hains, 1974)

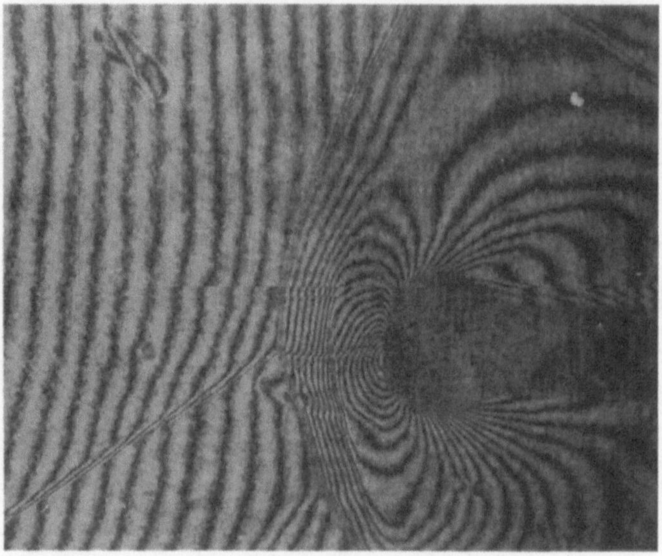

Figure 4 : Type III Interaction, P_0 = 2/3 atm

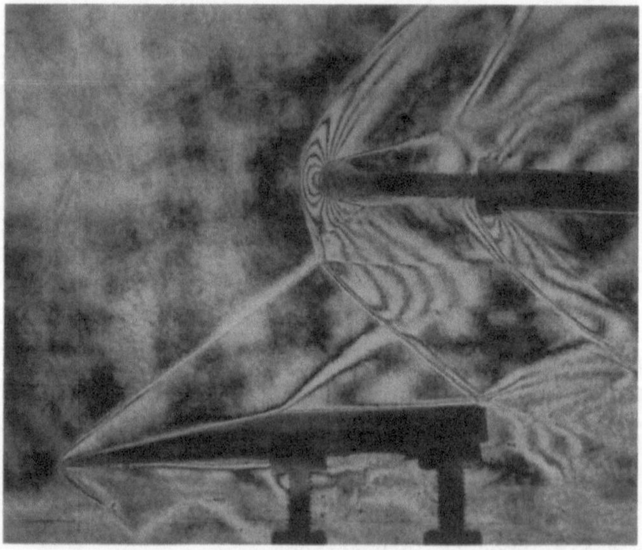

Figure 5 : Type III Interaction, P_0 = 1/3 atm

The Design of Super-Concorde and Space Vehicles Using Global Optimization Techniques

A. Nastase

Lehrgebiet Aerodynamik des Fluges, RWTH,
Templergraben 55, W-5100 Aachen, Fed. Rep. of Germany

Abstract : The optimum-optimorum configuration of the space vehicle is the configuration for which the shapes of its surface and also of its planprojection are simultaneously determined in such a manner that its drag attains its minimum at a given cruising Mach number M_∞ . The problem of the determination of the optimum-optimorum configuration of a space vehicle of variable geometry which presents a minimum drag at two cruising Mach numbers M_∞ and M_∞^* are here also considered.

1. THE OPTIMUM-OPTIMORUM THEORY.

The optimum-optimorum theory, introduced by the author in [1]-[3], was used for the determination of the optimum-optimorum shape of the wing alone (Fig. 1a) at a given cruising Mach number M_∞ as in [1]- [5], [8]. More recently, the variational problem concerning the determination of the optimum-optimorum shape of the integrated wing-fuselage configuration (Fig. 1b) (at a cruising Mach number M_∞) was considered by the author in [8]-[12]. For the integrated wing- fuselage configuration, all its geometrical parameters, i.e. the distributions of cambers, twists, thicknesses, and also the similarity parameters of the planprojections of the wing and of the fuselage, are optimized in order to obtain a minimum drag (at a given cruising Mach number M_∞). This is a proposal of the author for the design of the optimal shapes of the both vehicles of a two stage configuration (like Sänger and Horus) and for the Super-Concorde.

The numerical analysis performed by the author on about hundred optimized wings leads to the conclusion that the dimensionsless span ℓ_{opt} shows a strong dependence on cruising Mach number M_∞ , as it can be seen in the optimal hyperbola (Fig. 3c) (hereby, is $B = \sqrt{M_\infty^2 - 1}$).

Therefore the further step in the optimization of the entire configuration of the space vehicle is to determine, the optimum- optimorum shape of the integrated wing-fuselage configuration of variable geometry. This variable geometry can be realized with the help of movable leading edge flaps, as formerly proposed by the author in [1], [7], [13], for the delta wing alone.

The integrated wing-fuselage configuration with movable flaps (Fig. 1b, 1c) can be optimized at two supersonic cruising Mach numbers M_∞^* and M_∞ ($M_\infty^* < M_\infty$) .At the higher supersonic Mach number M_∞ , the integrated configuration of the space vehicle is flying with the flaps in retracted position. At the lower supersonic Mach numbers M_∞^* , the space vehicle is flying with the flaps in open position. This is the authors proposal for the design of the shape of a single-stage vehicle (like Hotol, Hermes, and NASP).

The determination of the optimum-optimorum configuration of the space vehicle with variable geometry leads to the solving of two three-dimensional boundary-value problems for the axial disturbance velocity u on the space vehicle with flaps (in retracted and open positions), and of two successive enlarged variational problems (with free boundary) for the space vehicle with flaps (in retracted and in open positions). The wing-fuselage configuration with flaps in retracted position is here considered as a wing alone, for which the surface is discontinuous along the junction lines between the wing and the fuselage. The wing-fuselage configuration with flaps in open position is considered also as a wing alone, for which the surface is now discontinuous along the junction lines between the wing and the fuselage, and between the wing and the leading edge flaps.

The author has proposed in [1]-[5] a method for the design of fully-optimized shape of the aircraft configurations and space vehicles, which she called it optimum-optimorum theory. This theory allows the simultaneous determination of the optimal shapes of the surface and of the plan-projection of the space vehicle in order to obtain a minimum drag. The determination of the shape of the optimum-optimorum space vehicle leads to an extended variational problem for the drag functional C_d , i.e. :

$$C_d \equiv \int_{S(x_1,x_2)} F[x_1,\ x_2,\ Z(x_1,x_2)]\ dx_1\ dx_2 = \text{min.} \tag{1}$$

Here, the function $Z(x_1,x_2)$ and also the boundary $S(x_1,x_2)$ of the integral are a priori unknown, and are determined by the solving of this extended variational problem. The optimum-optimorum space vehicle is chosen among a set of space vehicles, which are defined through some common properties. In the frame of the optimum- optimorum theory of the author, two space vehicles belong to the same set if : their surfaces can be piecewise approximated through a superposition of homogeneous polynomes of the same degree, their planprojections are polygons which can be related through affine transformations, and the shapes of the space vehicles of the set fulfill the same auxiliary conditions (of geometrical or aerodynamical nature).

The parameters of the optimization are the coefficients Z_{ij} of homogeneous polynomes of the equations of the surfaces and the similarity parameters $(\nu_1, \nu_2, \ldots, \nu_n)$ of the planprojections of the space vehicles of the set. In order to solve this enlarged variational problem for the determination of the extremum of the drag functional $C_d^{(t)}$ with free boundary, the author uses her hybrid, numerical-analytical method. This method starts with the remark, that the dependence of the drag functional $C_d^{(t)}$ versus the coefficients Z_{ij} of the polynomes, which piecewise approximate the surfaces of the space vehicles, is a quadratic form, while the dependence versus the similarity parameters of the planform are nonlinear and very complicated. The method presents two steps.

1. In the first step, the set of similarity parameters of the planform $(\nu_1, \nu_2, \ldots, \nu_n)$ are considered as given. The boundary of the drag functional $C_d^{(t)}$ is now a priori known. The optimal value of the coefficients of polynomial expansions of the surface of the space vehicle are obtained by solving a linear, algebraic system. These optimal coefficients determine uniquely the value of the drag functional $(C_d^{(t)})_{opt}$, for the prescribed set of similarity parameters of the planform. This value of $(C_d^{(t)})_{opt}$ represents a "point" of what is called here lower limit hypersurface of the drag functional $C_d^{(t)}$, i.e. :

$$(C_d^{(t)})_{opt} = f(\nu_1, \nu_2, \ldots, \nu_n) \tag{2}$$

Each of these points can be analytically determined.

2. In the second step, through systematical variation of the set of similarity parameters, the "position" of the minimum of this hypersurface is numerically (or graphically) determined, and gives the best set of similarity parameters $(\nu_1, \nu_2, \ldots, \nu_n)$ of the planform, as presented in (Fig. 2), for two similarity parameters. The optimal set of similarity parameters, together with a chosen area S_0 of the plan-projection, determine the shapes of the planform and of the surface of the optimum-optimorum space vehicle of a given set of space vehicles. The optimum-optimorum space vehicle is exactly the optimal space vehicle corresponding to this optimal set of similarity parameters. The minimum value of the "ordinate" of the hypersurface represents the drag coefficient of the optimum-optimorum space vehicle of the set. The above theory was successfully used by the author for the effective design of the shape of optimum-optimorum delta wing Adela [1]-[5], and of the optimum-optimorum shape of the integrated wing-fuselage

configuration (at cruising Mach number $M_\infty = 2$), as in [8]-[12], [21]. The shape of delta wing Adela is given in (Fig. 3a), and the modification in the shape of the wing Adela, due to the fuselage integration (in the section $\tilde{x}_1 = 0{,}6$), is given in (Fig. 1a,b). A further new application of the optimum- optimorum theory taken here into consideration is the determination of the shape of the entire space vehicle which is an integrated wing-fuselage-flaps configuration of minimum drag at two cruising Mach number M_∞ and M^*_∞. Two variational problems in cascade are here occurring; one for the determination of the optimum-optimorum shape of the space vehicle at the higher cruising Mach number M_∞ with the flaps in retracted position, and the second for the determination of the optimum-optimorum shapes of the flap-surface and its planprojection in such a manner, that the entire space vehicle (with the flaps in open position) is of minimum drag at the supersonic cruising Mach number M^*_∞.

2. DETERMINATION OF THE AXIAL DISTURBANCE VELOCITIES.

Let us refer the integrated thick, lifting delta wing to a three-orthogonal system of axes $Ox_1x_2x_3$ having the apex O of the wing as origin. The plane Ox_1x_2 is the plane of symmetry of the integrated wing, and the axis Ox_1 is the bisectrix of the angle of the integrated wing, in the plane Ox_1x_3, at its apex (the shock-free entry direction). The integrated thick, lifting delta wing surface, is supposed to be flattened in the plane Ox_1x_2 (Fig. 4a), and is considered in a parallel stream with the undisturbed velocity $\vec{V}_\infty$ at a moderate angle of attack α (measured between the Ox_1-axis and $\vec{V}_\infty$).

In the framework of linearised theory for flattened integrated thick, lifting delta wings at moderate angle of attack α, in the boundary value problem concerning the determination of the axial disturbance velocity u, the effect of lift can be separated from the effect of thickness. Further, the following two delta wing components will be separately considered. The <u>thin integrated delta wing</u> which is the skeleton surface of the thick, lifting integrated delta wing, and is considered at the same angle of attack α, and <u>the thick-symmetrical integrated delta wing</u> which has the same thickness distribution as the thick, lifting integrated delta wing, but its skeleton surface is a plane. This component is considered at zero angle of attack. The skeleton surface $Z(x_1,x_2)$ of the integrated delta wing is supposed to be continuous, but, for the sake of generality, the thickness distributions $Z^*(x_1,x_2)$ on the lateral sides OA_1C_1 and OA_2C_2 (corresponding to the wing) and $Z'^*(x_1,x_2)$ on the central part OC_1C_2 (corresponding to the fuselage) are supposed to be different. Further, this wing will be called initial integrated delta wing. The author introduced, as in [1]-[13], a well-suited affine transformation in order to obtain dimensionless coordinates :

$$\tilde{x}_1 = \frac{x_1}{h_1}\ , \quad \tilde{x}_2 = \frac{x_2}{\ell_1}\ , \quad \tilde{x}_3 = \frac{x_3}{h_1}\ , \quad B = \sqrt{M_\infty^2 - 1}$$

$$(\tilde{y} = \frac{y}{\ell}\ , \quad \ell = \frac{\ell_1}{h_1}\ , \quad \bar{k} = \frac{c'}{\ell})\ . \tag{3}$$

A transformed integrated delta wing is obtained, which has the maximal depth 1 and the half-span 1 (Fig. 4b). The traces $\tilde{C}_1$ and $\tilde{C}_2$ of the junction lines $\tilde{O}\tilde{C}_1$ and $\tilde{O}\tilde{C}_2$ (between the wing and the fuselage) have the following positions on the axis $\tilde{C}\tilde{y}$ (parallel to axis Ox_2) : $y_c = \pm \bar{k}$ ($\bar{k} = \bar{\nu}/\nu$ is here constant). The transformed integrated delta wing is placed in a supersonic flow with the cruising Mach number $\tilde{M}_\infty = \sqrt{1 + \nu^2}$. Between the dimensionless axial disturbance velocities u , u^* and $\tilde{u}$, $\tilde{u}^*$ and the dimensionless downwashes w , w^*, w'^* and $\tilde{w}$, $\tilde{w}^*$, $\bar{w}^*$ of the initial and transformed integrated delta wing components, there are the following relations :

$$u = \ell\tilde{u}\ , \quad w = \tilde{w}\ , \quad u^* = \ell\tilde{u}^*\ , \quad w^* = \tilde{w}^*\ , \quad w'^* = \bar{w}^*$$

$$(\nu = B\ell\ , \quad \bar{\nu} = Bc')\ . \tag{4}$$

Further, the assumption is made, that the downwashes $\tilde{w}$, $\tilde{w}^*$ and $\bar{w}^*$ are expressed in form of superpositions of homogeneous polynomes in $\tilde{x}_1$ and $\tilde{x}_2$, i.e. on the thin component of the transformed integrated wing :

$$\tilde{w} = \sum_{m=1}^{N} \tilde{x}_1^{m-1} \sum_{k=0}^{m-1} \tilde{w}_{m-k-1,k}\ |\tilde{y}|^k \tag{5}$$

and on the thick-symmetrical component of the transformed integrated wing :

$$\tilde{w}^* = \sum_{m=1}^{N} \tilde{x}_1^{m-1} \sum_{k=0}^{m-1} \tilde{w}^*_{m-k-1,k}\ |\tilde{y}|^k\ , \qquad (\text{if}\ \ \bar{k} < |\tilde{y}| < 1) \tag{6a}$$

$$\bar{w}^* = \sum_{m=1}^{N} \tilde{x}_1^{m-1} \sum_{k=0}^{m-1} \bar{w}^*_{m-k-1,k}\ |\tilde{y}|^k\ , \qquad (\text{if}\ \ 0 < |\tilde{y}| < \bar{k})\ . \tag{6b}$$

The coefficients $\tilde{w}_{ij}$, $\tilde{w}^*_{ij}$ and $\bar{w}^*_{ij}$ and the similarity parameter ν are unknown, and will be determined through the fully-optimization process. The axial disturbance velocity $\tilde{u}$ on the thin component of the transformed integrated thick-lifting delta wing with subsonic leading edges and a central ridge, is of the form :

$$\tilde{u} = \sum_{n=1}^{N} \tilde{x}_1^{n-1} \left\{ \sum_{q=0}^{E(\frac{n}{2})} \frac{\tilde{A}_{n,2q}\ \tilde{y}^{2q}}{\sqrt{1-\tilde{y}^2}} + \sum_{q=1}^{E(\frac{n-1}{2})} \tilde{C}_{n,2q}\ \tilde{y}^{2q} \cosh^{-1} \frac{1}{\sqrt{\tilde{y}^2}} \right\} \qquad (7)$$

If the integrated thin delta wing has supersonic leading edges it results in :

$$\tilde{u} = \sum_{n=1}^{N} \tilde{x}_1^{n-1} \left\{ \sum_{q=0}^{n-1} \tilde{H}_{nq}\ \tilde{y}^{q} [\cos^{-1} M_1 + (-1)^q \cos^{-1} M_2] + \sum_{q=0}^{E(\frac{n}{2})} \frac{\tilde{P}_{n,2q}\ \tilde{y}^{2q}}{\sqrt{1-\nu^2\tilde{y}^2}} + \sum_{q=1}^{E(\frac{n-1}{2})} \tilde{C}_{n,2q}\ \tilde{y}^{2q} \cosh^{-1} \frac{1}{\sqrt{\nu^2\tilde{y}^2}} \right\}. \qquad (8)$$

The axial disturbance velocity $\tilde{u}^*$ of the thick-symmetrical integrated delta wing component of the transformed integrated thick, lifting delta wing with subsonic leading edges, as in [6]-[12], is :

$$\tilde{u}^* = \sum_{n=1}^{N} \tilde{x}_1^{n-1} \left\{ \sum_{q=0}^{E(\frac{n}{2})} \frac{\tilde{P}^*_{n,2q}\ \tilde{y}^{2q}}{\sqrt{1-\nu^2\tilde{y}^2}} + \sum_{q=0}^{n-1} \tilde{H}^*_{nq}\ \tilde{y}^{q} [\cosh^{-1} M_1 + (-1)^q \cosh^{-1} M_2] \right.$$
$$\left. + \sum_{q=1}^{E(\frac{n-1}{2})} \tilde{C}^*_{n,2q}\ \tilde{y}^{2q} \cosh^{-1} \frac{1}{\sqrt{\nu^2\tilde{y}^2}} + \sum_{q=0}^{n-1} \tilde{G}^*_{nq}\ \tilde{y}^{q} [\cosh^{-1} R_1 + (-1)^q \cosh^{-1} R_2] \right\}. \qquad (9)$$

If the thick-symmetrical integrated delta wing has supersonic leading edges in the formula (9), the terms $\cosh^{-1} M_1$ and $\cosh^{-1} M_2$ are to be replaced with $\cos^{-1} M_1$ and $\cos^{-1} M_2$. Here, the following notations will be made :

$$R_1 = \sqrt{\frac{(1+\bar{\nu})(1-\nu\tilde{y})}{2(\bar{\nu}-\nu\tilde{y})}}, \qquad R_2 = \sqrt{\frac{(1+\bar{\nu})(1+\nu\tilde{y})}{2(\bar{\nu}+\nu\tilde{y})}} \qquad (10a)$$

$$M_1 = \sqrt{\frac{(1+\nu)(1-\nu\tilde{y})}{2\nu(1-\tilde{y})}}, \qquad M_2 = \sqrt{\frac{(1+\nu)(1+\nu\tilde{y})}{2\nu(1+\tilde{y})}}. \qquad (10b)$$

The coefficients of $\tilde{u}$ for the thin transformed integrated delta wing are related to the coefficients of the downwash $\tilde{w}$, and the coefficients of $\tilde{u}^*$ for the thick-symmetrical transformed integrated delta wing are related to the coefficients of the downwashes $\tilde{w}^*$ and $\bar{w}^*$ through the following linear and homogeneous relations :

$$\tilde{A}_{n,2q} = \sum_{j=0}^{n-1} \tilde{a}_{2q,j}^{(n)} \tilde{w}_{n-j-1,j} ,$$

$$\tilde{P}^*_{n,2q} = \sum_{j=0}^{n-1} (\tilde{p}_{2q,j}^{*(n)} \tilde{w}^*_{n-j-1,j} + \bar{p}_{2q,j}^{*(n)} \bar{w}^*_{n-j-1,j}) . \tag{11}$$

The constants $\tilde{a}_{2q,j}^{(n)}$, $\tilde{p}_{2q,j}^{*(n)}$, $\bar{p}_{2q,j}^{*(n)}$, etc. are functions only of the similarity parameter ν .

Let us now consider the open integrated delta wing (with flaps in open positions) (Fig. 1c), (Fig. 4c), at cruising Mach number M^*_∞ ($B^* = \sqrt{M^{*2}_\infty - 1}$) . The downwashes on the wing and fuselage are unchanged, and are given for the transformed configuration in the formulas (3), (4) and (5), and for the initial configuration as in the formulas (2). The downwashes $\tilde{\bar{w}}$ and $\tilde{\bar{w}}^*$ on the thin- and thick-symmetrical components on the flaps of the transformed open integrated wing (Fig. 4d) are supposed to be expressed in the form of superposition of homogeneous polynomes on the thin, and on the thick-symmetrical transformed flaps components.

$$\tilde{\bar{w}} = \sum_{n=1}^{N} \tilde{x}_1^{n-1} \sum_{k=0}^{n-1} \tilde{\bar{w}}_{n-k-1,k} |\tilde{y}|^k , \qquad \tilde{\bar{w}}^* = \sum_{n=1}^{N} \tilde{x}_1^{n-1} \sum_{k=0}^{n-1} \tilde{\bar{w}}^*_{n-k-1,k} |\tilde{y}|^k . \tag{12}$$

Between the downwashes and axial disturbance velocities on the initial and transformed flaps, there are the following relations :

$$\tilde{\bar{w}}' = \tilde{\bar{w}} , \qquad \tilde{\bar{w}}'^* = \tilde{\bar{w}}^* , \qquad \tilde{\bar{u}}' = \ell \, \tilde{\bar{u}} , \qquad \tilde{\bar{u}}'^* = \ell \, \tilde{\bar{u}}^* . \tag{13}$$

The following notations are further made :

$$\bar{\nu}^* = B^* c' , \qquad \tilde{\nu}^* = B^* \ell , \qquad \nu^* = B^* L , \qquad k^* = \frac{L}{\ell} . \tag{14}$$

The transformed open integrated delta wing is placed in a supersonic flow with the cruising Mach number $M^*_\infty = \sqrt{1 + \nu^{*2}}$. The axial disturbance velocities $\tilde{\bar{u}}$ and $\tilde{\bar{u}}^*$ on the thin and thick-symmetrical transformed open integrated delta wing with subsonic leading edges at the cruising Mach number M^*_∞ are obtained by the author here under the following forms :

$$\tilde{\tilde{u}} = \sum_{n=1}^{N} \tilde{x}_1^{n-1} \left\{ \sum_{q=0}^{n-1} \tilde{\tilde{K}}_{nq} \, \tilde{y}^q \left[\cosh^{-1} N_1 + (-1)^q \cosh^{-1} N_2\right] \right.$$

$$\left. + \sum_{q=0}^{E(\frac{n}{2})} \frac{\tilde{\tilde{A}}_{n,2q} \, \tilde{y}^{2q}}{\sqrt{k^{*2} - \tilde{y}^2}} + \sum_{q=1}^{E(\frac{n-1}{2})} \tilde{\tilde{C}}_{n,2q} \, \tilde{y}^{2q} \cosh^{-1} \sqrt{\frac{k^{*2}}{\tilde{y}^2}} \right\}, \qquad (15)$$

$$\tilde{\tilde{u}}^* = \sum_{n=1}^{N} \tilde{x}_1^{n-1} \left\{ \sum_{q=0}^{n-1} \tilde{\tilde{K}}_{nq} \, \tilde{y}^q \left[\cosh^{-1} N_1^* + (-1)^q \cosh^{-1} N_2^*\right] \right.$$

$$+ \sum_{q=0}^{E(\frac{n}{2})} \frac{\tilde{\tilde{P}}_{n,2q} \, \tilde{y}^{2q}}{\sqrt{1 - \nu^{*2} \, \tilde{y}^2}} + \sum_{q=1}^{E(\frac{n-1}{2})} \tilde{\tilde{C}}^*_{n,2q} \, \tilde{y}^{2q} \cosh^{-1} \sqrt{\frac{1}{\tilde{\nu}^{*2} \tilde{y}^2}}$$

$$+ \sum_{q=0}^{n-1} \tilde{\tilde{G}}^*_{nq} \, \tilde{y}^q \left[\cosh^{-1} R_1^* + (-1)^q \cosh^{-1} R_2^*\right]$$

$$\left. + \sum_{q=0}^{n-1} \tilde{\tilde{H}}^*_{nq} \, \tilde{y}^q \left[\cosh^{-1} M_1^* + (-1)^q \cosh^{-1} M_2^*\right] \right\}. \qquad (16)$$

In the formulas (15) and (16), the following notations have been made

$$N_1 = \sqrt{\frac{(1 + k^*)(k^* - \tilde{y})}{2k^*(1 - \tilde{y})}}, \qquad N_2 = \sqrt{\frac{(1 + k^*)(k^* + \tilde{y})}{2k^*(1 + \tilde{y})}}$$

$$R_1^* = \sqrt{\frac{(1 + \bar{\nu}^*)(1 - \tilde{\nu}^* \tilde{y})}{2\tilde{\nu}^*(\bar{k} - \tilde{y})}}, \qquad R_2^* = \sqrt{\frac{(1 + \bar{\nu}^*)(1 + \tilde{\nu}^* \tilde{y})}{2\tilde{\nu}^*(\bar{k} + \tilde{y})}}$$

$$M_1^* = \sqrt{\frac{(1 + \tilde{\nu}^*)(1 - \tilde{\nu}^* \tilde{y})}{2\tilde{\nu}^*(1 - \tilde{y})}}, \qquad M_2^* = \sqrt{\frac{(1 + \tilde{\nu}^*)(1 + \tilde{\nu}^* \tilde{y})}{2\tilde{\nu}^*(1 + \tilde{y})}}$$

$$N_1^* = \sqrt{\frac{(1 + \nu^*)(1 - \tilde{\nu}^* \tilde{y})}{2\nu^*(k^* - \tilde{y})}}, \qquad N_2^* = \sqrt{\frac{(1 + \nu^*)(1 + \tilde{\nu}^* \tilde{y})}{2\nu^*(k^* + \tilde{y})}}. \qquad (17)$$

These new formulas are obtained by the author by using the results of high conical flow theory of Germain [14], the hydrodynamic analogy of Carafoli [15], [16], and the principle of minimum singularities ([17], [18]). The coefficients of the axial velocities $\tilde{\tilde{u}}$ and $\tilde{\tilde{u}}^*$ are related to the coefficients of the downwashes $\tilde{w}$, $\tilde{\tilde{w}}$ and $\bar{w}^*$, $\tilde{w}^*$ and $\tilde{\tilde{w}}^*$ through linear and homogeneous relations of the form :

$$\tilde{\tilde{A}}_{n,2q} = \sum_{j=0}^{n-1} (\tilde{a}^{(n)}_{2q,j} \; \tilde{w}_{n-j-1,j} + \tilde{\tilde{a}}^{(n)}_{2q,j} \; \tilde{\tilde{w}}_{n-j-1,j}) \tag{18}$$

$$\tilde{\tilde{P}}^*_{n,2q} = \sum_{j=0}^{n-1} (\tilde{p}^{*(n)}_{2q,j} \; \tilde{w}^*_{n-j-1,j} + \bar{p}^{*(n)}_{2q,j} \; \bar{w}^*_{n-j-1,j} + \tilde{\tilde{p}}^{*(n)}_{2q,j} \; \tilde{\tilde{w}}^*_{n-j-1,j}) \; . \tag{19}$$

The coefficients $\tilde{a}^{(n)}_{2q,j}$, $\tilde{\tilde{a}}^{(n)}_{2q,j}$, $\bar{p}^{*(n)}_{2q,j}$, $\tilde{p}^{*(n)}_{2q,j}$, $\tilde{\tilde{p}}^{*(n)}_{2q,j}$, etc. are functions only on the similarity parameters $\bar{\nu}^*$, $\tilde{\nu}^*$ and ν^* .

The theoretical determined pressure coefficient C_p according to the present theory (i.e. by using the formulas (7) and (9) with $G^*_{nq} = 0$) are in good agreement with experimental results for a large range of Mach numbers ($M_\infty = 1{,}25 - 2{,}2$) and angles of attack α($|\alpha| < 10°$), as it can be seen in [26], for the longitudinal central section, and for the transversal section $\tilde{x}_1 = 0{,}599$ of the upper side of the optimum-optimorum delta wing Adela, and for the angles of attack $\alpha = -8°$, $+8°$. This agreement between theory and experiment is due to the accuracy of the solutions of the boundary value problems for the axial disturbance velocities $\tilde{u}$ and $\tilde{u}^*$ given in formulas (7) and (8). These solutions for $\tilde{u}$ and $\tilde{u}^*$ present the following advantages in comparison with the ones obtained in the frame of slender body theory [22],[23] :

a) They fulfil the full-linearised partial differential equation, which is hyperbolic, includes the influence of Mach number M_∞ , and needs no restrictions concerning the magnitude of span.

b) The boundary conditions along the characteristic surface, i.e. the Mach cone of the apex of the integrated wing, and at the infinity (forward) are satisfied.

c) According to the hydrodynamic analogy of Carafoli ([17], [18]), the singularities in these solutions of u and u^* are located only along the singular lines (i. e. along the leading edges of the wing, along the junction lines of the wing-fuselage configuration, etc.), and therefore are easier to be applied as the solutions for axial disturbance velocities given in [23], which are obtained by using singularities located on the whole wing surface.

d) These singularities are chosen according to the principle of minimum singularities ([17], [18]), and therefore the potential solutions for u and u^* given here are matched with a boundary layer solution, and are zonal solutions.

e) The solutions (8) and (9) for u and u^* can be also used for the calculation of pressure distribution and of aerodynamic characteristics of space vehicle which shape is given in discrete form. The surface of the integrated wing can be piecewise approximated in form of polynomial expansions which are obtained by using the two-dimensional minimal quadratic error similar, as in [5].

The optimization of the shape of the thin and thick-symmetrical and thick, lifting integrated wings with flaps in retracted position, can be treated as the optimization problem of the integrated wing-fuselage configuration, as in [8]-[12]. It results in the optimal values of the coefficients $\tilde{w}_{\theta\sigma}$, $\tilde{w}^*_{\theta\sigma}$ and $\bar{w}^*_{\theta\sigma}$ of the downwashes $\tilde{w}$, $\tilde{w}^*$ and $\bar{w}^*$, and the optimal values of the similarity parameters ν and $\bar{\nu}$. The optimization of the flap shapes are further considered.

3. OPTIMIZATION OF THE FLAPS SHAPE.

The optimization of the shape of the thin and thick-symmetrical open integrated wing (i.e. the wing-fuselage configuration with flaps in open position) at the second, lower Mach number M^*_∞ is here considered (Fig. 1c). The shape of the integrated wing-fuselage configuration (with flaps in retracted position) is determined by the solution of the precedent variational problem, and remains unchanged in this second variational problem, which consists in the determination of the shape of the flap in such a manner, that the entire space vehicle (with flaps in open position) is of minimum drag at the second cruising Mach number M^*_∞. The variational problem of the thin open integrated delta wing is firstly considered. The downwashes $\tilde{w}$ (on the transformed thin integrated wing-fuselage configuration) and $\tilde{\tilde{w}}$ on the flap are given as in formulas (3) and (12). The coefficients $\tilde{w}_{ij}$ of $\tilde{w}$ are previously determined by the precedent variational problem, and are here supposed known and constant. The optimization of the <u>thin flap component</u> (for a given value of ν) leads to the determination of the coefficient $\tilde{\tilde{w}}_{ij}$ of the downwash $\tilde{\tilde{w}}$ in such a manner, that the drag coefficient C'_d at the cruising Mach number M^*_∞ attains its minimum, i.e. :

$$C'_d \equiv \ell \sum_{n=1}^{N} \sum_{m=1}^{N} \sum_{k=0}^{m-1} \sum_{j=0}^{n-1} \Big\{ [\, \tilde{\Omega}_{nmkj}\, \tilde{w}_{n-j-1,j} + \tilde{\tilde{\Omega}}_{nmkj}\, \tilde{\tilde{w}}_{n-j-1,j}]\, \tilde{w}_{m-k-1,k} + [\tilde{\Omega}'_{nmkj}\, \tilde{w}_{n-j-1,j} + \tilde{\tilde{\Omega}}'_{nmkj}\, \tilde{\tilde{w}}_{n-j-1,j}]\, \tilde{\tilde{w}}_{m-k-1,k} \Big\} = \min. \tag{20}$$

with the following auxiliary condition (at the cruising Mach number M^*_∞).

- The lift coefficient C_ℓ' is given :

$$\tilde{\tilde{C}}_\ell \equiv \sum_{n=1}^{N} \sum_{j=0}^{n-1} (\tilde{\Lambda}'_{nj} \; \tilde{w}_{n-j-1,j} + \tilde{\tilde{\Lambda}}_{nj} \; \tilde{\tilde{w}}_{n-j-1,j}) = \frac{C'_{\ell_0}}{\ell} \tag{21}$$

- The pitching moment coefficient C_m' is given :

$$\tilde{\tilde{C}}_m \equiv \sum_{n=1}^{N} \sum_{j=0}^{n-1} (\tilde{\Gamma}'_{nj} \; \tilde{w}_{n-j-1,j} + \tilde{\tilde{\Gamma}}_{nj} \; \tilde{\tilde{w}}_{n-j-1,j}) = \frac{C'_{m_0}}{\ell} \tag{22}$$

- The axial disturbance velocity $\tilde{\tilde{u}}$ vanishes along the leading edge :

$$\tilde{\tilde{F}}_t \equiv \sum_{j=0}^{E(\frac{t}{2})} \tilde{\tilde{\Psi}}_{tj} \; \tilde{\tilde{w}}_{t-j-1,j} = 0 \qquad (t = 1,\ldots,N) \tag{23}$$

- The wing and the flap surfaces are continuous of class C_1 along the junction line between the wing and the flap. If

$$\Delta\tilde{\tilde{w}}_{m-k-1,k} = \tilde{\tilde{w}}_{m-k-1,k} - \tilde{w}_{m-k-1,k} \quad ,$$

then :

$$\tilde{\tilde{E}}_t \equiv \sum_{t=1}^{N} \sum_{k=0}^{m-1} \tilde{\tilde{c}}^{(t)}_{mk} \; \Delta \tilde{\tilde{w}}_{m-k-1,k} = 0 \; , \qquad \tilde{\tilde{G}}_t \equiv \sum_{t=1}^{N} \sum_{k=0}^{m-1} \tilde{\tilde{g}}^{(t)}_{mk} \; \Delta \tilde{\tilde{w}}_{m-k-1,k} = 0 \; ,$$

$$\tilde{\tilde{L}}_t \equiv \sum_{t=1}^{N} \sum_{k=0}^{m-1} \tilde{\tilde{\ell}}^{(t)}_{mk} \; \Delta \tilde{\tilde{w}}_{m-k-1,k} = 0 \; , \qquad (t = 0,1,2,\ldots,(N-1)) \; . \tag{24}$$

All the coefficients $\tilde{\Lambda}'_{nj}$, $\tilde{\tilde{\Lambda}}_{nj}$, etc., are depending only on the similarity parameters ν^* and $\tilde{\nu}^*$ ($\tilde{\nu}^*$ is here constant). The corresponding Hamilton's operator of this variational problem is :

$$\tilde{\tilde{H}} \equiv \ell \; \tilde{\tilde{H}}'$$

$$= \ell \left\{ \tilde{\tilde{C}}_d + \tilde{\tilde{\lambda}}^{(1)} \; \tilde{\tilde{C}}_\ell + \tilde{\tilde{\lambda}}^{(2)} \; \tilde{\tilde{C}}_m + \sum_{t=1}^{N} (\bar{\lambda}_t \; \tilde{\tilde{F}}_t + \tilde{\mu}_t \; \tilde{\tilde{E}}_t + \bar{\mu}_t \; \tilde{\tilde{G}}_t + \tilde{\tilde{\mu}}_t \; \tilde{\tilde{L}}_t) \right\} \tag{25}$$

<u>Remark</u> : The Lagrange's multipliers $\tilde{\tilde{\lambda}}^{(1)}$, $\tilde{\tilde{\lambda}}^{(2)}$, $\bar{\lambda}_t$, $\tilde{\mu}_t$, $\bar{\mu}_t$ and $\tilde{\tilde{\mu}}_t$ are only functions of the similarity parameter ν^* . If the first variation of the Hamilton's operator $\tilde{\tilde{H}}$ is cancelled ($\delta\tilde{H} = 0$) the following equations are obtained :

$$\sum_{n=1}^{N}\sum_{j=0}^{n-1}\Big\{[\tilde{\Omega}'_{n,\theta+\sigma+1,\sigma,j}+\tilde{\Omega}'_{\theta+\sigma+1,n,j,\sigma}]\,\tilde{\tilde{w}}_{n-j-1,j}$$

$$+[\tilde{\Omega}'_{n,\theta+\sigma+1,\sigma,j}+\tilde{\tilde{\Omega}}_{\theta+\sigma+1,n,j,\sigma}]\,\tilde{w}_{n-j-1,j}\Big\}$$

$$+\tilde{\tilde{\lambda}}^{(1)}\,\tilde{\tilde{\Lambda}}_{\theta+\sigma+1,\sigma}+\tilde{\tilde{\lambda}}^{(2)}\,\tilde{\Gamma}_{\theta+\sigma+1,\sigma}+\tilde{\tilde{\lambda}}_{\theta+\sigma+1,\sigma}\,\tilde{\tilde{\Psi}}_{\theta+\sigma+1,\sigma}$$

$$+\sum_{t=1}^{N}[\tilde{\mu}_t\,\tilde{\tilde{c}}^{(t)}_{\theta\sigma}+\bar{\mu}_t\,\tilde{\tilde{g}}^{(t)}_{\theta\sigma}+\tilde{\tilde{\mu}}_t\,\tilde{\tilde{\ell}}^{(t)}_{\theta\sigma}]=0$$

$$(1\leqslant\theta+\sigma+1\leqslant N\,,\quad\theta=0,1,\dots,(N-1))\tag{26}$$

This algebraic system, together with the auxiliary conditions (21)-(24) determine uniquely the coefficients $\tilde{\tilde{w}}_{\theta\sigma}$ of the downwashes and the Lagrange's multipliers $\tilde{\tilde{\lambda}}^{(1)}$, $\tilde{\tilde{\lambda}}^{(2)}$, $\bar{\lambda}_t$, $\tilde{\mu}_t$, $\bar{\mu}_t$ and $\tilde{\tilde{\mu}}_t$ (as functions of the similarity parameter ν^*).

Let us consider now the second variational problem concerning the optimization of the <u>thick-symmetrical component of the flap</u> by cruising Mach number M^*_∞. The downwashes coefficients $\tilde{\tilde{w}}^*_{ij}$ are determined in such a manner that the drag coefficient $C_d'^*$ of the thick-symmetrical component of the open integrated wing attains its minimum (at cruising Mach number M^*_∞), i.e. :

$$C_d'^*\equiv\ell\sum_{n=1}^{N}\sum_{m=1}^{N}\sum_{k=0}^{m-1}\sum_{j=0}^{n-1}\Big\{[\tilde{\Omega}^{**}_{nmkj}\,\tilde{w}^*_{n-j-1,j}+\bar{\Omega}^{**}_{nmkj}\,\bar{w}^*_{n-j-1,j}$$

$$+\tilde{\tilde{\Omega}}^{**}_{nmkj}\,\tilde{\tilde{w}}^*_{n-j-1,j}]\,\tilde{w}^*_{m-k-1,k}$$

$$+[\tilde{\Omega}'^{**}_{nmkj}\,\tilde{w}^*_{n-j-1,j}+\bar{\Omega}'^{**}_{nmkj}\,\bar{w}^*_{n-j-1,j}+\tilde{\tilde{\Omega}}'^{**}_{nmkj}\,\tilde{\tilde{w}}^*_{n-j-1,j}]\,\bar{w}^*_{m-k-1,k}\tag{27}$$

$$+[\tilde{\Omega}^{(**)}_{nmkj}\,\tilde{w}^*_{n-j-1,j}+\bar{\Omega}^{(**)}_{nmkj}\,\bar{w}^*_{n-j-1,j}+\tilde{\tilde{\Omega}}^{(**)}_{nmkj}\,\tilde{\tilde{w}}^*_{n-j-1,j}]\,\tilde{\tilde{w}}^*_{m-k-1,k}\Big\}=\min.$$

Additionally auxiliary conditions are considered :

- The flap is of null-thickness along its leading edges

$$\tilde{\tilde{F}}^*_t=\sum_{m=t+1}^{N}\tilde{\tilde{d}}^{*(t)}_{mk}\,\tilde{\tilde{w}}^*_{m-k-1,k}=0\qquad(t=0,1,\dots,(N-1))\tag{28}$$

- The surfaces of thick-symmetrical wing and flap are continuous of class C_1 along the junction line between wing and flap. If

$$\Delta \tilde{\tilde{w}}^*_{m-k-1,k} = \tilde{\tilde{w}}^*_{m-k-1,k} - \tilde{w}^*_{m-k-1,k} \quad ,$$

then :

$$\tilde{\tilde{E}}^*_t \equiv \sum_{m=t+1}^{N} \sum_{k=0}^{m-1} \tilde{\tilde{c}}^{*(t)}_{mk} \Delta \tilde{\tilde{w}}^*_{m-k-1,k} = 0 \ , \ \tilde{\tilde{G}}^*_t \equiv \sum_{m=t+1}^{N} \sum_{k=0}^{m-1} \tilde{\tilde{g}}^{*(t)}_{mk} \Delta \tilde{\tilde{w}}^*_{m-k-1,k} = 0 \ ,$$

$$\tilde{\tilde{L}}^*_t \equiv \sum_{m=t+1}^{N} \sum_{k=0}^{m-1} \tilde{\tilde{\ell}}^{*(t)}_{mk} \Delta \tilde{\tilde{w}}^*_{m-k-1,k} = 0 \ , \qquad (t = 0,1,\ldots,(N-1)) \qquad (29)$$

- The relative volume τ' of the flap is given :

$$\tilde{\tilde{\tau}}^*_t \equiv \sum_{m=1}^{N} \sum_{k=0}^{m-1} \tilde{\tilde{\tau}}^*_{mk} \ \tilde{\tilde{w}}^*_{m-k-1,k} = \tau'_0 \sqrt{\ell} \quad , \qquad (30)$$

The Hamilton's operator $\tilde{\tilde{H}}$ of this variational problem is :

$$\tilde{\tilde{H}}^* = \ell \, \tilde{\tilde{H}}'^* = \ell \Big\{ \tilde{\tilde{C}}^*_d + \mu^* \, \tilde{\tilde{\tau}}^* + \sum_{t=1}^{N} (\tilde{\mu}^*_t \, \tilde{\tilde{F}}^*_t + \bar{\mu}^*_t \, \tilde{\tilde{E}}^*_t + \tilde{\eta}^*_t \, \tilde{\tilde{G}}^*_t + \bar{\eta}^*_t \, \tilde{\tilde{L}}^*_t) \Big\} \ . \quad (31)$$

Here, μ^* , $\tilde{\mu}^*_t$, $\bar{\mu}^*_t$, $\tilde{\eta}^*_t$ and $\bar{\eta}^*_t$ are the Lagrange's multipliers. If the first variation of $\tilde{\tilde{H}}^*$ is cancelled, the following equations are obtained :

$$\sum_{n=1}^{N} \sum_{j=0}^{n-1} \Big\{ [\tilde{\tilde{\Omega}}^{(**)}_{n,\theta+\sigma+1,\sigma,j} + \tilde{\tilde{\Omega}}^{(**)}_{\theta+\sigma+1,n,j,\sigma}] \ \tilde{\tilde{w}}^*_{n-j-1,j}$$

$$+ [\tilde{\Omega}^{(**)}_{n,\theta+\sigma+1,\sigma,j} + \tilde{\tilde{\Omega}}^{**}_{\theta+\sigma+1,n,j,\sigma}] \ \tilde{w}^*_{n-j-1,j}$$

$$+ [\bar{\Omega}^{(**)}_{n,\theta+\sigma+1,\sigma,j} + \tilde{\tilde{\Omega}}'^{**}_{\theta+\sigma+1,n,j,\sigma}] \ \bar{w}^*_{n-j-1,j} \ \Big\}$$

$$+ \mu^* \ \tilde{\tilde{\tau}}^*_{\theta+\sigma+1,\sigma} + \sum_{t=1}^{N} [\tilde{\mu}^*_t \ \tilde{\tilde{d}}^{*(t)}_{\theta+\sigma+1,\sigma} + \bar{\mu}^*_t \ \tilde{\tilde{c}}^{*(t)}_{\theta+\sigma+1,\sigma}$$

$$+ \tilde{\eta}^*_t \ \tilde{\tilde{g}}^{*(t)}_{\theta+\sigma+1,\sigma} + \bar{\eta}^*_t \ \tilde{\tilde{\ell}}^{*(t)}_{\theta+\sigma+1,\sigma}] = 0$$

$$(1 \leqslant \theta + \sigma + 1 \leqslant N \ , \quad \theta = 0,1,\ldots,(N-1)) \qquad (32)$$

These equations, together with the auxiliary conditions (30)-(32), form a linear algebraic system which determines uniquely the values of the coefficients $\tilde{\tilde{w}}^*_{\theta\sigma}$ of the downwash $\tilde{\tilde{w}}^*$ and the values of the Lagrange's multipliers μ^* , $\tilde{\mu}^*_t$, $\bar{\mu}^*_t$, $\tilde{\eta}^*_t$ and $\bar{\eta}^*_t$ as function of the similarity parameter ν^* . The best value of the similarity parameter ν^* for the thick-lifting flap (at cruising Mach number M^*_∞ can be also determined by using the hybrid numerical-analytical method of the author, as in [1]-[4]. The optimal value of $\nu(\nu = \nu_{opt})$ is the position of the minimum of the lower limit line of the drag functional $(C_d'^{(t)})_{opt}$, similarly as in (Fig. 3) :

$$(C_d'^{(t)})_{opt} = f(\nu) \qquad (C_d'^{(t)} = C_d' + C_d'^{*}) \tag{33}$$

Remark : The lift and pitching moment coefficients C_ℓ and C_m on the optimum-optimorum wing model Adela (Fig. 3) measured by the author and collaborators, in trisonic wind tunnel (section 60 x 60 cm²) of the DFVLR-Köln (and supported by DFG) are in very good agreement with the theoretically predicted values, according to the above theory as in [5], [19], for the all range of Mach numbers (M_∞ = 1,25 - 2,2) and angles of attack α ($|\alpha| < 14°$) . The measurements of C_ℓ and C_m on wedged delta wing model, performed at higher supersonic Mach numbers (M_∞ = 2,4 - 4,0) , are in good agreement with the values of C_ℓ and C_m , predicted by the present theory, for the ranges of angles of attack $|\alpha| \leqslant 12°$, as in [27] !

4. CONCLUSIONS.

The optimum-optimorum theory of the author can be successfully applied for the global-optimization of the entire configuration of the space vehicle, which shape presents the following advantages :

a) It is total integrated (i. e. wing-fuselage and wing-flap integration), and therefore has no drag due to corners.

b) It is of minimum drag for two different supersonic/hypersonic cruising Mach numbers, and therefore is usefull for space vehicle.

c) It is of high lift due to Kutta auxiliary conditions along the leading edges.

d) It presents a reduced drag and increased lift for a large range of Mach numbers and angles of attack.

The hybrid numerical-analytical method of the author which allows the effective determination of the optimum-optimorum shape of the space vehicle presents the following advantages :

a) It is accurate, because it allows the simultaneous optimization of all geometrical parameters of its shape.

b) It is flexible, while it can be applied to the optimization of complex shape of space vehicle and allows, to add or to suppress some auxiliary conditions and to change the cruising Mach number chosen for the optimization.

c) It is able to determine the shape of the space vehicle of variable geometry in order to obtain a minimum drag at two, very different, supersonic/hypersonic cruising Mach numbers.

d) <u>It is fast</u> (6 sec. computer time at Cyber 175) for the full- optimization of the space vehicle shape !

REFERENCES

[1] A. NASTASE. - *"Use of computers in the optimization of aerodynamic shapes"*, Ed. Acad. Romania, 1973, 280 p.

[2] A. NASTASE. - "Eine graphisch-analytische Methode zur Bestimmung der Optimum-Optimorum Form dünner Deltaflügel in Uberschallströmungen", RRST-MA 1, 19, Romania, 1974.

[3] A. NASTASE. - "Die Theorie des Optimum-Optimorum Tragflügels im Uberschall", ZAMM 57, 1977.

[4] A. NASTASE. - "New concepts for design of fully-optimized configurations for future supersonic aircraft", ICAS-Proceedings Munich, 1980.

[5] A.NASTASE. - "Optimierte Tragflügelformen in Uberschallströmungen", Herchen Ed., Frankfurt, 1990, 300 p.

[6] A. NASTASE. - "Contribution à l'étude des formes aérodynamiques optimales", Thèse, Fac. Sc. Paris (Sorbonne), 1970, 150 p.

[7] A. NASTASE. - "The thin delta wing with variable geometry, optimum for two supersonic cruising speeds", RRST-MA 3, 14, 1969.

[8] A. NASTASE. - "Wing optimization and fuselage integration for future generation of supersonic aircraft", Israel Journal of Technology, Jerusalem, 1985.

[9] A. NASTASE . - "Computation of wing-fuselage configuration for supersonic aircraft", Numerical methods in fluid mechanics II, K. Oshima Ed., Tokyo, 1987.

[10] A. NASTASE. - "Computation of fully-optimized wing-fuselage configuration for future generation of supersonic aircraft", Integral methods in science and engineering, F. Payne, C. Corduneanu, A. Haji--Sheikh and T. Huang Ed., Hemisphere Corp., Washington D.C., 1986, 650 p.

[11] A. NASTASE. - "Optimum-optimorum wing-fuselage integration in transonic-supersonic flow. Proceedings of high speed aerodynamics I", A. Nastase Ed., Herchen Ed., Frankfurt, 1987, 220 p.

[12] A. NASTASE. - "Optimum-optimorum integrated wing-fuselage configuration for supersonic transport aircraft of second generation", ICAS-Proceedings, London, 1986.

[13] A. NASTASE. - *"Optimum aerodynamic shape by means of variational method"*, Ed. Acad. Romania, 1969, 240 p.

[14] P. GERMAIN. - "La théorie des mouvements homogènes et son application au calcul de certaines ailes Delta en régime supersonique", Rech. Aero., 7, France, 1949, p. 3-16.

[15] E. CARAFOLI. - "About the hydrodynamic character of the solutions of conicol flow used in the theory of polygonal wings II", Com. Acad. of Romania, 1952.

[16] E. CARAFOLI, D. MATEESCU, A. NASTASE. - *"Wing theory in supersonic flow"*, Pergamon Press, London, 1969, 500 p.

[17] M. VAN DYKE. - *"Perturbation methods in fluid mechanics"*, Acad. Press, New York, 1964.

[18] A. NASTASE. - "L'étude du comportement asymptotique des vitesses axiales de perturbation au voisinage des singularités", RRST-MA 4, 17, Romania, 1972.

[19] A. NASTASE, A. SCHEICH. - "Theoretical prediction of aerodynamic characteristics of wings in transonic-supersonic flow at higher angles of attack and its agreement with experimental results", ZAMM 67, 5, 1987.

[20] R. L. STOLLINGS, M. LAMB. - "Wing alone aerodynamic characteristics for high angles of attack at supersonic speeds", NASA Technical paper 1889, 1981.

[21] A. NASTASE. - "The optimum-optimorum theory and its application to the optimization of the entire supersonic transport aircraft", Comput. Fluid Dynamics, G. Davis and C. Fletcher, Ed., North-Holland, 1988.

[22] D. KUCHEMANN. - "The aerodynamic design of aircraft", Pergamon Press, London, 1978, Chap. 5.

[23] R. JONES, D. COHEN. - *"High speed wing theory"*, Princeton Press 1960.

[24] A. NASTASE. - "The design of optimum-optimorum shape of space vehicle", Proc. of the first internat. Conf. on Hypersonic flight in the 21st Century, M. Higbea, and J. Vedda Ed., Univ. of North Dakota, 1988.

[25] A. NASTASE. - "The space vehicle of variable geometry, optimum for two supersonic cruising speeds", ZAMM, 69, 1989.

[26] A. NASTASE, E. STANISAV. - "Prediction of pressure distribution on optimum-optimorum Delta wing at higher angles of attack in supersonic flow and its agreement with experimental results", ZAMM, 70, 1990.

[27] A. NASTASE, A. HONERMANN. - "Theoretical prediction of aerodynamic characteristics of Delta wings with supersonic leading edges, in supersonic-hypersonic flow and its agreement with theoretical results", ZAMM, 71, 1991.

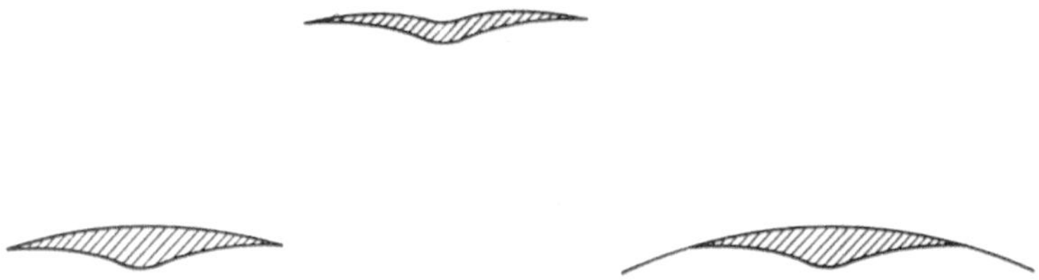

Fig. 1a-c

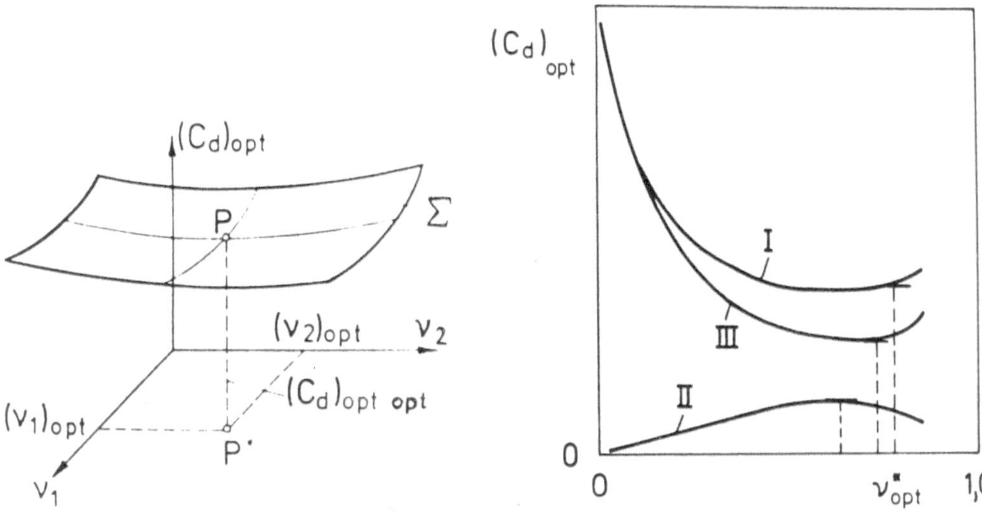

Fig. 2a,b

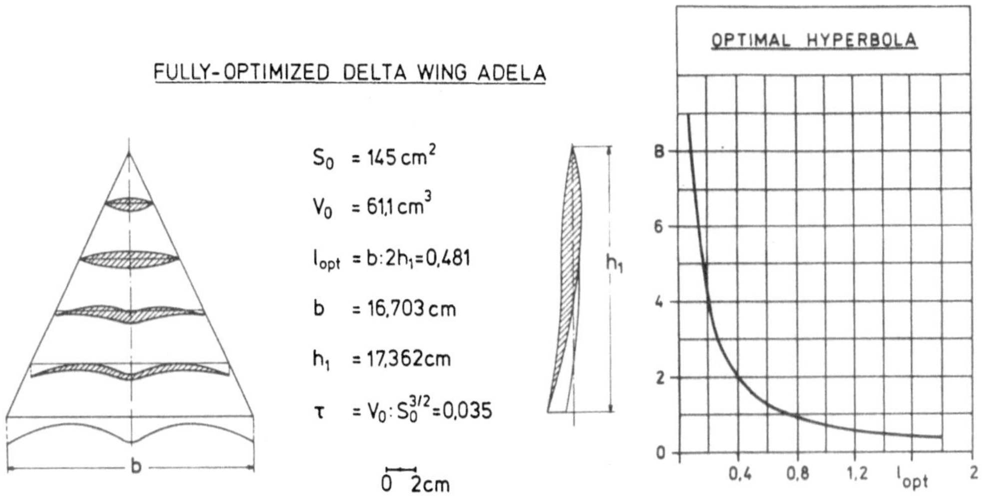

Fig. 3a,b

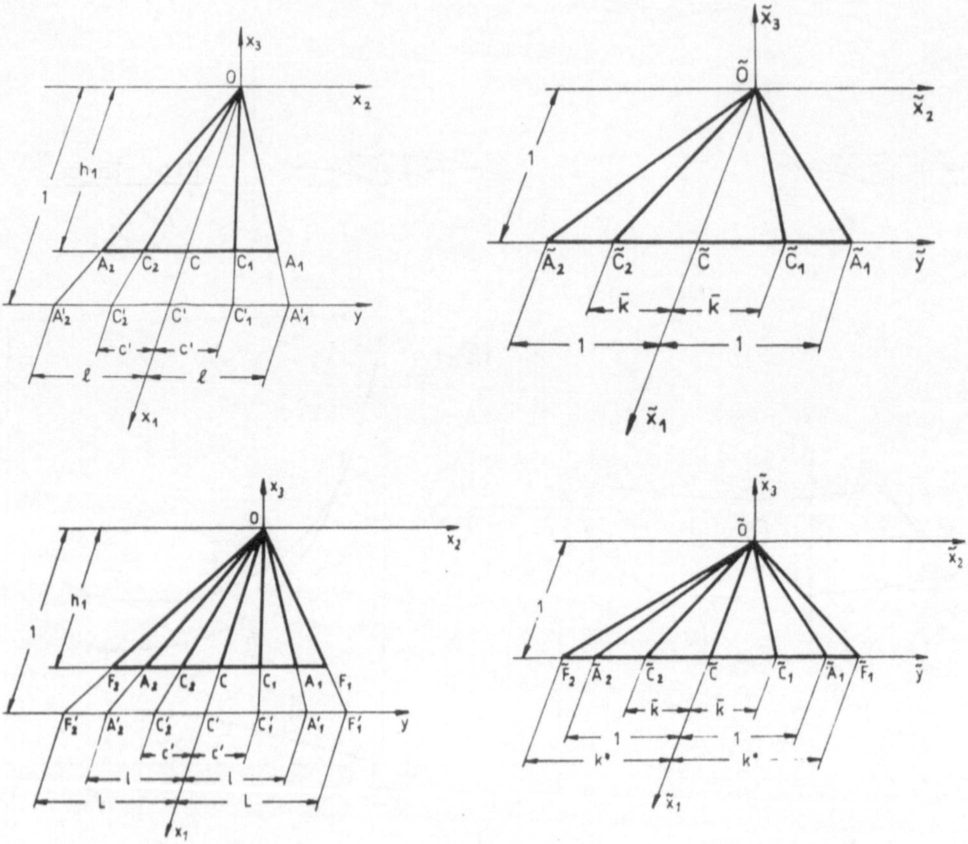

x3
O
x2
h1
1
A2 C2 C C1 A1
A'2 C'2 C' C'1 A'1 y
c' c'
ℓ ℓ
x1
x̃3
Õ
x̃2
1
Ã2 C̃2 C̃ C̃1 Ã1 ỹ
k̄ k̄
1 1
x̃1
x3
O
x2
h1
1
F2 A2 C2 C C1 A1 F1
F'2 A'2 C'2 C' C'1 A'1 F'1 y
c' c'
l l
L L
x1
x̃3
Õ
x̃2
1
F̃2 Ã2 C̃2 C̃ C̃1 Ã1 F̃1 ỹ
k̄ k̄
1 1
k* k*
x̃1

Fig. 4a-d

Part IV

Continuum Mechanics

Vibrating Strings with Obstacles: The Analytic Study

C. Citrini

Dipartimento di Matematica, Politecnico di Milano,
Piazza Leonardo da Vinci 32, I-20133 Milano, Italy

It is a very pleasant task for me to be here in Paris to pay honour to Professor Henri Cabannes on the occasion of his retirement. I remember with a great pleasure many opportunities of scientific contact with him in the last years, and above all the period I spent in this University three years ago. I thank him for his kindness, and I wish him to take again many satisfactions in continuing his activity.

Unilateral obstacle problems for vibrating strings are at same time charming and difficult. Their interest results from their own nature : the problem of the rebound of a string against an obstacle is the simplest one with an infinite number of degrees of freedom. Hence, it is astonishing that this problem has been firstly solved only in 1975 (by Amerio and Prouse). But this delay is obviously caused by the difficulties hidding in such a simple frame, because most of the methods running in elliptic and parabolic cases do not apply in hyperbolic ones.

These reasons explain, in my opinion, the small number of researchers who devote themselves to these problems. It is however a pleasure for me to recall that they friendly met in Colloquium Euromech 209 "Vibrations with unilateral constraints", I had the honour to coordinate with H. Cabannes, and which took place in Como (Villa Olmo) from 5 to 7 june 1986.

Nevertheless, in the last fifteen years, many results have been obtained ; and whereas, only few problems have been completely solved, many others are well established, and the general frame has been enlightened enough. The reference list at the end of this exposition proves it clearly.

Most of the unilateral problems for vibrating strings may be written in the form :

$$\Box\, y(x,t) = f + J$$

$$J \geqslant 0 , \quad y \geqslant \varphi$$

$$\text{supp } J \subseteq \{(x,t) : y = \varphi\}$$

where $f = f(x,t)$ are the active forces, and J is an unknown measure giving the reaction of the obstacle φ .

I will limit my talk to a synthesis about problems and analytic methods; the numeric questions are treated by M. Glowinski. Friction problems, where the inequality on the d'Alembertian is controlled by the value of the speed and not by the position of the string, have a different nature, and we will not speak about them, though they are the object of many works of H. Cabannes.

We can divide the works about these topics according to the type of obstacle in three groups :

I - Continuous obstacle (wall),
II - Pointshaped obstacle (nail, ring),
III - Mixed cases.

I. CONTINUOUS OBSTACLE.

We must distinguish between the following cases :

I.1 - Rectilinear obstacle, no external forces,
I.2 - Curvilinear obstacle, or non vanishing external forces.

I.1. Rectilinear obstacle, no external forces.

This case has been deeply investigated since the first work of Amerio and Prouse [1]. They gave the explicit solution, on the ground of the notion of influence line of the obstacle.

We can prove that the support of J is constituted by at most countably many space-like "impact arcs", which are joined by characteristic segments to form the so called "influence lines of the obstacle" $t = t_i(x)$ (Amerio and Prouse). If Z_P is the characteristic backward semicone having the point P as vertex, and if $z(x,t)$ is the free solution corresponding to the given initial-boundary value data, we define :

$$t_1(x) = \sup \{t : \forall\ (\xi,\tau) \in Z_{(x,t)} \Rightarrow z(\xi,\tau) \geqslant \varphi\} .$$

For $t \leqslant t_1(x)$, the solution agrees with the free one. The computation of the solution for $t > t_1(x)$ is simple : on every impact arc, we change the sign of the speed, and we solve the corresponding Cauchy problem in the mixtilinear triangle bounded by the same arc and the forward characteristic segments issuing from its boundary points. Hence, we have the solution on a polygonal line, from which we calculate a new free solution, as far as we find a second influence line, and so on.

As there are at most countably many influence lines, which are well separated, the solution exists for any $t > 0$, and then we obtain the solution by a suitable partitioning of the half-strip $\{(x,t) : 0 \leqslant x \leqslant 1 , 0 \leqslant t)\}$ by solving a sequence of elementary problems (of Cauchy, Darboux and Goursat).

Many questions have been treated in this simple case. We list them shortly :

I.1.a. **Regularity** of the solutions ([4], [9]), also in non linear cases ([40]).

I.1.b. **Inelastic impact** and **energy** relation, evaluation of the **reaction** ([27], [28], [41], [9], [26]).

I.1.c. **Periodicity** or **almost periodicity** of solutions. Very beautiful results are due to H. Cabannes (the first ones with A. Haraux) ([24], [25], [26], [13], [16], [17], [22]).

Most of these works make use of the explicit expression of the solution, which is known in the simplest cases. Firstly, the case of the "symmetric pinched string" (corresponding to the initial data $v(0,t') = 1 - 2|x'|$, $v_t(0,t') = 0$) is studied, whose solution $v(x',t')$ has the period $T = 1 + h$ (if $\varphi(t') = - h$). The change of variables :

$$2x' = F(x + t) + F(x - t) , \quad 2t' = F(x + t) - F(x - t)$$

(F satisfying some simple conditions), under which the rebound condition holds too, reduces to this case all Cauchy problems with initial position having only a maximum and no initial speed. Hence, the solution will be periodic (if $h = p/q$, we have the period$T = 2(p+q)$ or $T = p+q$ if this number is even) or almost periodic if h is irrational.

The Fourier expansion of the solution corresponding to the symmetric pinched string is given in [15].

Also the cases where the obstacle agrees or is near enough to the equilibrium position, or with special initial data (for instance, piecewise linear initial position) are studied ([17], [22]).

The symmetric pinched string between two walls $(h \leqslant y \leqslant H)$ has period $T = H - h$ ([38]).

I.1.d. **Penalization.** The convergence of this method has been investigated by M. Schatzman and A. Bamberger ([53], [9]) and by A. Motet ([46], [47], [48]). In this last article also the case of a membrane is treated. She proves that the solutions of the penalized equation $\Box u_\varepsilon = u_\varepsilon^- / \varepsilon$ converge in $H^{3/2-\eta}_{loc}$ to the solution of the inequality $\Box u \geqslant 0$ which conserves the energy in a neighborhood of the non characteristic portion of the influence line (surface). She makes use of the theorem of propagation of singularities, and, if the surface is not C^∞, of the paradifferential calculus.

I.1.e. Impact of **two strings** with the **same** propagation speed ([33], [34], [35]).

I.2. Curvilinear obstacle or non vanishing external forces.

Both conditions (which are easily proved to be equivalent) are difficult to treat if the force is directed toward the obstacle($f \leqslant 0$), or if the obstacle is convex ($\varphi'' \leqslant 0$). In the "good" case, all the above arguments work well ([51]).

I.2.a. A first difficulty is the **loss of uniqueness** ([32]) : there are forces (also C^∞) for which also a non vanishing solution exists (as for the basket ball pushed against the floor). Sufficient hypotheses for uniqueness are yet to be found.

I.2.b. The greatest difficulty of this case lies however in the fact that the contact set is no longer formed by space-like arcs, obtained from the free solution, as in the case I.1, but in general it is a domain in the (x,t) plane, bounded by time-like lines $(x = x(t))$ to be found. Hence, in general we have a **free boundary problem.**

It has been solved in the case of the string of sitar (an ancient Indian instrument) by R. Burridge (and others) [11], and, in the general case, by Amerio [6]. See also [21].

On the line separating the contact domain Ω from the region where the string is detached from the obstacle, we must impose that the solution and its derivatives in the outward directions vanish. It turns into an equation of the implicit type $G(\xi,\eta) = 0$, where (ξ,η) are the characteristic coordinates.

Other results on this subject are in [37] (external force depending on the unknown and on its derivatives) and in [7].

I.2.c. A problem similar to the previous one is the problem of the string vibrating against a **gluing obstacle**, which has been treated in [12] and [39].

II. THE POINTSHAPED CASE.

The pointshaped case is easier, because we have more a priori informations about the support of the reaction J , that is we know the longitudinal position $x = \lambda(t)$ of the obstacle, which can be represented by inequalities of the type :

$$a(t) \leqslant y(\lambda(t),t) \leqslant b(t)$$

(nail, moving ring with variable diameter). We have therefore only to find the instants of contact and the intensity of the reaction.

II.1. The general expression of the solution.

It is obtained explicitly from the free solution in the case $f = f(x,t)$ ([2], [3]), and by a fixed point theorem in the case $f = f(x,t,y)$ ([5]) and $f = f(x,t,y,y_x,y_t)$ ([36]).

There is always conservation of the energy ([29]) for fixed obstacle.

Irregular cases (with discontinuous solutions) have been studied in [31] and [43]. The mechanic meaning is obviously lost, so that we have to consider the electric one. The telegraph equation has been studied in [10].

For convergence problems and for first examples of periodicity, see also [30], [50], [52]. In [30], there is a simple example of asymptotically resonant behaviour.

The case of two strings vibrating in two intersecting planes is treated in [38].

II.2. Periodicity and almost-periodicity results.

If $a(t) = a = \text{Const}$, and the string is symmetric with only a maximum, the period is $T = 3q - p$ (or $(3q - p)/2$ if this value is even) ([42], [20]), p/q being the value $a/z(0,0)$, supposed rational.

If $a(t) = 0$ and the free solution is $z(x,t) = \cos m\pi x \cos m\pi t$ $(m = 2k + 1)$, the period is $T = 1 + m + 1/m$. If the string is symmetric and has two minima of value -1 , and a maximum at the origin of value $M = p/q$, we have : $T = 12q^2 + 3pq - 2p^2$ if $p \leqslant 2q$, $T = 3p^2 + 3pq - 8q^2$ if $p > 2q$,([14], [18]).

The almost-periodicity (initial position with only a maximum) is proved again by Cabannes and Haraux [25].

Periodicity or almost-periodicity results for the case of two strings can be found in [38].

III. THE TWO TYPES OF PROBLEM MIX IN SOME SITUATIONS.

III.1. When there are effectively **two obstacles** ([38]) : we can give examples of periodic solutions (the period being $(1 + H)/2 - h$ if the conditions are $h \leqslant y(0,t)$, $y(x,t) \leqslant H$, in the case of the symmetric pinched string) or almost-periodic.

III.2. When we have an obstacle with convex but **irregular** parts (as by instance a piecewise linear obstacle or a rectilinear obstacle having a length less than the length of the string). A string rebounding against a roof can behave as in the continuous case (if the slope is small) or as in the pointshaped case (if the roof is sharp). Obviously, this depends on the speed at the instant of contact. See Marchionna [44], [45].

III.3. The **impact of two strings**, in the general case where the **two propagation speeds** are **different**, gives a very complex situation, where the two models mix in astonishing fashion. This problem has been recently solved by Amerio [8]. Let us consider two strings $y_1(x,t)$ and $y_2(x,t)$, and suppose that their speeds are $V_1 < V_2$. The problem can be then schematized by the inequalities :

$$\Box_1\, y_1(x,t) = \mu_1\, y_{1tt} - T_1\, y_{1xx} = -J$$

$$\Box_2\, y_2(x,t) = \mu_2\, y_{2tt} - T_2\, y_{2xx} = J$$

$$J \geqslant 0\,, \qquad y_2 \geqslant y_1$$

$$\operatorname{supp} J \subseteq \{(x,t) : y_2 = y_1\}\,.$$

The characteristic cone Z_2 of the "quick" string contains the cone of the "slow" string. Hence, the first influence line satisfies the most restrictive Lipschitz condition, i.e. $|t'(x)| \leqslant 1/V_2$. However, it will no more be constituted by a true impact arc : supposing that the first contact point is at the origin of the axes, in a neighborhood of this point the influence line will be constituted by two forward characteristic segments, belonging to the straight lines $x = \pm V_2\, t$. But the reaction J between the two strings is not supported by this line, except at the origin. Its support, on the contrary, will be a pair of unknown lines, $\Lambda^+ : x = v(t)$ and $\Lambda^- : x = -u(t)$. May be they can reduce to a single one. We can foresee that they will be space-like for the slow string, and time-like for the quick one, i.e. that :

$$V_1 \leqslant v'(t) \leqslant V_2$$

(and the same for $u(t)$). This hypothesis derives from the remark that in the quick string the effect of the tension prevails, and for T_2 tending to infinity the string tends to a rigid wall : hence, with respect to the slow string it behaves like a wall, which implies that the support of the reaction is space-like for string 1 . On the contrary, for the slow string inertial effects prevail : for μ_1 tending to infinity, the string tends to a fixed mass, i.e. to a pointshaped obstacle. Hence, for the quick string, the support of J must be time-like. This insight turns out to be winning, because we obtain effectively existence and uniqueness of the solution.

If we trace from a point $P_1(v(t),t)$ of Λ^+ the V_2-backward characteristic straight line, we meet line Λ^- at a point $Q_2(-u(t),\varphi(t))$. The function $\varphi(t)$ is connected to $u(t)$ and $v(t)$ by an integral (or differential) equation. Likewise, a V_2 - backward characteristic links the points $P_2(-u(t),t)$ of Λ^- and $Q_1(v(t),\psi(t))$ of Λ^+ .

After introducing the functions $k(t)$ and $h(t)$ giving the intensity of the reactions (which are measures supported by the two lines Λ^+ and Λ^-), we can compute the shifting of both strings due to these measures. Finally, we impose the conservation of the energy, by writing that the total work of reactions vanishes (on both Λ^+ and Λ^-).

We obtain a system of six differential-functional equations in the six unknowns $v(t)$, $u(t)$, $\varphi(t)$, $\psi(t)$, $k(t)$, $h(t)$. The functions $k(t)$ and $h(t)$ appear linearly, and we can eliminate them, and obtain a system of four equations in four unknowns. By the contraction mapping theorem, we deduce local existence and uniqueness of solutions.

This argument applies if the values of some parameters (depending on the initial data) satisfy simple conditions. If these conditions do not hold, one of the lines supporting the reaction (e.g. Λ^-) disappears, the expression of $\varphi(t)$ is immediate, and the system reduces to a simple differential equation in the remaining unknown $v(t)$.

OPEN PROBLEMS.

The list could be very long. I limit myself to the more significant problems.

- Almost periodicity of solutions for fixed obstacle. There is no evidence that the hypotheses made until now (e.g. no initial speed) are truly necessary.
- Fourier expansion of the solution in the periodic case.
- Uniqueness of the solution in the case of the convex obstacle.
- Resonance in presence of external periodic forces.
- Completion of the study of convergence methods.
- Generalization to the case of general hyperbolic operators.

- Generalization to the case of the membrane and of waves in more space dimensions.
- Generalization to the case of the beam (hyperbolic case = beam of Timoshenko).

REFERENCES

[1] L. AMERIO and G. PROUSE. - "Study of the motion of a string vibrating against an obstacle", Rend. di Mat., Ser. 6, 8, n° 2, 1975, p. 563-585.

[2] L. AMERIO. - "Su un problema di vincoli unilaterali per l'equazione non omogenea della corda vibrante", Pubbl. I.A.C., Ser. III, 109, 1976, p. 1-11.

[3] L. AMERIO. - "On the motion of a string vibrating through a moving ring with a continuously variable diameter", Atti Accad. Naz. Lincei Rend. Cl. Sci. Fis. Natur, Ser. VIII, 62, n° 2, 1977, p. 134-142.

[4] L. AMERIO. - "Continuous solutions of the problem of a string vibrating against an obstacle", Rend. Sem. Mat. Univ. Padova, 59, 1978, p. 67-96.

[5] L. AMERIO. -"A unilateral problem for a nonlinear vibrating string equation", Atti Accad. Naz. Lincei Rend. Cl. Sci. Fis. Natur., Ser. VIII, 64, n° 1, 1978, p. 8-21.

[6] L. AMERIO. - "Studio del moto di una corda vibrante contro una parete di forma qualsiasi, sotto l'azione di una forza esterna arbitraria : domini di appoggio : un problema unilaterale di frontiera libera", Rend. Accad. Naz. XL, 102, Vol. VIII, n° 10, 1984, p. 185-246.

[7] L. AMERIO. - "Sulla connessione tra i fenomeni di rimbalzo elastico e di appoggio per una corda vibrante", Rend. Accad. Naz. XL, 105, Vol. XI, n° 2, 1987, p. 43-52.

[8] L. AMERIO. - "On the elastic impact of two vibrating strings with different characteristic velocities : study of a free- boundary unilateral problem", Rend. Accad. Naz. XL, 107, Vol. XIII, n° 22, 1989, p. 341-380.

[9] A. BAMBERGER and M. SCHATZMAN. - "New results on the vibrating string with a continuous obstacle", SIAM J. Math. Anal., Vol. 14, 1983, p. 560-595.

[10] "Analisi di un problema di propagazione con vincolo unilaterale", Quad. I.A.C., Ser. III, n° 136, 1981, p. 1-19.

[11] R. BURRIDGE, J. KAPPRAFF and C. MORSHEDI. - "The sitar string, a vibrating string with a one-sided inelastic constraint", SIAM J. Appl. Math., Vol. 42, n° 6, 1982, p. 1231-1251.

[12] R. BURRIDGE and J. B. KELLER. - "Peeling, slipping and cracking. Some one-dimensional free-boundary problems in mechanics", SIAM Review, Vol. 20, n° 1, 1978, p. 31-61.

[13] H. CABANNES. - "Mouvements périodiques d'une corde vibrante en présence d'un obstacle rectiligne", ZAMP, 31, 1980, p. 473-482.

[14] H. CABANNES. - "Mouvements périodiques d'une corde vibrante en présence d'un obstacle ponctuel", J. Mécanique, 20, n° 1, 1981, p. 41-58.

[15] H. CABANNES. - "Mouvements d'une corde vibrante en présence d'un obstacle rectiligne fixe. II", C. R. Acad. Sc. Paris, A 295, 1982, p. 637-640.

[16] H. CABANNES. - "Mouvements d'une corde vibrante en présence d'un obstacle rectiligne", C. R. Acad. Sc. Paris A 296, 1983, p. 1367--1371.

[17] H. CABANNES. - "Mouvement d'une corde vibrante en présence d'un obstacle rectiligne", Journ. Mécanique Théor. Appl., 3, n° 3, 1984, p. 397-414.

[18] H. CABANNES. - "Periodic motions of a string vibrating against a fixed point-mass obstacle : II", Math. Meth. in the Appl. Sc., 6, 1984, p. 55-67.

[19] H. CABANNES. - "Cordes vibrantes avec obstacles", Acustica, 55, 1984, p. 14-20.

[20] H. CABANNES. - "Mouvements d'une corde vibrante en présence d'un obstacle ponctuel fixe", C. R. Acad. Sc. Paris, A 298, 1984, p. 613-616.

[21] H. CABANNES. - "Mouvements d'une corde vibrante en présence d'un obstacle convexe : un problème à frontière libre", C. R. Acad. Sc. Paris, A 301, 1985, p. 125-129.

[22] H. CABANNES. - "Mouvements d'une corde vibrante qui oscille en présence d'un obstacle voisin de la position d'équilibre", C. R. Acad. Sc. Paris, A 301, 1985, p. 1273-1276.

[23] H. CABANNES and C. CITRINI, Editors. - *"Vibrations with unilateral constraints"*, Proceedings Colloquium Euromech. 209, Como (Villa Olmo), Italy, june 5-7, 1986.

[24] H. CABANNES and A. HARAUX. - "Mouvements presque périodiques d'une corde vibrante en présence d'un obstacle rectiligne", C. R. Acad. Paris, A 291, 1980, p. 563-565.

[25] H. CABANNES and A. HARAUX. - "Mouvements presque périodiques d'une corde vibrante en présence d'un obstacle fixe, rectiligne ou ponctuel", Int. J. Nonlinear Mechanics, 16, n° 5/6, 1981, p. 449-458.

[26] H. CABANNES and A. HARAUX. - "Almost periodic motion of a string vibrating against a straight fixed obstacle", Nonlinear Analysis, 7, 1983, p. 129-141.

[27] C. CITRINI. - "Sull'urto parzialmente elastico o anelastico di una corda vibrante contro un ostacolo", Atti Accad. Naz. Lincei Rend. Cl. Sci. Fis. Natur., Ser. VIII, 59, n° 5, 1975, p. 368-376 and n° 6, p. 667-676.

[28] C. CITRINI. - "Energia ed impulso nell'urto parzialmente elastico o anelastico di una corda vibrante contro un ostacolo", Ist. Lombardo Accad. Sc. Lett. Rend., A 110, 1976, p. 271-280.

[29] C. CITRINI. - "The energy theorem in the impact of a string vibrating against a pointshaped obstacle", Atti Accad. Naz. Lincei Rend. Cl. Sc. Fis. Natur., Ser. VIII, 62, n° 2, 1977, p. 143-149.

[30] C. CITRINI. - "Risultati tipici sul problema della corda vibrante con ostacolo puntiforme", Pubbl. I.A.C., Ser. III, n° 134, 1978, p. 1-24.

[31] C. CITRINI. - "Discontinuous solutions of a nonlinear hyperbolic equation with unilateral constraints", Manuscripta Math., 29, 1979, p. 323-352.

[32] C. CITRINI. - "Controesempi all'unicità del moto di una corda in presenza di una parete", Atti Accad. Naz. Lincei Rend. Cl. Sci. Fis. Natur., Ser. VIII, 67, n° 3-4, 1979, p. 79-85.

[33] C. CITRINI and B. D'ACUNTO. - "Sull'urto tra due corde", Ricerche Mat., 28, n° 2, 1978, p. 375-398.

[34] C. CITRINI and B. D'ACUNTO. - "Sur le choc de deux cordes", C. R. Acad. Sc. Paris, A 289, 1979, p. 5-7.

[35] C. CITRINI and B. D'ACUNTO. - "A remark on the elastic impact of two vibrating strings", Rend. Accad. Naz. XL, 108, Vol. XIV (to appear).

[36] C. CITRINI and C. MARCHIONNA. - "Sul problema dell'ostacolo puntiforme per l'equazione iperbolica $\square y = f(x,t,y,y_x,y_t)$ ", Rend. Accad. Naz., XL, 99, Vol. V, n° 5, 1981-1982, p. 53-72.

[37] C. CITRINI and C. MARCHIONNA. - "Support domains for a quasi-linear string vibrating against a wall : a unilateral free boundary problem", Rend. Accad. Naz. XL, 103, Vol. IX, n° 6, 1985, p. 61-86.

[38] C. CITRINI and C. MARCHIONNA. - "Some unilateral problems for the vibrating string equation", Europ. J. Mech. A (Solids), Vol. 8, n°1, 1989, p. 73-85.

[39] C. CITRINI and C. MARCHIONNA. - "On the motion of a string vibrating against a gluing obstacle", Rend. Accad. Naz. XL, 107, Vol. XIII, n°8, 1989, p. 141-164.

[40] B. D'ACUNTO. - "Sull'urto elastico di una corda in un caso non lineare", Ricerche Mat., 27, n° 2, 1978, p. 301-317.

[41] B. D'ACUNTO. - "Teorema dell'energia nell'urto elastico di una corda in un caso non lineare", Rend. Accad. Sc. Fis. Mat. Napoli, Ser. IV, 46, 1979, p. 27-41.

[42] M. FRONTINI and L. GOTUSSO. - "Risultati analitici dedotti da una sperimentazione numerica per l'equazione della corda vibrante con ostacolo puntiforme", Quad. I.A.C., Ser. III, n° 67, 1978, p. 1-20.

[43] C. MARCHIONNA. - "Sul problema unilaterale della corda vibrante con ostacolo puntiforme dotato di velocità qualsiasi", Rend. Accad. Naz., XL, 102, Vol. VIII, n° 1, 1984, p. 1-26.

[44] C. MARCHIONNA. - "Studio del moto di una corda vibrante contro una parete con spigoli : domini e segmenti di appoggio", Rend. Accad. Naz., XL, 103, Vol. IX, n° 13, 1985, p. 275-312.

[45] C. MARCHIONNA. - "On the motion of a string vibrating against an interrupted wall", Rend. Accad. Naz. XL, 104, Vol. X, n° 19, 1986, p. 213-222.

[46] A. BACHELOT-MOTET. - "Sur les pénalisées de l'équation des ondes", Thèse de 3e cycle, Univ. Bordeaux I, 1982.

[47] A. BACHELOT-MOTET. - "Une estimation a priori sur les solutions de l'équation des ondes pénalisée, en dimension 1 d'espace", C. R. Acad. Sc. Paris, A 299, 1984, p. 659-662.

[48] A. BACHELOT-MOTET. - "Une estimation a priori sur les solutions de l'équation des ondes pénalisée, en dimension 1 et 2 d'espace", Proc. Euromech. 209, 1986, p. 56-68.

[49] L. NARRATONE. - "Su un modello matematico della scrica elettrica in un condensatore piano e illimitato", Quad. I.A.C., Ser. III, n° 113, 1977, p. 1-15.

[50] C. REDER. - "Etude qualitative d'un problème hyperbolique avec contrainte unilatérale", Thèse de 3e cycle, Univ. Bordeaux I, 1979.

[51] M. SCHATZMAN. - "A hyperbolic problem of second order with unilateral constraints : the vibrating string with a concave obstacle", Journ. Math. Anal. Appl., 73, 1980, p. 138-191.

[52] M. SCHATZMAN. - "Un problème hyperbolique du 2e ordre avec contrainte unilatérale : La corde vibrante avec obstacle ponctuel", Journ. Diff. Eq., 36, 1980, p. 295-334.

[53] M. SCHATZMAN. - "The penalty method for the vibrating string with an obstacle", in "Analytical and numerical approaches to asymptotic problems in analysis", Axelsson, Frank, van der Sluis Ed., North-Holland Publ. Co., 1981, p. 345-357.

Vibrations of Euler-Bernoulli Beams with Pointwise Obstacles

H. Carlsson[1] *and R. Glowinski*[2]

[1]Royal Institute of Technology, Stockholm, Sweden, and
Department of Mechanical Engineering, University of Houston,
Houston, TX 77004, USA

[2]Department of Mathematics, University of Houston,
Houston, TX 77004, USA, and
Université Pierre et Marie Curie, 4, place Jussieu,
F-75252 Paris Cedex 05, France, and
INRIA, B.P. 105, F-78150 Le Chesnay, France

Abstract : In this paper, we discuss the numerical simulation of the vibrations of a beam in the presence of pointwise obstacles. We suppose that these vibrations are modeled by the Euler-Bernoulli equation for linear beams and that one extremity of the beam is clamped while the other may be rigidly attached to a rigid body. The numerical methodology is based on the following techniques : Hermite cubic finite elements for the space discretization, an energy preserving finite difference time discretization scheme, and a penalty treatment of the inequalities associated to the obstacles. The resulting methodology is robust, seems to be accurate and is easy to implement. The results of numerical experiments show the possibilities of the methods discussed here.

INTRODUCTION

In addition to his interest in Kinetic Theory and Boltzman equations, Henri Cabannes has investigated with success various mathematical problems associated to the modeling of the vibrations of elastic strings with friction and/or obstacles (see [1]-[6] for some H. Cabannes articles on the above topic ; see also the references therein). In a recent paper [7], we have investigated the numerical solution of some of the problems addressed in [1]-[6], taking advantage of test problems with known exact solutions given in these references.

In this paper, which is partly motivated by studying the vibrations of *flexible structures*, we essentially apply the methods described in [7] to the numerical simulation of the oscillations of *elastic beams* in the presence of *pointwise obstacles*. We are particularly interested in those situations where one extremity of the beam is clamped, while the other is possibly rigidly attached to a rigid body (a pay-load) of given mass and moment of inertia.

In Section 1 we shall discuss the mathematical modeling of the problem including the *penalty* treatment of the obstacles ; through penalty the problem is reduced to a *nonlinear wave equation.*

In Section 2, we discuss the *time discretization* of the above nonlinear wave equation by an *energy preserving time discretization scheme.*

In Section 3, we discuss the *space discretization* of the above nonlinear wave equation using *Hermite cubic finite element approximations.*

In Section 4, we discuss the solution, by *Newton's method,* of the systems of nonlinear algebraic equations resulting of the space and time discretization methods described in Sections 2 and 3.

Finally, the results of numerical experiments validating the above methodology are shown in Section 5.

It is worth mentioning that a very basic reference concerning *contact problems* and their mathematical and numerical analysis is [8] (cf. also the numerous references given there).

1. PHYSICAL PROBLEM AND MATHEMATICAL FORMULATION

In this paper, which is strongly influenced by the work of Littman and Markus (cf. [9], [10]) on *hybrid structures*, we consider the oscillations of a specific such hybrid system consisting of an *Euler-Bernoulli beam, clamped* at one extremity, while at the other it is *rigidly attached to a rigid body* of mass M and moment of inertia J.

We suppose that the beam is *clamped* at $x = 0$, that its length is L, that its stiffness is EI, that its mass per unit length is ρ, and that it is horizontal if no forces (external or contact) act on it. We denote by y the vertical displacement of the beam and we assume that y is sufficiently small so that the *Euler-Bernoulli* model holds. In such a case the motion of the beam in the absence of obstacles and external forces is modeled by

$$\rho y_{tt} + EIy_{xxxx} = 0 \qquad \text{on } (0,L), \tag{1.1}$$

$$y(x,0 = y_0(x) \quad \text{and} \quad y_t(x,0) = y_1(x) \quad \text{on } (0,L)\ , \tag{1.2}$$

$$y_t(L,0) = z_0 \quad , \quad y_{xt}(L,0) = z_1, \tag{1.3}$$

$$y(0,t) = 0 \quad , \quad y_x(0,t) = 0 \ , \tag{1.4}$$

$$EIy_{xxxx}(L,t) = My_{tt}(L,t), \tag{1.5}$$

$$-\ EIy_{xx}(L,t) = Jy_{xtt}(L,t). \tag{1.6}$$

In (1.1)-(1.6), we have used the following notation :

$$y_x = \frac{\partial y}{\partial x}\ ,\ y_t = \frac{\partial y}{\partial t}\ ,\ y_{tt} = \frac{\partial^2 y}{\partial t^2}\ ,\ y_{xx} = \frac{\partial^2 y}{\partial x^2}\ ,\ y_{xt} = \frac{\partial^2 y}{\partial x \partial t}\ ,\ y_{xxxx} = \frac{\partial^4 y}{\partial x^4}\ ,$$

and so on.

In order to approximate the dynamical problem (1.1)-(1.6) it is quite convenient to write it in *variational form*, which is nothing but an application of the *virtual work* principle. Let's introduce first the *functional space* (of Sobolev type) :

$$V = \{v \mid v \in \mathbf{H}^2(0,L)\ ,\ v(0) = v_x(0) = 0\}\ . \tag{1.7}$$

Take now $v \in V$, and multiply both sides of (1.1) by v. We obtain then after integrating over $(0, L)$.

$$\int_0^L (\rho y_{tt} + EI y_{xxxx})\, v dx = 0, \qquad \forall\ v \in V. \tag{1.8}$$

Integrating now by parts in (1.8), and using the boundary conditions (1.4)-(1.6), we obtain

$$\rho \int_0^L y_{tt} v dx + EI \int_0^L y_{xx} v_{xx} dx$$

$$+ M y_{tt}(L,t)\ v(L) + J y_{xtt}(L,t)\ v_x(L) = 0, \forall\ v \in V. \tag{1.9}$$

We shall complete (1.9) by requiring the function y to satisfy the initial conditions (1.2), (1.3) and also

$$y(t) \in V \text{ for almost every } t\ ; \tag{1.10}$$

(in (1.10), $y(t)$ denotes the function $x \longrightarrow y(x,t)$).

We shall not discuss here the existence and uniqueness properties of the above dynamical problem. Before discussing the obstacle variant of the above problem, observe that if one takes $v = y_t$ in (1.9) we obtain the following energy relation :

$$\frac{d}{dt} E(t) = 0\ , \tag{1.11}$$

where the *energy* $E(t)$ is given by

$$\begin{aligned} E(t) &= \frac{1}{2} \int_0^L (\rho \mid y_t(x,t) \mid^2\ + EI \mid y_{xx}(x,t) \mid^2)\, dx \\ &\quad + \frac{1}{2}\, M \mid y_t(L,t) \mid^2\ + \frac{1}{2}\, J \mid y_{xt}(L,t) \mid^2 . \end{aligned} \tag{1.12}$$

We consider now the case where pointwise obstacles, say N_{obs}, are perturbating the motion of the above beam ; we suppose that contact with these obstacles is *non dissipative*.

For simplicity, we shall suppose that these obstacles are located above the beam and that their locations are defined by the pairs $\{x_k, H_k\}$, $k = 1, ..., N_{obs}$, with $0 < x_k < L$, and H_k the height of these obstacles. There would be no difficulty to have a mixture of obstacles with some above the beam and some others below.

We introduce now the following *closed convex subset* of V

$$K = \{v \mid v \in V, \quad v(x_k) \leq H_k, \quad \forall\, k = 1, ..., N_{obs}\}. \tag{1.13}$$

Following [11] we shall model the motion of the beam in the presence of the above obstacles by the following time dependent *variational inequality*

$$\left.\begin{aligned} & y(t) \in K \text{ for almost every } t, \\ & \rho \int_0^L y_{tt}(v - y(t))dx + EI \int_0^L y_{xx}(v - y(t))_{xx}dx \\ & + My_{tt}(L,t)(v(L) - y(L,t)) \\ & + Jy_{xtt}(L,t)(v_x(L) - y_x(L,t)) \geq 0, \quad \forall\, v \in K. \end{aligned}\right\} \tag{1.14}$$

The variational inequality (1.14) has to be completed by the initial conditions

$$y(0) = y_0 \in K \quad , \quad y_t(0) = y_1, \tag{1.15}$$

$$y_t(L,0) = z_0 \quad , \quad y_{xt}(L,0) = z_1 \,. \tag{1.16}$$

We shall not discuss here the existence and uniqueness properties of the solution of problem (1.14)-(1.16), since it is a quite nontrivial problem. Instead, we shalldiscuss, following e.g., [12], the *penalty approximation* of problem (1.14)-(1.15).

From a *purely formal* point of view (1.14) can also be written

$$\left.\begin{aligned} & y(t) \in V \text{ for almost every } t, \\ & \rho \int_0^L y_{tt}v dx + EI \int_0^L y_{xx}v_{xx}dx + My_{tt}(L,t)\, v(L) \\ & + Jy_{xtt}(L,t)\, v_x(L) + \langle \partial j(y), v\rangle = 0, \quad \forall\, v \in V, \end{aligned}\right\} \tag{1.17}$$

where $j : V \longrightarrow \mathbb{R} \cup \{+\infty\}$ is the *indicator functional* of K, i.e.

$$j(v) = 0 \quad \text{if} \quad v \in K \quad , \quad j(v) = +\infty \quad \text{if} \quad v \notin K\,; \tag{1.18}$$

functional $j(.)$ is *convex, proper* and *lower semi-continous*. In (1.17), $\partial j(y)$ is the (generalized) gradient of j at $v = y(t)$. The usual penalty treatments of the constraint $v \in K$ essentially consists of *regularizing* the functional $j(.)$, i.e., of approximating it by a *differentiable* functional *vanishing* on K and *taking very large values* on $V \backslash K$.

An obvious candidate is j_ϵ defined by

$$j_\epsilon(v) = \frac{1}{2\epsilon} \sum_{k=1}^{N_{obs}} \left((v(x_k) - H_k)^+ \right)^2 . \tag{1.19}$$

This penalty functional leads to approximate (1.14), (1.17) by the following *nonlinear beam equation* (written in *variational form*) :

$$\left. \begin{aligned} & y_\epsilon(t) \in V \text{ for almost every } t \text{ ; for every } v \in V \text{ we have} \\ & \rho \int_0^L y_{\epsilon tt}\, v dx + EI \int_0^L y_{\epsilon xx}\, v_{xx} dx + M y_{\epsilon tt}(L,t)\, v(L) \\ & + J y_{\epsilon xtt}(L,t)\, v_x(L) + \frac{1}{\epsilon} \sum_{k=1}^{N_{obs}} (y_\epsilon(x_k,t) - H_k)^+ v(x_k) = 0 , \end{aligned} \right\} \tag{1.20}$$

to be completed by the initial conditions (1.15), (1.16).

In the following sections we shall discuss the numerical solution of the dynamical problem (1.15), (1.16), (1.20).

Remark 1.1 : Using penalty to handle the constraints associated to the obstacles implies that a *boundary layer* is created when the beam hits an obstacle. According to [7, Section 8.1.1.], we can expect these boundary layers to have a thickness of order $\sqrt{\epsilon}$ at most.

2. TIME DISCRETIZATION OF THE PENALIZED DYNAMICAL PROBLEM (1.15), (1.16), (1.20)

2.1. - Generalities

We consider the following class of dynamical problems

$$M\ddot{u} + Au + \varphi(u) = 0 , \tag{2.1}$$

$$u(0) = u_0 \quad , \quad \dot{u}(0) = u_1 , \tag{2.2}$$

where in (2.1), (2.2) :

(i) u is a function of t taking its values in $\mathbb{R}^N$;

(ii) M and A are $N \times N$, symmetric and positive definite matrices ;

(iii) $\dot{u} = \dfrac{du}{dt} \quad , \quad \ddot{u} = \dfrac{d^2u}{dt^2}$;

(iv) $\varphi : \mathbb{R}^N \longrightarrow \mathbb{R}^N$ **is a nonlinear operator such that**

$$\varphi(v) = \{\varphi_j(v_j)\}_{j=1}^N \quad , \ \forall\, v = \{v_j\}_{j=1}^N \in \mathbb{R}^N ,$$

the functions φ_j being continuous from $\mathbb{R}$ into itselft ;
(v) u_0 and u_1 belong both to $\mathbb{R}^N$.

There are many methods to discretize the initial value problem (2.1), (2.2) ; in this article, following [7], we shall consider a particular one which is *second order accurate* and *energy preserving*. Our guideline here is the following observation :

Define $\mathcal{E}(t)$ by

$$\mathcal{E}(t) = \frac{1}{2}\,(M\dot{u},\dot{u}) + \frac{1}{2}\,(Au,u) + \sum_{j=1}^{N} \phi_j(u_j) \ , \tag{2.3}$$

where $(.,.)$ denotes the usual scalar product of $\mathbb{R}^N$, and where $\phi_j(x) = \int_0^x \varphi_j(\xi)d\xi$; then if u is solution of (2.1), (2.2) we have

$$\frac{d}{dt}\,\mathcal{E}(t) = 0 \ . \tag{2.4}$$

To prove (2.4) we just take the scalar product of $\dot{u}$ with both sides of (2.1).

In order to have a scheme which preserves -in some sense- the energy relation (2.4), we introduce a *time discretization step* $\Delta t (> 0)$, and denoting by u^n the approximation of u at $t = n\Delta t$, we approximate (2.1), (2.2) by :

Assuming that u^{n-1} and u^n *are known, for* $n \geq 1$, *we approximate equation* (2.1) *at* $t = n\Delta t$ by

$$M\left(\frac{u^{n+1} + u^{n-1} - 2u^n}{|\,\Delta t\,|^2}\right) + A\left(\frac{u^{n+1} + 2u^n + u^{n-1}}{4}\right)$$
$$+ 2\left\{\frac{\phi_j\left(\dfrac{u_j^{n+1} + u_j^n}{2}\right) - \phi_j\left(\dfrac{u_j^n + u_j^{n-1}}{2}\right)}{u_j^{n+1} - u_j^{n-1}}\right\}_{j=1}^{N} = 0 \ . \tag{2.5}$$

To use scheme (2.5) we need to know u^0 and u^1. There is no problem with u^0 since we shall take

$$u^0 = u_0 \ . \tag{2.6}$$

Concerning the calculation of u^1 we shall approximate $\dot{u}(0) = u_1$ by

$$u^1 - u^{-1} = 2\,\Delta t\, u_1 \ , \tag{2.7}$$

and (2.1), at $t = 0$, by (for example)

$$M\left(\frac{u^1 + u^{-1} - 2u^0}{|\Delta t|^2}\right) + A\left(\frac{u^1 + u^{-1} + 2u^0}{4}\right) + \varphi(u^0) = 0\,; \qquad (2.8)$$

eliminating u^{-1} between (2.7), (2.8) we obtain a linear system whose solution provides u^1. The stability and energy preserving properties of the above scheme are proved in [7, Section 6]. The *discrete energy* which is preserved is

$$\begin{aligned}\mathcal{E}_{n+1/2} = \; & \frac{1}{2}\left\{\left(M\left(\frac{u^{n+1}-u^n}{\Delta t}\right), \frac{u^{n+1}-u^n}{\Delta t}\right)\right. \\ & \left. + \left(A\left(\frac{u^{n+1}+u^n}{2}\right), \frac{u^{n+1}+u^n}{2}\right)\right\} \\ & + \sum_{j=1}^{N} \phi_j\left(\frac{u_j^{n+1}+u_j^n}{2}\right).\end{aligned} \qquad (2.9)$$

Concerning the practical solution of the nonlinear problem (2.5), we advocate *Newton's method* to compute u^{n+1}.

2.2. - Application to the time discretization of problem (1.15), (1.16), (1.20)

First of all, we drop the subscript ϵ in (1.20) ; applying then the discretization principles discussed in Section 2.1, we approximate the evolution problem (1.15), (1.16), (1.20) by the following sequence of fourth order elliptic problems :

Assuming that y^{n-1} and y^n are known for $n \geq 1$, we approximate (1.20) at $t = n\Delta t$ by :

$$\left.\begin{aligned} & y^{n+1} \in V\,;\ \forall v \in V, \text{ we have} \\ & \rho\int_0^L \frac{y^{n+1} + y^{n-1} - 2y^n}{|\Delta t|^2} v\,dx \\ & + \frac{EI}{4}\int_0^L \left(y^{n+1} + 2y^n + y^{n-1}\right)_{xx} v_{xx}\,dx \\ & + M\,\frac{y^{n+1}(L) - 2y^n(L) + y^{n-1}(L)}{|\Delta t|^2} v(L) \\ & + J\,\frac{y_x^{n+1}(L) - 2y_x^n(L) + y_x^{n-1}(L)}{|\Delta t|^2} v_x(L) \\ & + \frac{2}{\epsilon}\sum_{k=1}^{N_{obs}} \frac{\phi_k\left(\dfrac{y^{n+1}(x_k)+y^n(x_k)}{2}\right) - \phi_k\left(\dfrac{y^n(x_k)+y^{n-1}(x_k)}{2}\right)}{y^{n+1}(x_k) - y^{n-1}(x_k)} v(x_k) = 0 \end{aligned}\right\} \qquad (2.10)$$

with

$$\phi_k(\eta) = \frac{1}{2}\left((\eta - H_k)^+\right)^2. \qquad (2.11)$$

The *space discretization* of (2.10) will be discussed in Section 3 ; the main point is that after space discretization the nonlinear elliptic problem (2.10) is reduced to a nonlinear finite dimensional system well suited to a solution by *Newton's method.* To use (2.10), we need to know y^0 and y^1 ; taking (1.15), (1.16) into account we define y^0 by

$$y^0 = y_0 \ , \tag{2.12}$$

and concerning the calculation of y^1, we approximate the other initial conditions in (1.15), (1.16) by

$$y^1 - y^{-1} = 2\,\Delta t\; y_1 \ , \tag{2.13}$$

$$y^1(L) - y^{-1}(L) = 2\,\Delta t\; z_0 \quad , \quad y^1_x(L) - y^{-1}_x(L) = 2\,\Delta t\; z_1 \ , \tag{2.14}$$

and (1.20) at $t = 0$ by

$$\left.\begin{aligned} &\rho\int_0^L \frac{y^1 + y^{-1} - 2y^0}{|\,\Delta t\,|^2} v\,dx + \frac{EI}{4}\int_0^L \left(y^1 + 2y^0 + y^{-1}\right)_{xx} v_{xx}\,dx \\ &+ M\frac{y^1(L) - 2y^0(L) + y^{-1}(L)}{|\,\Delta t\,|^2} v(L) \\ &+ J\frac{y^1_x(L) - 2y^0_x(L) + y^{-1}_x(L)}{|\,\Delta t\,|^2} v_x(L) = 0, \quad \forall v \in V, \end{aligned}\right\} \tag{2.15}$$

with $y^1 \in V$; since $y_0 \in K$, the nonlinear term occurring in (1.20) vanishes at $t = 0$, justifying therefore (2.15). Eliminating y^{-1} between (2.13)-(2.15) we obtain a linear problem whose solution provides y^1.

3. SPACE DISCRETIZATION OF THE SEMI-DISCRETE PROBLEM (2.10)-(2.15)

In order to approximate the elliptic problems (2.10) and (2.15) we shall use a *finite element method* based on *Hermite cubic* approximations.

3.1. - The fundamental discrete space

We introduce first $I + 1$ grid points x_i, $i = 0, ..., I$, such that

$$x_0 = 0, \quad x_I = L \ ; \quad x_i < x_{i+1}, \qquad \forall i = 0, 1, ... I - 1 \ . \tag{3.1}$$

We denote by e_i the closed interval $[x_{i-1}, x_i]$. The N_{obs} obstacles will be located at grid points (this is not necessary, but it improves accuracy and simplifies the programming).

We approximate V by

$$V_h = \left\{ v_h \mid v_h \in C^1[0, L],\ v_h(0) = \frac{d}{dx}\, v_h(0) = 0,\ v_h \mid_{e_i} \in P_3,\ \forall i = 1, ... I \right\}. \tag{3.2}$$

In (3.2), P_3 is the space of the polynomials in one variable of degree ≤ 3. An easy way to define $v_h \in V_h$, is to prescribe

$$v_h(x_i) \ , \quad \frac{d}{dx} v_h(x_i), \qquad i = 1, ..., I, \tag{3.3}$$

(we necessarily have $v_h(0) = \frac{d}{dx} v_h(0) = 0$). A typical v_h is the one shown in Figure 3.1., below

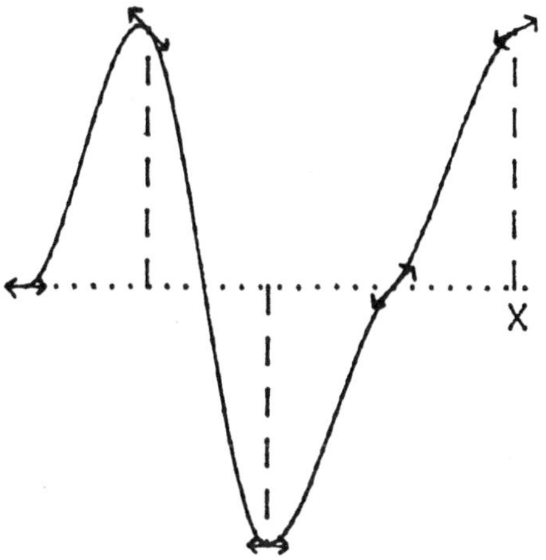

Figure 3.1 : *A typical element of* $V_h(I = 4)$.

At the points x_i we have continuity of v_h and $\frac{d}{dx} v_h$; the space V_h is therefore of *Hermite cubic* type. In order to approximate problems (2.10), (2.15) by finite dimensional ones, it is quite convenient to introduce a vector basis of V_h. In order to construct such basis, we consider first the functions $\hat{w}_1$ and $\hat{w}_2$ defined by

$$\hat{w}_1(\xi) = (1- \mid \xi \mid)^2(1 + 2 \mid \xi \mid) \text{ if } \mid \xi \mid \leq 1 \ , \quad \hat{w}_1(\xi) = 0 \text{ if } \mid \xi \mid \geq 1 \ , \tag{3.4}$$

$$\hat{w}_2(\xi) = \xi(1- \mid \xi \mid)^2 \text{ if } \mid \xi \mid \leq 1 \ , \quad \hat{w}_2(\xi) = 0 \quad \text{if } \mid \xi \mid \geq 1 \ , \tag{3.5}$$

We observe that $\hat{w}_1$ and $\hat{w}_2$ are C^1, piecewise cubic functions. Next, to each grid point x_i, $i = 1, ...I$, we associate a pair of functions $\{w_{1i}, w_{2i}\}$ defined as follows

If $i = 1, ..., I-1$,

$$\left.\begin{aligned} w_{1i}(x) &= \hat{w}_1\left(\frac{x-x_i}{x_{i+1}-x_i}\right) \qquad if\ x \in [x_i, x_{i+1}], \\ w_{1i}(x) &= \hat{w}_1\left(\frac{x-x_i}{x_i-x_{i-1}}\right) \qquad if\ x \in [x_{i-1}, x_i], \\ w_{1i}(x) &= 0 \qquad on\ [0,L]\backslash(x_{i-1}, x_{i+1}), \end{aligned}\right\} \tag{3.6}$$

$$\left.\begin{aligned} w_{2i}(x) &= (x_{i+1}-x_i)\hat{w}_2\left(\frac{x-x_i}{x_{i+1}-x_i}\right) \qquad if\ x \in [x_i, x_{i+1}], \\ w_{2i}(x) &= (x_i-x_{i-1})\,\hat{w}_2\left(\frac{x-x_i}{x_i-x_{i-1}}\right) \qquad if\ x \in [x_{i-1}, x_i], \\ w_{2i}(x) &= 0 \qquad on\ [0,L]\backslash(x_{i-1}, x_{i+1}), \end{aligned}\right\} \tag{3.7}$$

and, *if* $i = I$,

$$\begin{aligned} w_{1I}(x) &= \hat{w}_1\left(\frac{x-x_I}{x_I-x_{I-1}}\right) \qquad if\ x \in [x_{I-1}, x_I], \\ w_{1I}(x) &= 0 \qquad on\ [0, x_{I-1}], \end{aligned} \tag{3.8}$$

$$\begin{aligned} w_{2I}(x) &= (x_I-x_{I-1})\,\hat{w}_2\left(\frac{x-x_I}{x_I-x_{I-1}}\right) \qquad if\ x \in [x_{I-1}, x_I], \\ w_{2I}(x) &= 0 \qquad on\ [0, x_{I-1}]. \end{aligned} \tag{3.9}$$

The set $\mathcal{B}_h = \{w_{1i}\}_{i=1}^{I} \cup \{w_{2i}\}_{i=1}^{I}$ is clearly a basis of V_h and we have

$$V_h(x) = \sum_{i=1}^{I} (v_h(x_i)w_{1i}(x) \ + \ v_h'(x_i)w_{2i}(x))\,, \forall v_h \ \in \ V_h. \tag{3.10}$$

3.2. - The fully discrete problem

Using the space V_h defined in the above Section 3.1, we approximate problem (2.10) by the following (nonlinear) variational problem in V_h :

$y_h^{n+1} \in V_h \; ; \qquad \forall \, v \in V_h$ *we have*

$$\begin{aligned} &\rho \int_0^L \frac{y_h^{n+1} - 2y_h^n + y_h^{n-1}}{|\,\Delta t\,|^2} v dx + \frac{EI}{4} \int_0^L \left(y_h^{n+1} + 2y_h^n + y_h^{n-1}\right)_{xx} v_{xx} dx \\ &+ M \, \frac{y_h^{n+1}(L) - 2y_h^n(L) + y_h^{n-1}(L)}{|\,\Delta t\,|^2} \, v(L) \\ &+ J \frac{y_{hx}^{n+1}(L) - 2y_{hx}^n(L) + y_{hx}^{n-1}(L)}{|\,\Delta t\,|^2} \, v_x(L) \\ &+ \frac{2}{\epsilon} \sum_{k=1}^{N_{obs}} \frac{\phi_k \left(\frac{y_h^{n+1}(x_k) + y_h^n(x_k)}{2}\right) - \phi_k \left(\frac{y_h^n(x_k) + y_h^{n-1}(x_k)}{2}\right)}{y_h^{n+1}(x_k) \; - \; y_h^{n-1}(x_k)} \, v(x_k) \; = \; 0. \end{aligned} \tag{3.11}$$

We should space discretize (2.15) in a similar fashion.

In practice, we shall expand y_h^{n+1} over the basis $\mathcal{B}_h$ as follows

$$y_h = \sum_{j=1}^{I} y_j w_{1j} \; + \; \sum_{j=1}^{I} p_j w_{2j} \; , \tag{3.12}$$

and, in (3.11), take for v all the elements of the basis $\mathcal{B}_h$. We obtain thus a nonlinear system of $2I$ equations for the $2I$ unknowns $\{y_j\}_{j=1}^I \, , \, \{p_j\}_{j=1}^I \, .$

4. SOLUTION OF THE NONLINEAR PROBLEM (3.11) BY NEWTON'S METHOD

Problem (3.11) is a nonlinear variational problem in V_h equivalent to a finite dimensional system of nonlinear equations. Indeed the above system has the following form

$$AY + F(Y) \; = B, \tag{4.1}$$

where $Y, \, B \, \in \, \mathbb{R}^{2I}$. where A is a $2I \, \times \, 2I$ symmetric and positive definite matrix, and where F is a nonlinear operator. Operator F is essentially C^1 and the corresponding linearized operator is quite sparse. Indeed matrix A is also sparse making Newton's method an interesting option to solve problem (4.1). Newton's method clearly yields the following algorithm

$$Y^0 \; \in \; \mathbb{R}^{2I} \; \textit{is given}, \tag{4.2}$$

then for $m \geq 0$,

$$Y^{m+1} \; = Y^m - \left(A + F'(Y^m)\right)^{-1} \left(AY^m + F(Y^m) - B\right) \; . \tag{4.3}$$

The initialization of algorithm (4.2), (4.3) is easy in practice, since if the above algorithm is used to solve (3.11), then y_h^n and y_h^{n-1} can be used to construct good initial guesses for algorithm (4.2), (4.3). An obvious choice is to use y_h^n to define the initial guess in (4.2) ; a more sophisticated one is to define Y^0 from $2y_h^n - y_h^{n-1}$, i.e., by *linear extrapolation* between y_h^n and y_h^{n-1}. Actually the first choice is already quite good since in practice, with the values of Δt and ϵ commonly used in (3.11), the Newton's method (4.2), (4.3) applied to the solution of (3.11) and initialized using just y_h^n converges in 3 iterations at most.

5. NUMERICAL EXPERIMENTS

The numerical methods described in the preceding sections have been applied to the test problem defined by :

$$L = 1 \quad , \quad \rho = 1 \quad , \quad EI = 100 \quad , \quad M = 10 \quad , \quad J = 0.1 \; .$$

The initial velocity is zero and the initial position y_0 has been taken as the solution of the elasto-static problem defined by

$$y_{xxxx} = 0 \quad on \; (0,1) \; , \tag{5.1}$$

$$y(0) = y_x(0) = 0 \; , \tag{5.2}$$

$$y(L) \;=\; -0.5 \quad , \quad y_{xx}(L) \;=\; 0 \; . \tag{5.3}$$

There is only one point obstacle (i.e. $N_{obs} = 1$) at $x = 1/2$, and the corresponding height iz zero (i.e. $H = 0$).

Concerning the approximation of problem (1.14)-(1.16), we have taken $\epsilon = 10^{-5}$ as *penalty parameter*, $\Delta t = 10^{-3}$ *as time discretization step* and finally for the *space discretization* we have been using a *uniform mesh* of step size $h = 1/20$ (i.e. $I = 20$).

On Figure 5.1 we have shown the initial position of the beam (i.e. the solution of problem (5.1)-(5.3)) ; we have also visualized the obstacle at $x = 1/2$. On Figures 5.2 to 5.7 we have shown the position of the beam for various values of t. We have been particularly interested by those values of t at which the beam hits the obstacle and Figures 5.3, 5.5, 5.6 and 5.7 show that the beam slightly penetrates into the obstacle ; the penetration is in the order of $\sqrt{\epsilon}$, at most.

Finally, we have shown on Figure 5.8 the variation versus time of the various energies associated to the system, namely the *kinetic energy*, the *elastic energy of the beam*, the *elastic energy of the obstacle* and finally, the *total energy* of the system, i.e. the sum of all the above ones. We observe that the total energy is remarkably *constant*, as it should be. Figure 5.8 strongly suggests that the beam motion is periodic.

6. CONCLUSION

We have discussed in this paper a methodology for the numerical simulation of the oscillations of an Euler-Bernoulli beam, clamped at one extremity, and rigidly coupled to a rigid body at the other extremity ; we have also supposed that the oscillations are constrained by pointwise obstacles. Using *penalization*, the time dependent differential inequalities modeling the beam oscillations with obstacle, are approximated by a nonlinear partial differential equation which can be solved by fairly classical numerical methods.

Numerical experiments show that the method is quite robust and reproduces accurately various aspects of the behavior of the system which are qualitatively known such as total energy conservation, thickness of boundary layers, etc...

Acknowledgments : The authors would like to thank Professor G. Duvaut from Paris VI University and ONERA for helpful discussions and comments concerning the problems treated in this article. They acknowledge also the support of the National Science Foundation, via Grant INT 8612680, and they thank Juana A. Wilson for her processing of this paper.

References

[1] H. CABANNES, Mouvement d'une corde vibrante en présence d'un obstacle convexe : un problème à frontière libre, *C. R. Acad. Sc. Paris,* t. 301, Série II, (1985), p.125-129.

[2] H. CABANNES, Motion of a vibrating string in the presence of a convex obstacle : a free boundary problem, *Math. Meth. in the Applied Sciences,* 9, (1987), p. 276-297.

[3] H. CABANNES, Cordes vibrantes avec obstacle, cordes vibrantes avec frottement, in *Vibrations with Unilateral Constraints*, H. Cabannes, C. Citrini eds. , EUROMECH 209, 1987, p. 33-43.

[4] H. CABANNES, Mouvement d'une corde vibrante soumise à un frottement solide, *C. R. Acad. Sc. Paris,* t. 287, Série A, (1978), p. 671-673.

[5] H. CABANNES, Propagation of discontinuities in vibrating strings subject to solid friction, *Meccanica*, 14, (1979), p. 175-179.

[6] H. CABANNES, Study of the motions of a vibrating string subject to solid friction, *Math. Meth. in the Appl. sci.*, 3, (1981), p. 287-300.

[7] E.J. DEAN, R. GLOWINSKI, Y.M. KUO, M.G. NASSER, On the discretization of some second order in time differential equations. Application to nonlinear wave problems, University of Houston Research Report UH/MD80, February 1990 (to appear).

[8] N. KIKUCHI, J.T. ODEN, *Contact Problems in Elasticity ; A Study of Variational Inequalities and Finite Element Methods*, SIAM, Philadelphia, 1988.

[9] W. LITTMAN, L. MARKUS, Stabilization of a hybrid system of elasticity by feedback damping, *Annali Mat. Pura Appl.*, 52, (1988), p. 281-330.

[10] W. LITTMAN, L. MARKUS, Exact boundary controllability of a hybrid system of elasticity, *Arch. Rat. Mech. Anal.*, 103, (1988), p. 193-236.

[11] G. DUVAUT, J.L. LIONS, *Inequalities in Mechanics and Physics*, Springer-Verlag, Berlin, 1976.

[12] J.L. LIONS, *Quelques méthodes de résolution des problèmes aux limites non linéaires*, Dunod, Paris, 1969.

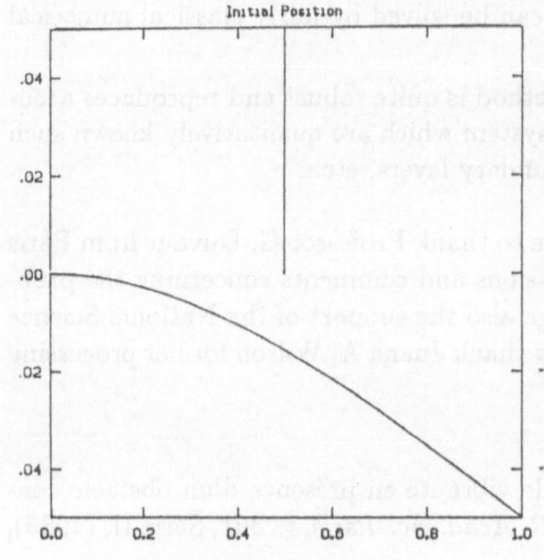

Figure 5.1

Initial Position of the Beam

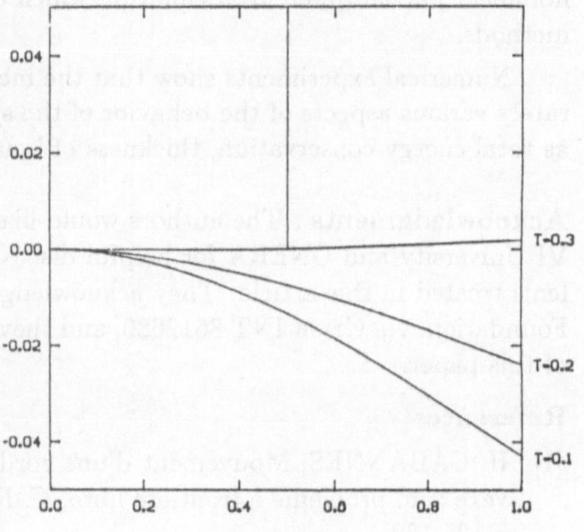

Figure 5.2

Position of the Beam at t=.1, .2, .3.

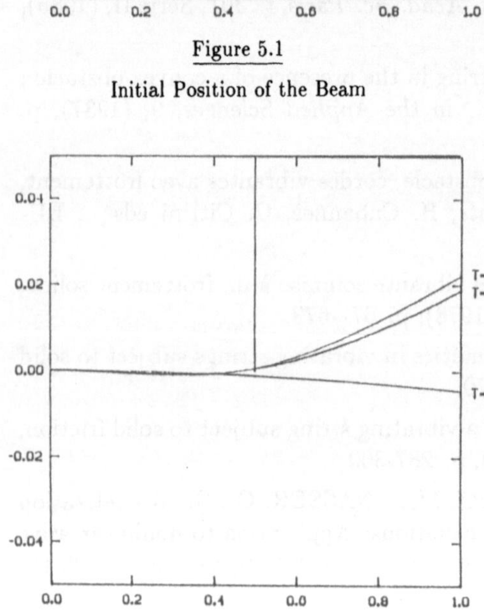

Figure 5.3

Position of the Beam at t=.4, .5, .6.

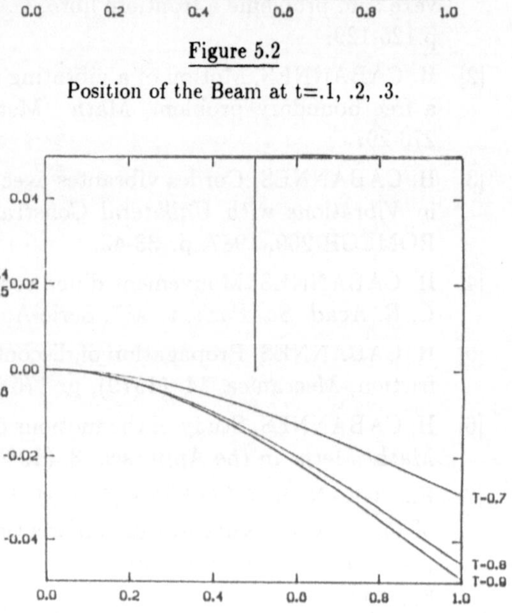

Figure 5.4

Position of the Beam at t=.7, .8, .9.

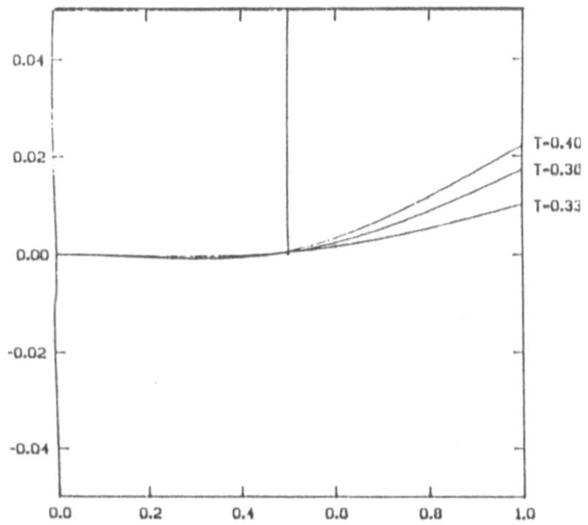

Figure 5.5

Position of the Beam at t=.333, .367, .400

(the Beam stays in contact with the obstacle).

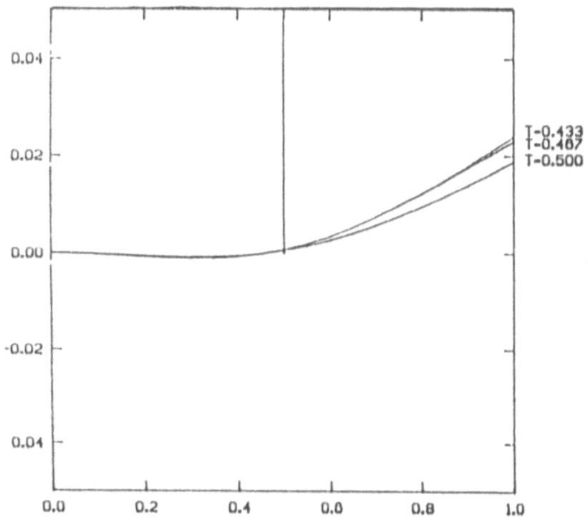

Figure 5.6

Position of the Beam at t=.433, .467, .500.

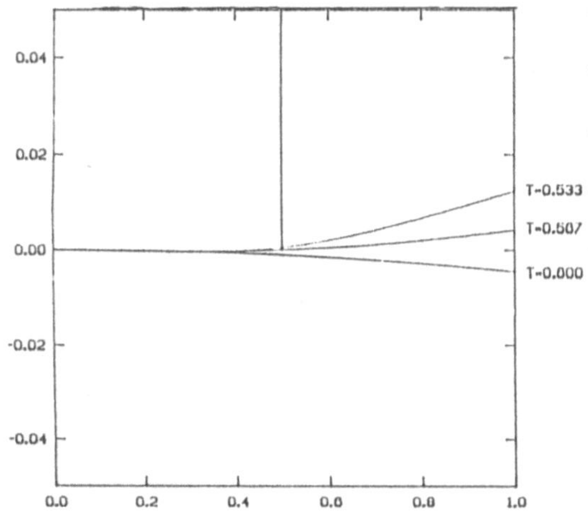

Figure 5.7

Position of the Beam at t= .533, .567, .6.

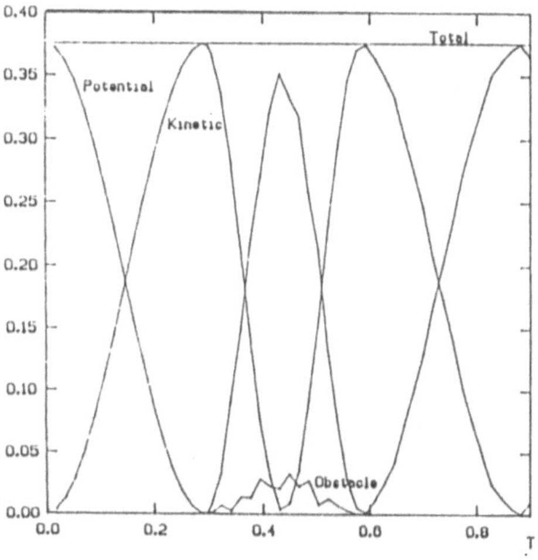

Figure 5.8

Variations of the various energies associated to the system.

Contribution to the Fracture Analysis of Composite Materials

D. Leguillon and E. Sanchez-Palencia

Laboratoire de Modélisation en Mécanique, associé au CNRS,
Université Pierre et Marie Curie, 4, place Jussieu,
F-75252 Paris Cedex 05, France

Abstract : Within the generalized plane elasticity framework, the computation of singularities (stress concentration), together with a matched asymptotic expansion method, allow to build solutions to various problems involving a small perturbation : rigid inclusion, micro-void, ... Taking a micro-crack as small perturbation brings to the analysis of brittle fracture and to the asymptotics of the energy release rate. A Griffith criterion divides the singularities into two classes, namely the strong and weak singularities. As an example, the case of a cracked ply embedded in a stratified composite is examined. With given assumptions on the toughness of the components and of the interface, the crack can propagate to become a delamination crack (i.e. lying at the interface between the components).

INTRODUCTION

Stress concentration phenomena play an important role in the analysis of structures. Composite materials are especially sensitive to that since the multiplicity of components is itself a cause of stress concentration. The mechanical and mathematical approaches bring to seek singular solutions to elliptic problems. Many works have been devoted to that problem since the pioneering works by Kondratiev (1967). A list of references can be found in Leguillon & Sanchez-Palencia (1987a) (this reference will be denoted L. & S., 1987a in the following). These solutions are based on a separation of variables :

$$\underline{U}_\alpha(x) = r^\alpha \underline{u}_\alpha(\theta, \varphi); \alpha \in \mathbf{C}, 0 < \mathcal{R}e(\alpha) < 1, \ \underline{u}_\alpha \in \underline{H}^1(\Omega) \tag{1}$$

where $x = (x_1, x_2)$ denotes the cartesian coordinates and (r, θ) the polar coordinates. The condition $0 < \mathcal{R}e(\alpha)$ ensures the solution to have a finite energy in the vicinity of the origin, while $\mathcal{R}e(\alpha) < 1$ leads to singular solutions : strains and stresses tend to infinity in a vicinity of the origin.

One can also derive from these solutions global informations such as the amount of energy released during the onset or the growth of micro-flaws (Leguillon, 1989, 1990, L. & S. 1990), which allow to define propagation criteria.

COMPUTATION OF SOLUTIONS $k\, r^\alpha\, \underline{u}(\theta)$

Computation of α and $\underline{u}(\theta)$

This section briefly recalls results presented in L. & S. (1987a). The presentation is based mainly on generalized plane elasticity problems (Pipes & Pagano, 1970) and a computation software has been developed supported by the GRECO "Calcul des Structures" (L. & S., 1989). The generalization to three dimensional problems, which is slightly different, requires essentially most powerful computation devices (the extensions are mentioned in L. & S., 1987b).

Equations, defined on a subdomain of $\mathbb{R}^2$, can be written out :

$$\left.\begin{aligned} -\frac{\partial \sigma_{i\beta}}{\partial x_\beta} &= 0 \text{ (momentum)} \\ \sigma_{i\beta} &= a_{i\beta j\gamma} e_{j\gamma}(\underline{U}) \text{ (constitutive equation)} \\ e_{j\gamma}(\underline{U}) &= \frac{1}{2}\left(\frac{\partial U_j}{\partial x_\gamma} + \frac{\partial U_\gamma}{\partial x_j}\right) \\ &+ \text{ boundary conditions} \end{aligned}\right\} \tag{2}$$

where $\beta, \gamma = 1, 2$ and $i, j = 1, 2, 3$ (resp. $i, j = 1, 2$) in generalized plane elasticity (resp. plane elasticity). σ denotes the stress tensor, e the strain tensor, $\underline{U}$ the displacement field and a the elasticity tensor.

The method we proposed allows to compute α and $\underline{u}_\alpha$ in a wide range of situations in which analytical methods become quickly unwieldy (Dempsey & Sinclair, 1979). It is based on a variational formulation which takes into account the change of variable by a judicious choice of test functions. Homogeneous boundary conditions and applied external loads in the vicinity of the origin bring out the one dimensional problem (see L. & S., 1987a and 1987b for local non-homogeneous problems) :

$$\left.\begin{aligned} &\text{Find } \underline{u}_\alpha(\theta) \in \underline{H}^1(0, \omega) \quad \text{and} \quad \alpha \in \mathbf{C} \text{ such that :} \\ &0 < \mathcal{R}e(\alpha) < 1 \\ &A(\alpha)\underline{u}_\alpha = -\alpha^2 A_1 \underline{u}_\alpha + \alpha A_2 \underline{u}_\alpha + A_3 \underline{u}_\alpha = 0 \end{aligned}\right\} \tag{3}$$

where $A_1$and A_2 are hermitian operators and A_3 is skew-hermitian (L. & S., 1987a, p.89). Problem (3) can be also expressed as a classical eigenvalue problem :

$$\mathcal{A}\vec{U}_\alpha = \alpha \vec{U}_\alpha \quad \text{with} \quad \vec{U}_\alpha = \begin{pmatrix} \underline{U}_\alpha \\ r\dfrac{\partial \underline{U}_\alpha}{\partial r} \end{pmatrix} \quad \text{and} \quad \mathcal{A} = \begin{bmatrix} 0 & I \\ A_1^{-1} A_3 & A_1^{-1} A_2 \end{bmatrix} \tag{4}$$

Moreover, from (4) and ignoring the condition $\mathcal{R}e(\alpha) < 1$, it is possible to derive an expansion of the solution to the elastic problem in a corner, involving classical power terms $r^{\alpha_i} \underline{u}_{\alpha_i}$ and logarithmic terms $r^{\alpha_j}(\text{Log} r\ \underline{u}_{\alpha_j} + \underline{v}_{\alpha_j})$, $O < \mathcal{R}e(\alpha_1) \leq \mathcal{R}e(\alpha_2) \leq \ldots$ in the case of Jordan blocks (see also Maz'ya and Nazarov, 1988).

In a very similar way, replacing the condition $\mathcal{R}e(\alpha) > 0$ by $\mathcal{R}e(\alpha) < 0$, allows to study the local behavior at infinity and an analogous expansion can be derived. This kind of duality can be expressed with the following property (L. & S., 1987a, p.91) :

Property 1 : *If α is a solution, then $-\alpha$ (as well as $\bar{\alpha}$ and $-\bar{\alpha}$ when complex exponents are involved) is also a solution, these solutions have the same geometric and algebraic multiplicities.*

Numerically, after discretization of (3) by one dimensional finite elements of first order, one can solve the eigenvalue problem (4). This procedure requires a powerful software since Jordan blocks and complex eigenvalues are involved. The code "SL" (L. & S., 1989) solves directly the equation (3) ($A^h(\alpha)$ is the discrete version of $A(\alpha)$) by the determinant method :

$$\det(A^h(\alpha)) = 0 \qquad (5)$$

Then the $\underline{u}_\alpha$ and $\underline{v}_\alpha$ are obtained, by a Gauss method (computation of the Jordan basis) or by inverse iterations (which provides only the eigenvectors but not the root vectors). The accuracy of the results is strictly the same whatever the approach (eigenvalue problem or determinant method), and the determinant method is of course by far cheaper an can be implemented on a PC computer (L. & S., 1987a, p.107, L. & S., 1989). It provides only one singular value at each computation instead of the whole set (or a large part of it), but generally, as seen further, just two or three values are required for the analysis.

Numerical results

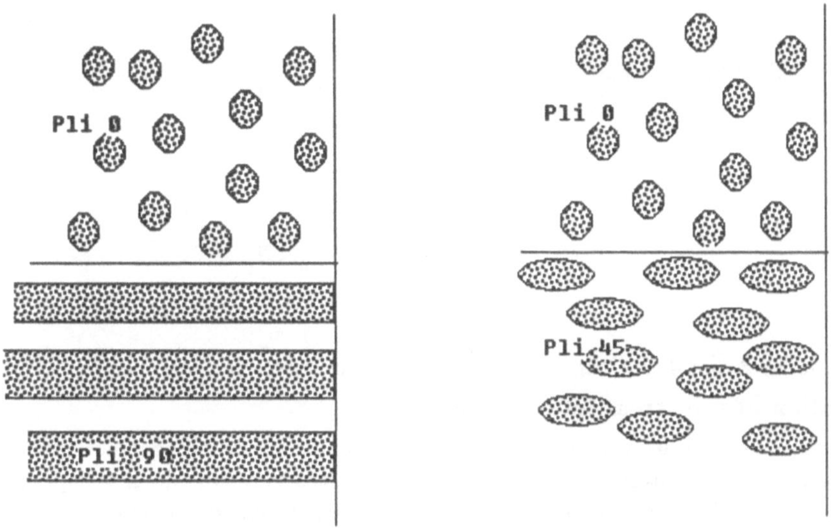

Figures 1 *and* 2 : *Stress free boundary singularities* (0/90 – 0/45).

Results exhibited hereafter cannot be obtained using analytical approaches. They estimate singularities occuring in a multi-layers carbon/epoxy fiber reinforced material. Properties of one ply are as follow (fibers are assumed to be parallel to the Ox_1 axis) :

$a_{1111} = 145000.$, $a_{1122} = 6720.$, $a_{1212} = 4850.$, $a_{1313} = 4850.$, $a_{2222} = 12700.$
$a_{2323} = 3240.$, $a_{1133} = 6720.$, $a_{2233} = 6220.$, $a_{3333} = 12700.$(Mpa).

The two first cases (figures 1 and 2) involve edge singularities: 1- between a 0° ply and a 90° ply $\alpha = 0.892$, 2- between a 0° ply and a 45° ply $\alpha = 0.963$. The third example involves a delamination crack between a 0° ply and a 90° ply $\alpha_1 = 0.500$, $\alpha_2 = 0.500 \pm i0.035$. It exhibits a real and a complex value (both with the same real part 0.5). The last example is more surprising, it concerns a reinforcing of the multilayered plate by a covering on the edge of the plate. The covering is made of the same material, the fibers being in the stacking up direction (figure 4). This entails a stronger singularity than those existing previously (without the covering) $\alpha = 0.852$. Of course this result does not mean that such a reinforcing is unuseful or even dangerous since the values of the stress intensity factors must be taken into account to have a full description of the stress concentration.

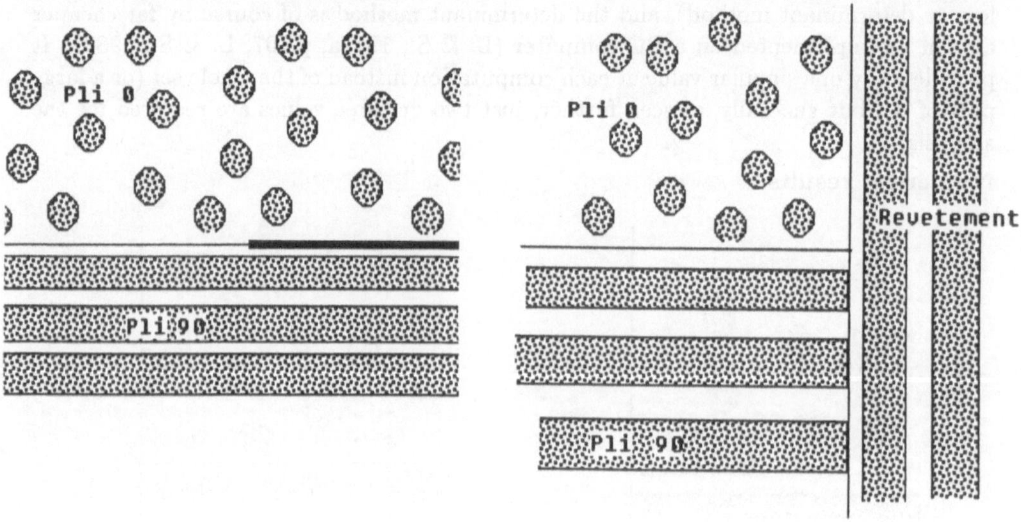

Figures 3 and 4 : Other singularities (Delamination – Covering)

Computation of the stress intensity factor k

Exponents and modes previously computed depend only on the geometry and on the properties of the materials involved in the local structure but do not depend on the external applied loads. On the opposite, the stress intensity factors k obviously do depend on the applied loads, therefore their computations require the knowledge of the exact(!) solution or at least a finite element approximation of it. The computation is based on an extraction method (Babuska & Miller, 1984) with a judicious choice of the extraction function which is $\underline{u}_{-\alpha}$ as a consequence of property 1 (L. & S., 1987a, p.122) :

$$k = \psi(\underline{U}^{fe}, r^{-\alpha}\underline{u}_{-\alpha})/\psi(r^{\alpha}\underline{u}_{\alpha}, r^{-\alpha}\underline{u}_{-\alpha}) \tag{6}$$

where $\underline{U}^{fe}$ is the finite element solution taking into account the external applied loads and where :

$$\psi(\underline{U}, \underline{V}) = \int_{\Gamma} [\sigma_{ij}(\underline{U})n_j V_i - \sigma_{ij}(\underline{V})n_j U_i \, ds] \tag{7}$$

Γ is any contour including the origin and starting and finishing at the straight edges of the corner, $\underline{n}$ is the normal to Γ toward the origin. The contour integral ψ defined in (7) is independent of Γ provided $\underline{U}$ and $\underline{V}$ satisfy the momentum equation and provided the Green's formula holds (Betti's theorem).

All these results can be extended to the case of multiple and complex singularities for homogeneous and non-homogeneous problems. The numerical calculations must be cautiously performed as well as for any contour integral (L. & S., 1987a, p.121).

MATCHED ASYMPTOTIC EXPANSIONS

Matched asymptotic expansions

In this section we study the change in the elastic solution due to a small perturbation located in the vicinity of a singular point. A micro-crack, a micro-void, a damage area, a rigid inclusion, a plastic zone, ... can form this small perturbation whose diameter ℓ must be small with respect to a characteristic length L of the structure : $\varepsilon = \ell/L << 1$. Let $\underline{U}^\varepsilon$ be the solution to the perturbed problem, it is defined in the perturbed domain Ω^ε. We assume that there exists an asymptotic expansion :

$$\underline{U}^\varepsilon(x) = \underline{U}^0(x) + f_1(\varepsilon)\underline{U}^1(x) + ... \qquad \text{with} \qquad \lim_{\varepsilon \to 0} f_1(\varepsilon) = 0 \tag{8}$$

This is the so-called "outer" expansion, and $\underline{U}^0$, the leading term, is solution to the problem without perturbation (settled on the unperturbed domain Ω^{out}). $\underline{U}^0$ is valid out of a vicinity of the perturbation as well as each term of the expansion, and this is the reason for the name "outer".

Let us consider now the domain $\Omega^\varepsilon/\varepsilon$, homothetic to Ω^ε with ratio ε^{-1}. This domain becomes the infinite domain Ω^{in} as $\varepsilon \to 0$. The change of variable $y = x/\varepsilon$ brings to assume the existence of an "inner" expansion in Ω^{in} :

$$\underline{U}^\varepsilon(\varepsilon y) = F_0(\varepsilon)\underline{V}^0(y) + F_1(\varepsilon)\underline{V}^1(y) + ... \qquad \text{with} \qquad \lim_{\varepsilon \to 0} \frac{F_1(\varepsilon)}{F_0(\varepsilon)} = 0 \,. \tag{9}$$

Replacing (9) into (2), gives problems in the unknown functions $\underline{V}^0$, $\underline{V}^1$, ..., defined in Ω^{in}, but there are no conditions at infinity. These conditions (needed to have well posed problems) are a consequence of the matching rules: the behavior of terms involved in the outer expansion in the vicinity of the origin must coincides with the behavior of terms involved in the inner expansion, in the vicinity of infinity.
From the previous section it can be derived :

$$\left.\begin{array}{rcll} \underline{U}^0(x) & = & kr^\alpha \underline{u}_\alpha(\theta) + ... & as \quad r \longrightarrow 0 \\ F_0(\varepsilon)\underline{V}^0(y) & = & F_0(\varepsilon)[k'\rho^\beta \underline{u}_\beta(\theta) + ... & as \quad \rho \longrightarrow \infty \end{array}\right\} \tag{10}$$

where $r = \mid x \mid$ and $\rho = \mid y \mid = r/\varepsilon$. The matching rules can be expressed in an intermediate variable (L. & S., 1990), which means in other words that it exists an intermediate area where both expansions (8) and (9) are valid, and it upholds :

$$\alpha = \beta \qquad ; \qquad k'F_0(\varepsilon) = k\varepsilon^\alpha \tag{11}$$

Computation of the energy release rate

In the framework of brittle fracture mechanics, the energy release rate $G = -\partial W/\partial h$ (W potential energy of the structure, h crack length) is an important tool which allows to define a criterion for crack motion (Griffith's criterion). Unfortunately, in composite materials, many situations are such that $G \to 0$ or $G \to \infty$. These properties are connected with the existence of singularities in the vicinity of the crack tip (Leguillon, 1989). Let us consider a micro-crack located at a singular point of a two- dimensional structure (the result can be extended to the generalized plane elasticity as well as to the three dimensional elasticity). ℓ is the micro-crack length: $\ell/L = \varepsilon << 1$. The micro-crack is a particular choice, and the results extend to other kinds of perturbation and it can be simply a small extension to an existing macro-crack.

The potential energies of the unperturbed and perturbed structures can be written out :

$$W^0 = \frac{1}{2}\int_{\Omega^{out}} a_{ijkh} e_{kh}(\underline{U}^0) e_{ij}(\underline{U}^0) dx - \int_{\partial\Omega_2} \underline{H}\ \underline{U}^0 ds \tag{12}$$

$$W^\varepsilon = \frac{1}{2}\int_{\Omega^\varepsilon} a_{ijkh} e_{kh}(\underline{U}^\varepsilon) e_{ij}(\underline{U}^\varepsilon) dx - \int_{\partial\Omega_2} \underline{H}\ \underline{U}^\varepsilon ds \tag{13}$$

where $\partial\Omega_2$ is (unperturbed) subdomain of the boundary where the applied loads $\underline{H}$ act. Then, it is easy to show that :

$$\delta W^\varepsilon = W^\varepsilon - W^0 = \frac{1}{2}\int_\Gamma \left[\sigma_{ij}(\underline{U}^0)\, n_j\, U_i^\varepsilon - \sigma(\underline{U}^\varepsilon)\, n_j\, U_i^0\right] ds \tag{14}$$

Γ is any contour (with normal $\underline{n}$ toward O) including the perturbation and starting and finishing at the stress free edges of the origin. (14) does not depend on the choice of the contour.

For the sake of simplicity, we assume that the singularity located at O is real and single. To have an explicit expression of the leading term in (14), it is necessary, first, to have a more precise expression of the behavior of $\underline{V}^0$ in the vicinity of infinity. From property 1, we have :

$$\underline{V}^0(y) \simeq \rho^\alpha \underline{u}_\alpha(\theta) + \kappa\rho^{-\alpha}\underline{u}_{-\alpha}(\theta) \qquad \text{as} \qquad |\, y \,| \to \infty \tag{15}$$

then, using once more the matching conditions, we derive the second term of the outer expansion (8) :

$$\underline{U}^\varepsilon(x) = kr^\alpha \underline{u}_\alpha(\theta) + \widehat{\underline{U}}^0(x) + k\kappa\varepsilon^{2\varepsilon}\left[r^{-\alpha}\underline{u}_{-\alpha(\theta)} + \widehat{\underline{U}}^1(x)\right] + \ldots \tag{16}$$

Replacing (16) into (14) allows to exhibit the leading term :

$$\delta W^\varepsilon \simeq \frac{1}{2}k^2\kappa\varepsilon^{2\alpha} \leq 0 \tag{17}$$

after a judicious normalization of $\underline{u}_{-\alpha}$ (Leguillon, 1989). The energy release rate is derived immediately from (17) :

$$G = -\frac{\partial W}{\partial \ell} = -\lim_{\delta\ell \to 0} \frac{\delta W}{\delta \ell} = -\frac{\alpha}{L} k^2 \kappa \varepsilon^{2\alpha-1} \geq 0 \tag{18}$$

Similar results can be obtained in more entangled cases as multiple or complex values (Leguillon, 1989).

It can be seen from (18) that the limit G may not exist as $\varepsilon \to 0$ but it allows essentially to make a distinction between "strong" singularities when $\lim_{\varepsilon \to 0} G = +\infty$, and "weak" singularities when $\lim_{\varepsilon \to 0} G = 0$ holds. The boundary between these two classes is made of singularities such that $\alpha = 0.5$ (or $\mathcal{R}e(\alpha) = 0.5$). It is within this boundary that cracks in homogeneous media ($\alpha = 0.5$) and cracks lying at the interface between two materials ($\mathcal{R}e(\alpha) = 0.5$) take place.

Weak singularities have been shown in the previous section (figures 1, 2 and 4). Strong singularities occur for instance at a crack tip, when it meets an interface between two isotropic components provided the crack is lying in the stiffer material (Erdogan & Biricicoglu, 1973, L. & S., 1987a, p. 94 and 1987b).

Strong singularity and crack propagation

We consider a crack approaching an interface and triggering a strong, real, single singularity α (figures 5 and 6). If one consider the small perturbation as the small remaining ligament in the stiff material (the small distance between the crack tip and the interface, figure 5), then a similar reasoning to the previous one allows to build matched asymptotic expansions. The energy released during the break of this ligament is given by (17) where k is the stress intensity factor of the singularity α (at 0) in the outer problem and κ the stress intensity factor of the singularity $-\alpha$ (at infinity) in the inner problem. From (18) $G \to \infty$ as $\varepsilon \to 0$:

$$G = \frac{\alpha}{L}\, k^2\, \kappa\, \varepsilon^{2\alpha-1} \tag{19}$$

The existence of a critic value G^c_{st} of G (Griffith's criterion) in the stiff material implies an instable propagation of the crack tip if :

$$\varepsilon \leq \varepsilon_1^c = \left(\frac{LG^c_{st}}{\alpha k^2 \kappa}\right)^{1/(2\alpha-1)} \tag{20}$$

Let us consider now the final step, after propagation, new matched asymptotic expansions can be written out, considering the small kink as the perturbation (figure 6). A similar result to (19) can be obtained where only κ has been changed (the outer problem is the same, but the geometry of the inner problem has been modified) :

$$G = -\frac{\alpha}{L} k^2 \kappa' \varepsilon^{2\alpha-1} \tag{21}$$

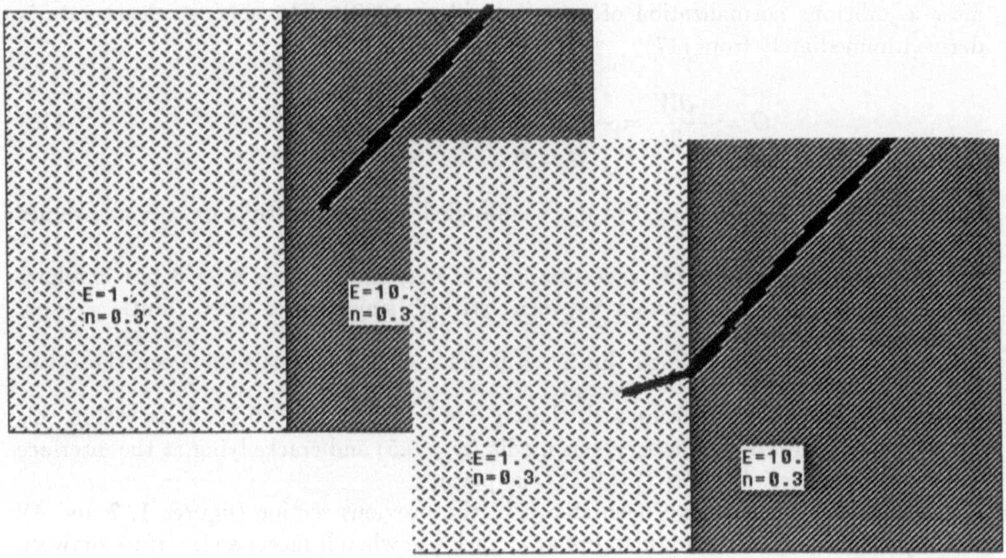

Figures 5 and 6 : The crack before and after propagation

As a consequence, following the propagation (19)-(20), the crack tip stops with a kink length :

$$\varepsilon \geq \varepsilon_2^c = \left(-\frac{LG^c}{\alpha k^2 \kappa'} \right)^{1/(2\alpha-1)} \tag{22}$$

where G^c is the critical value of G in the direction of propagation (in the soft material or along the interface or even back in the stiff material). One can see in (22) that the result depends on the angle φ made by the initial crack and the kink. More precisely, when the critical value is reached, we have :

$$G^c(\varphi)/\kappa'(\varphi) = -\frac{\alpha}{L} k^2 \varepsilon^{2\alpha-1} \tag{23}$$

Such an expression allows to compute the critic length (22), but also and previously to determine the angle of propagation φ_0 which satisfies :

$$G^c(\varphi_0)/\kappa'(\varphi_0) \leq G^c(\varphi)/\kappa'(\varphi) \qquad \forall \varphi, \qquad 0 \leq \varphi \leq 2\pi \tag{24}$$

In the case of a single singularity, this direction φ_0 is independent of the applied loads and depends only on the geometry and on the properties of the materials bonded together.

A numerical result

Computations have been performed on the geometry described in figure 5 and 6, within the plane elasticity framework for the sake of simplicity. The singularity is single :

$\alpha = 0.207$, and the inner problem is the only one to have been considered (the result is independent of the external applied loads). The stress intensity factor $\kappa'(\varphi)$ is computed for various values of φ between 0 and 2π. Results are plotted on figure 7.

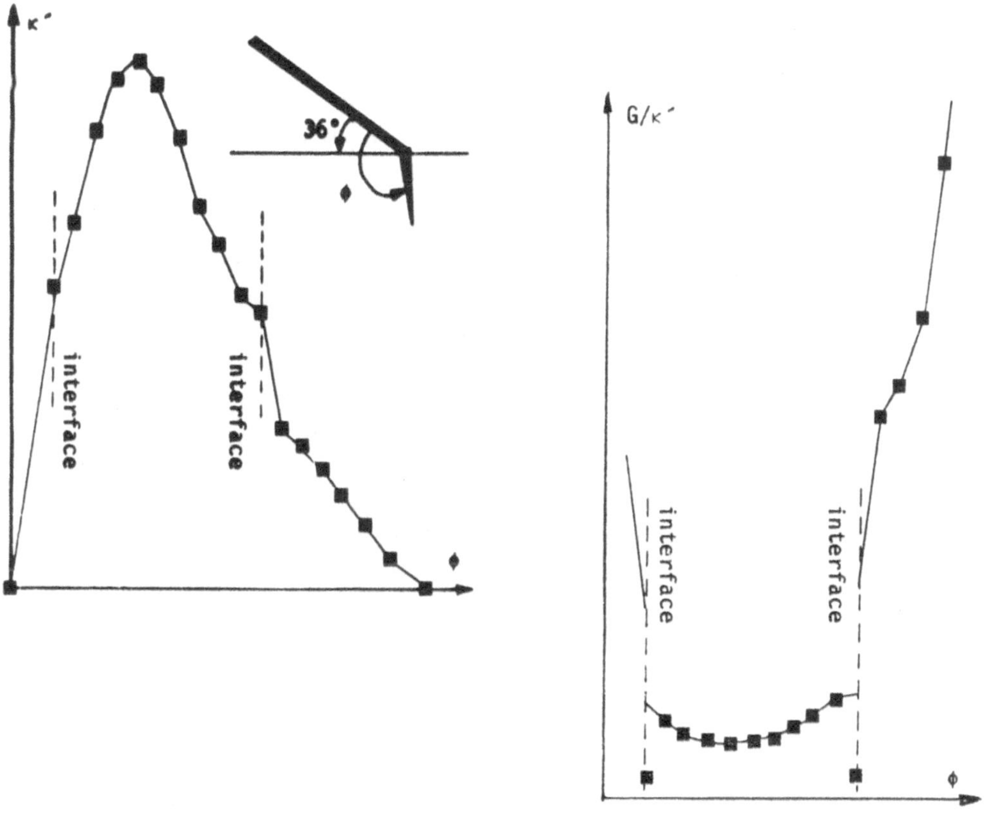

Figures 7 and 8 : κ' and G/κ' versus φ.

In order to use the criterion (24) to determine the direction of propagation of the kink, we have assumed the following data: $G^c_{st} = 8.$, $G^c_{so} = 4.$ and $G^c_i = 1.$, where G^c_{st}, G^c_{so} and G^c_i are respectively the critical values of G within the stiff material, within the soft material and along the interface (we emphasize that only the relative values are involved). The curve $G(\varphi)/\kappa'(\varphi)$ vs. φ is plotted on figure 8 and one can see that two obvious directions of propagation are priviledged along the interface which can brings to the situation shown on figure 9.

Of course, the solution to the outer problem would be required if one needs to know the kink length (22) (the stress intensity factor k is involved).

CONCLUSION

The singular solutions computed in the first section are the leading terms of the outer expansion. They contain global informations which can be used in some phenomena of crack propagation in brittle composite materials and which coincide with the classical fracture mechanics in the simplified case of elastic homogeneous brittle materials (Bui, 1978). In the case of a strong singularity, one can obtain informations on the length and on the direction of kinks. In the case of weak singularities, the above mentioned reasoning allows to determine only the direction of propagation. However, these results derive of a very recent work and we hope that many mechanical interpretations remain to do.

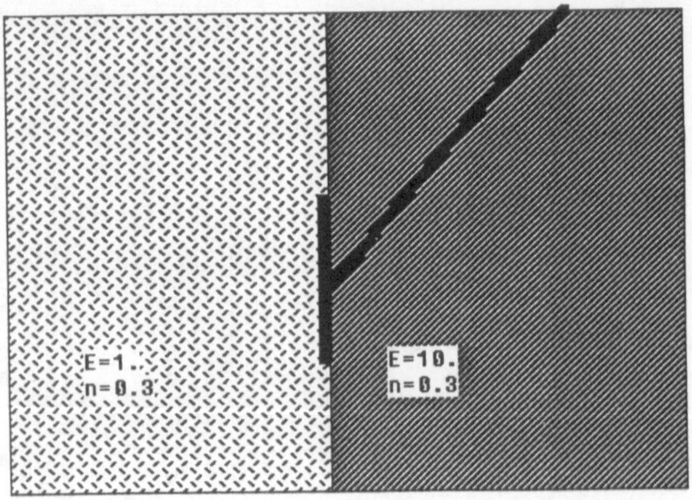

Figure 9 : *Delamination as a consequence of a strong singularity.*

References

BABUSKA I. & MILLER A., 1984, The post processing approach in the finite element method. Part 2 : the calculation of stress intensity factors, *Int. J. for Num. Met. in Eng.*, 20, p. 1111-1129.

BUI H.D., 1978, *Mécanique de la rupture fragile*, Masson, Paris.

DEMPSEY J.P. & SINCLAIR G.B., 1979, On the stress singularities in the plane elasticity of the composite wedge, *J. Elast.*, 9, 373-391.

ERDOGAN F. & BIRICICOGLU V., 1973, Two bonded half planes with a crack going through the interface, *Int. J. Eng. Sci.*, 11, 745-766.

KONDRATIEV V.A., 1967, Boundary value problems for elliptic equations in domains with conical or angular points, *Trans. Moscow Math. Soc.*, 16, 227-313.

LEGUILLON D., 1989, Calcul du taux de restitution de l'énergie au voisinage d'une singularité, *C. R. Acad. Sci. Paris*, t309, série II, 945-950.

LEGUILLON D., 1990, Comportement asymptotique du taux de restitution de l'énergie en fin de fracture, *C. R. Acad. Sci. Paris*, t310, série II, 155-160.

LEGUILLON D. & SANCHEZ-PALENCIA E., 1987a, *Computation of singular solutions in elliptic problems and elasticity*, Masson, Paris, John Wiley, New-York.

LEGUILLON D. & SANCHEZ-PALENCIA E., 1987b, Effets de bords et singularités dans les matériaux composites, *Annales des composites*, 1, 7-20.

LEGUILLON D. & SANCHEZ-PALENCIA E., 1989, rapport DRET 88/1454, Paris.

LEGUILLON D. & SANCHEZ-PALENCIA E., 1990, Solutions locales en élasticité, effets de concentration de contraintes, Colloque GRECO n°93 "Calcul des structures", Giens, France Mai 1990. A paraître dans "Calcul des strctures et intelligence artificielle", vol. 4, 1990, J.M. Fouet *et al* éditeurs, Pluralis, Paris.

MAZ'YA V.G. & NAZAROV S.A., 1988, The asymptotic behavior of energy integrals under small perturbations of the boundary near corner points and conical points, Trans. Moscow Math. Soc., 77-127.

PIPES R.B. & PAGANO N.J., 1970, Interlaminar stresses in composite laminates under uniform axial extension, *J. Comp. Materials*, 4, 538-548.

Homogenization Method Applied to Porous Media

T. Lévy

Laboratoire de Modélisation en Mécanique, associé au CNRS,
Université Pierre et Marie Curie, 4, place Jussieu, F-75252
Paris Cedex 05, France

1. INTRODUCTION

Homogenization method ([1], [2]) is used in the study of media with a microstructure on a scale which is very much smaller than the macroscopic scale of interest, the dimension of a specimen for instance. Under the assumption that the spacial distribution of the heterogeneities is, in some sense, periodic, it gives the passage from a microscopic description to a macroscopic description of a problem. Periodicity is an hypothesis which is convenient in order to obtain results in a precise mathematical form, and which may be used as a simplified model for more general situations. The fine periodic structure of the medium is associated with a small parameter ε. The homogenization process is an asymptotic study, as ε tends to zero, which gives firstly the local variations, in the period, of the field quantities and then leads to a rigourous deductive procedure for obtaining the macroscopic equations of the bulk behaviour.

This method is applied now to the linear stationary filtration of a viscous incompressible fluid in a porous rigid medium. The fluid flow (velocity $\vec{V}^\varepsilon$, pressure P^ε) satisfies

$$\mathrm{div}\vec{V}^\varepsilon = 0 \tag{1}$$

$$0 = -\,\mathrm{grad}P^\varepsilon + \mu \,\Delta\, \vec{V}^\varepsilon + \vec{f} \tag{2}$$

in the cavities ($\vec{f}$ is a constant density of forces, which may be the gravity), and the non-slip condition

$$\vec{V}^\varepsilon = 0 \tag{3}$$

on the solid boundaries.

In Section 2, the medium is supposed to be periodic with period dimensions of order ε, $\varepsilon \ll 1$. In the next sections the medium is doubly periodic, with periods of size ε and ε^2. Section 3 is devoted to the study of a medium where the solid obstacles have a doubly periodic structure : it contains small solids with characteristic dimensions of order ε and very small solids with dimensions of order ε^2. In Section 4, the cavities of the medium are distributed with a doubly periodic structure : pores of order ε separate for each other porous blocks where the pores are of order ε^2.

In this paper we are only concerned by the macroscopic equations and not by the macroscopic boundary conditions on the porous medium, consequently we may consider an infinite porous medium.

2. PERIODIC POROUS MEDIUM

2.1. - Formulation of the problem

We consider a periodic fixed porous medium, the period is a parallelepiped cell homothetic with the small ratio ε ($\varepsilon \ll 1$) of a basic period Y in which the fluid domain Y_F and the solid one Y_S have a smooth boundary Γ_y. The medium configuration is such that the fluid part is of one piece, so all the εY_F parts are connected.

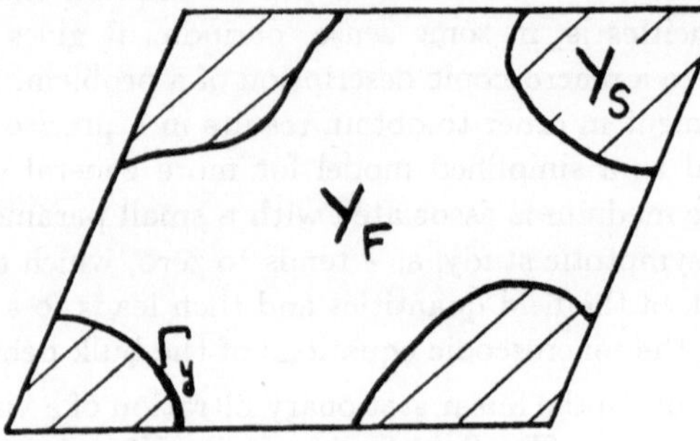

As usual in homogenization, to emphasize the two length scales : the scale of a specimen L and the scale of the pore $\ell = \varepsilon L$, the field quantities $\vec{V}^\varepsilon$, P^ε are sought in the form of double scale asymptotic expansions in power of the small parameter ε with coefficients depending both on the variable x ($x = (x_1,\, x_2,\, x_3)$ the position vector of a point in cartesian coordinates) which is the macroscopic variable, and $y = x/\varepsilon$ which is the microscopic variable. The macroscopic variable corresponds to the global structure of the fields and the microscopic variable to their local structure. To take into account the periodic distribution of the pores, we postulate that the terms of the asymptotic expansions are periodic functions of y with period Y. These expansions are substitued in equations (1)-(3) and by equating the terms of same orders in ε to zero (taking care that applied to a function of x and $y = x/\varepsilon$ the operator $\partial/\partial x_j$ becomes $\partial/\partial x_j + \varepsilon^{-1}\partial/\partial y_j$), we obtain problems with respect to the y variable which must be solved in the basic period Y, x being considered as a parameter. In order to obtain non trivial problems in Y, we choice asymptotic expansions in the form ([2], [3]) :

$$\begin{aligned} \vec{V}^\varepsilon(x) &= \varepsilon^2\, \vec{V}^0(x,y) \,+\, \varepsilon^3\, \vec{V}^1(x,y) \,+\, \ldots \\ P^\varepsilon(x) &= P^0(x) \,+\, \varepsilon P^1(x,y) \,+\, \ldots \end{aligned}$$

with $y = x/\varepsilon$, $y \in Y_F$ and functions $\vec{V}^i$, P^i Y-periodic in y.

2.2. - Local solution

In sections 2.2 and 2.3 we follow [3], to which we refer for further details.

The differential problem governing the local behaviour of the fluid is :

$$\left.\begin{aligned} \operatorname{div}_y \vec{V}^0 &= 0 \\ 0 = -\operatorname{grad}_y P_1 + \mu\, \Delta_y \vec{V}^0 &+ \vec{f} - \operatorname{grad}_x P^0 \end{aligned}\right\} \text{ in } Y_F$$
$$\vec{V}^0 = 0 \quad \text{on } \Gamma_y,$$

(the subindex y or x specifies the partial derivative). The solution is

$$\vec{V}^0(x,y) = \left(f_i - \frac{\partial P^0}{\partial x_i}\right) \vec{u}^i(y) \tag{4}$$

where $\vec{u}^i(y)$ is the unique solution of the variational problem

$$\begin{cases} \vec{u}^i \in V_y = \left\{\vec{w}, \vec{w} \in \left[H^1(Y_F)\right]^3, \ \vec{w}\ \mathrm{Y-period.},\ \vec{w} = 0 \text{ on } \Gamma_y,\ \operatorname{div}_y \vec{w} = 0\right\} \\ \text{such that} \\ \mu\left(\vec{u}^i, \vec{w}\right)_{V_y} = \int_{Y_F} w_i\, dy \qquad \forall\, \vec{w} \in V_y, \end{cases}$$

with
$$(\vec{u}, \vec{w})_{V_y} = \int_{Y_F} \frac{\partial u_j}{\partial y_k}\frac{\partial w_j}{\partial y_k} dy$$

the scalar product of the Hilbert space V_y.

And, up to a constant which may depend on x, the pressure P^1 is given by

$$P^1(x,y) = \left(f_i - \frac{\partial P^0}{\partial x_i}\right) q^i(y) \tag{5}$$

where $q^i(y)$ is the pressure associated with the velocity $\vec{u}^i(y)$ in the Stokes problem.

2.3. - Macroscopic laws

The macroscopic filtration velocity is physically defined by a surface mean value. It is small as ε^2 and determined at the first order by $\vec{V}^0$. As $\vec{V}^0(x,y)$ satisfies $\operatorname{div}_y \vec{V}^0 = 0$ in Y_F and $\vec{V}^0.\vec{n} = 0$ on Γ_y, it may be proved that its surface mean value is equal to its volume mean value.

$$\tilde{\vec{V}}^0 = \frac{1}{|Y|}\int_{Y_F} \vec{V}^0(x,y)dy$$

easier to compute.

At first order the filtration velocity is $\varepsilon^2 \vec{\tilde{V}}^0$ and satisfies the Darcy's law

$$\tilde{V}_i^0 = K_{ij}^Y \left(f_j - \frac{\partial P^0}{\partial x_j} \right) \tag{6}$$

with
$$K_{ij}^Y = \frac{1}{\mid Y \mid} \int_{Y_F} u_i^j(y) dy \; .$$

The permeability tensor K_{ij}^Y depends on the viscosity coefficient as μ^{-1} and only on the geometry of the period Y. It is symmetric and positive definite.

Another macroscopic law :

$$\mathrm{div} \vec{\tilde{V}}^0 = 0 \tag{7}$$

must be added to the Darcy's law, it expresses the macroscopic continuity equation and is obtained from the second approximation of (1)

$$\mathrm{div}_y \; \vec{V}^1 \; + \; \mathrm{div}_x \; \vec{V}^0 \; = 0$$

integrated in Y_F.

The macroscopic equations (6)-(7) determine, with suitable boundary conditions on the specimen, $P^0(x)$ and $\vec{\tilde{V}}^0(x)$. The microscopic velocity is then completly determined by (4).

3. POROUS MEDIUM MADE OF OBSTACLES WITH A DOUBLY PERIODIC STRUCTURE

The matrix of the porous medium admits two characteristic length scales : as a matter of fact it is the mixture of a porous medium and an impervious solid medium.

3.1. - Formulation of the problem

We consider two basic periods Y and Z, the solid parts are denoted Y_S and Z_S respectively, $Y_F = Y \backslash Y_S$ and $Z_F = Z \backslash Z_S$ are delimited by smooth boundaries Γ_y and Γ_z in Y and Zrespectively. The solid matrix is made by all the small parts εY_S distributed with the period εY and all the very small parts $\varepsilon^2 Z_S$ distributed with the period $\varepsilon^2 Z$ in the parts εY_F. We suppose that all the εY_F are connected and also all the $\varepsilon^2 Z_F$. This study was first investigated by J.L. Lions [4]. It is performed now in a form analogous to section 2 and which may be applied in section 4. Taking into account the geometry of the structure, we seek $\vec{V}^\varepsilon$ and P^ε in the form of triple scale asymptotic expansions. We postulate the following expressions :

$$\begin{aligned} \vec{V}^\varepsilon(x) &= \varepsilon^4 \, \vec{V}^2(x,y,z) \, + \, \varepsilon^5 \, \vec{V}^3(x,y,z) \, + \, ... \\ P^\varepsilon(x) &= P^0(x,y,z) \, + \, \varepsilon P^1(x,y,z) \, + \, \varepsilon^2 \, P^2(x,y,z) \, + \, ... \end{aligned}$$

with $y = x/\varepsilon$, $z = x/\varepsilon^2$, $y \in Y_F$ and $z \in Z_F$, the terms $\vec{V}^i$ and P^i being Y−periodic in y and Z−periodic in z.

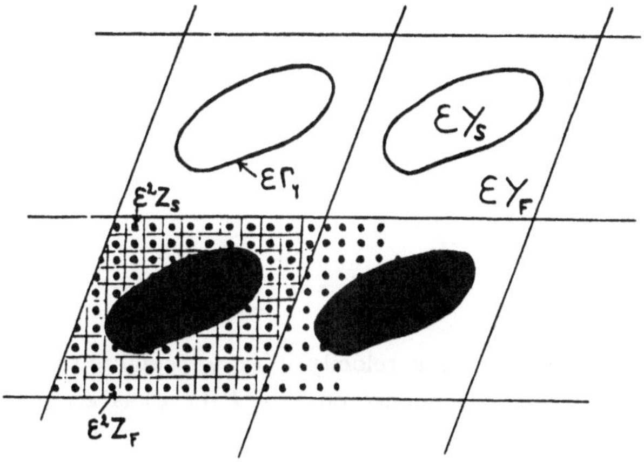

3.2. - Local solution

Substituting these expansions in problem (1)-(3), we obtain successively

$$P^0(x,y,z) = P^0(x) \ , \ P^1(x,y,z) = P^1(x,y) \ .$$

The first non trivial problem arising is :

$$\left.\begin{array}{l} \operatorname{div}_z \vec{V}^2 = 0 \\ 0 = -\operatorname{grad}_z P^2 + \mu \, \Delta_z \vec{V}^2 + \vec{f} - \operatorname{grad}_x P^0 - \operatorname{grad}_y P^1 \end{array}\right\} \quad \text{in } Z_F$$

$$\vec{V}^2 = 0 \quad \text{on } \Gamma_z.$$

As in section 2, the unique solution may be written

$$\vec{V}^2(x,y,z) = \left(f_i - \frac{\partial P^0}{\partial x_i} - \frac{\partial P^1}{\partial y_i} \right) \vec{v}^i(z) \tag{8}$$

with $\vec{v}^i(z)$ satisfying in $Z_F$the same problem than $\vec{u}^i(y)$ in Y_F (see section 2.2). This determines the variation of $\vec{V}^2$ with respect to z.

The filtration in the εY_F parts is the volume mean value in Z_F : $\varepsilon^4 \tilde{\vec{V}}^2$, it obeys the Darcy's law :

$$\tilde{V}_i^2 = K_{ij}^Z \left(f_j - \frac{\partial P^0}{\partial x_j} - \frac{\partial P^1}{\partial y_j} \right) \tag{9}$$

with $$K_{ij}^Z = \frac{1}{|Z|} \int_{Z_F} v_i^j(z) dz \ ,$$

and the continuity equation, obtained as (7) in section 2.3 :

$$\mathrm{div}_y \tilde{\vec{V}}^2 = 0. \tag{10}$$

This result may be understood in the following sense : the medium is an Y-periodic mixture of porous blocks, the εY_F, and rigid impervious solids, the εY_S. The local problem with respect to the y variable is determined by equations (9) and (10) in Y_F and the suitable boundary condition on Γ_y between a porous medium and an impervious solid [2]

$$\tilde{\vec{V}}^2 . \vec{n} = 0 \, . \tag{11}$$

This condition involving the filtration velocity $\tilde{\vec{V}}^2(x,y)$ replaces the non-slip condition on Γ_y for $\vec{V}^2(x,y,z)$. So the dependence on y of $P^1(x,y)$ is given by the Neumann problem :

$$\begin{cases} K^Z_{ij} \dfrac{\partial^2 P^1}{\partial y_i \, \partial y_j} = 0 & \text{in } Y_F \\ K^Z_{ij} \dfrac{\partial P^1}{\partial y_j} n_i = K^Z_{ij} \left(f_j - \dfrac{\partial P^0}{\partial x_j} \right) n_i & \text{on } \Gamma_y \, . \end{cases}$$

Up to a constant which may depends on x, the unique solution is

$$P^1(x,y) = \left(f_k - \frac{\partial P^0}{\partial x_k} \right) \Pi^k(y) \tag{12}$$

with $\Pi^k(y) \in H^1(Y_F)$, Y-periodic and such that

$$\int_{Y_F} K^Z_{ij} \left(\delta_{ki} - \frac{\partial \Pi^k}{\partial y_i} \right) \frac{\partial q}{\partial y_j} \, dy = 0 \quad \forall \, q \in H^1(Y_F) \text{ and } Y - \text{periodic} \, .$$

Taking (12) into account, the dependence on y of $\vec{V}^2(x,y,z)$ and $\tilde{\vec{V}}^2(x,y)$ are given by (8) and (9).

3.3. - Macroscopic laws

According to (10) and (11), as in section 2, we can defined the macroscopic filtration velocity by the volume mean value. At the first order, it is $\varepsilon^4 \tilde{\tilde{\vec{V}}}^2$ with

$$\tilde{\tilde{\vec{V}}}^2 = \frac{1}{|Y|} \int_{Y_F} \left[\frac{1}{|Z|} \int_{Z_F} \vec{V}^2(x,y,z) dz \right] dy = \frac{1}{|Y|} \int_{Y_F} \tilde{\vec{V}}^2(x,y) dy \, .$$

It obeys the Darcy's law

$$\tilde{\tilde{V}}^2_i = N_{ij} \left(f_j - \frac{\partial P^0}{\partial x_j} \right) \tag{13}$$

with

$$N_{ij} = \frac{1}{|Y|} \int_{Y_F} K_{i\ell}^{Z} \left(\delta_{j\ell} - \frac{\partial \Pi^j}{\partial y_\ell} \right) dy \; .$$

Furthermore, by integrating in Z_F the next approximation of the continuity equation (1)

$$\mathrm{div}_z \vec{V}^4 + \mathrm{div}_y \vec{V}^3 + \mathrm{div}_x \vec{V}^2 = 0$$

taking into account the $Z-$peridicity of $\vec{V}^4$, the $Y-$periodicity of $\tilde{\vec{V}}^3$ and the boundary conditions on $\Gamma_z$and Γ_y, we obtain

$$\mathrm{div}\; \tilde{\tilde{\vec{V}}}^2 = 0 \; . \tag{14}$$

The macroscopic filtration satisfies (13) and (14). As in section 2, this gives the variation with respect to x of $P^0$and $\tilde{\tilde{\vec{V}}}^2$ and then the variation of $\vec{V}^2(x, y, z)$ is entirely determined.

As a matter of fact the bulk behaviour it that of a porous medium with the permeability tensor N_{ij}, which can be written :

$$N_{ij} = \frac{1}{|Y|} \int_{Y_F} K_{\ell m}^{Z} \left(\frac{\partial y_i}{\partial y_\ell} - \frac{\partial \Pi^i}{\partial y_\ell} \right) \left(\frac{\partial y_j}{\partial y_m} - \frac{\partial \Pi^j}{\partial y_m} \right) dy \; .$$

It is symmetric and positive definite. In the case of macroscopic isotropy of the medium, we can prove that the permeability of the very small solids, the $\varepsilon^2 Z_S$, is reduced by the small solids εY_S.

4. POROUS MEDIUM MADE OF CAVITIES WITH A DOUBLY PERIODIC STRUCTURE

We study a porous medium in which the cavities introduce two characteristic length scales : for example a porous medium formed of porous blocks separated from each other by a system of fissures. In this section we follow [5] to which we refere for further details.

4.1. - Formulation of the problem

As a model, a porous medium with a doubly periodic distribution of the cavities, of order ε for what we call fissures, and of order $\varepsilon^2$for the pores in the porous blocs, is considered. As in section 3.1, we introduce the two basic periods Y and Z. The porous blocks and the fissures are distributed with the period εY and in the porous blocks εY_S the solids and the pores are distributed with the period $\varepsilon^2 Z$. All the $\varepsilon^2 Z_F$ parts are connected so the blocks εY_S are porous. In section 4.2, the fissures εY_F are connected (Fig. a), in section 4.3., they are closed (Fig. b).

Taking into account the periodicity of the medium, different forms of multiple scale asymptotic expansions of the solution are sought in the fissures and in the pores. Triple scale expansions are postulated in the pores of the porous blocks :

$$\begin{aligned} \vec{V}^{\varepsilon} &= \varepsilon^2 \vec{V}^{0S}(x,y,z) + \varepsilon^3 \vec{V}^{1S}(x,y,z) + \varepsilon^4 \vec{V}^{2S}(x,y,z) + ... \\ P^{\varepsilon} &= P^{0S}(x,y,z) + \varepsilon P^{1S}(x,y,z) + \varepsilon^2 P^{2S}(x,y,z) + ... \end{aligned}$$

with $y = x/\varepsilon$, $z = x/\varepsilon^2$, $y \in Y_S$ and $z \in Z_F$, the different terms being Y-periodic in y and Z-periodic in z. Double scale expansions are postulated in the fissures :

$$\begin{aligned} \vec{V}^{\varepsilon} &= \varepsilon^2 \vec{V}^{0F}(x,y) + \varepsilon^3 \vec{V}^{1F}(x,y) + \varepsilon^4 \vec{V}^{2F}(x,y) + ... \\ P^{\varepsilon} &= P^{0F}(x,y) + \varepsilon P^{1F}(x,y) + \varepsilon^2 P^{2F}(x,y) + ... \end{aligned}$$

with $y = x/\varepsilon$, $y \in Y_F$, the different terms being $Y-$periodic in y. The expansions are connected by continuity conditions on the blocks boundaries for the flow in the pores and in the fissures.

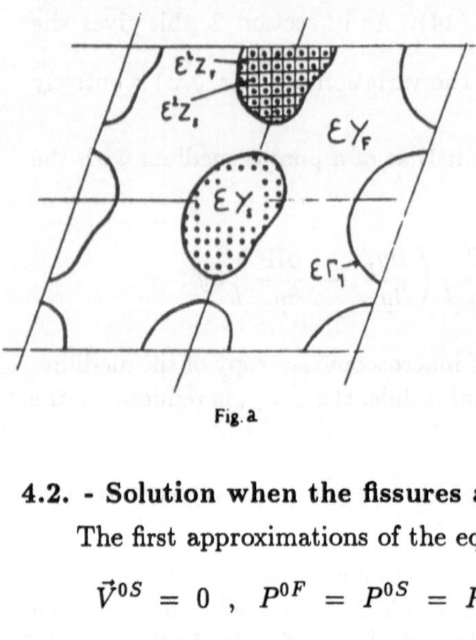

Fig. a

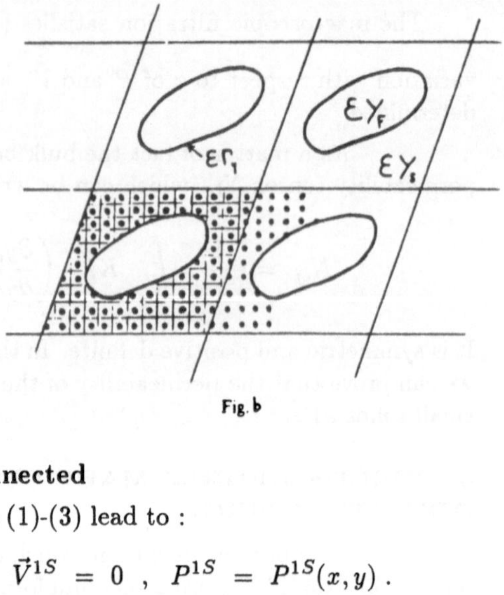

Fig. b

4.2. - Solution when the fissures are connected

The first approximations of the equations (1)-(3) lead to :

$$\vec{V}^{0S} = 0 \ , \ P^{0F} = P^{0S} = P^0(x) \ , \ \vec{V}^{1S} = 0 \ , \ P^{1S} = P^{1S}(x,y) \, .$$

Then we find that the z variation of $\vec{V}^{2S}$ is given by (8) as in section 3.2 with P^{1S} instead of P^1, the filtration velocity in the εY_S is $\varepsilon^4 \vec{V}^{2S}$ given by (9) and verifies (10). In the fissures, the y variation of $\vec{V}^{0F}$ is given by (4) and P^{1F} by (5) as when Y_S is impervious. Consequently the dependence on y of P^{1S} in the pores is determined : it results of the continuity equation (10) with the Darcy's law (9) in the porous blocks and the boundary condition

$$P^{1S}(x,y) = P^{1F}(x,y) \quad \text{on } \Gamma_y$$

which expresses the continuity of the pressure between pores and fissures. It may be written :

$$P^{1S}(x,y) = \left(f_i - \frac{\partial P^0}{\partial x_i}\right) r^i(y) \tag{15}$$

with $r^i(y)$ satisfying

$$K^Z_{\ell m} \frac{\partial^2 r^i}{\partial y_\ell \partial y_m} = 0 \quad \text{in } Y_S \ , \quad r^i = q^i \quad \text{on } \Gamma_y \ .$$

The mean filtration velocity in the porous blocks $\varepsilon^4 \tilde{\vec{V}}^{2S}$ is such that

$$\tilde{V}_i^{2S} = H_{ij}(y) \left(f_j - \frac{\partial P^0}{\partial x_j}\right)$$

with
$$H_{ij} = K^Z_{ik} \left(\delta_{jk} - \frac{\partial r^j}{\partial y_k}\right) .$$

It is related to the mean pressure gradient by a Darcy's law with a permeability tensor H_{ij} which is not that of the pores K^Z_{ij}, but is modified by the flow in the adjacent fissures.

The macroscopic filtration in the rock, according to the expansions of $\vec{V}^\varepsilon$ in the pores and in the fissures, is not influenced in the first approximation by the pores of the blocks : it is, as in section 2, $\varepsilon^2 \tilde{\vec{V}}^{0F}$ with the Darcy's law (6) and the continuity equation (7). As in section 2, this determines the variation with respect to the macroscopic variable x of the local solutions.

4.3. - Solution when the fissures are closed

In the pores, as in section 4.2

$$\vec{V}^{0S} = 0 \ , \ P^{0F} = P^{0S} = P^0(x) \ , \ \vec{V}^{1S} = 0 \ , \ P^{1S} = P^{1S}(x,y) \ ,$$

$\vec{V}^{2S}$ is given by (8), the mean filtration velocity in the porous blocks is $\varepsilon^4 \tilde{\vec{V}}^{2S}$ with (9) and (10). In the fissures, $\vec{V}^{0F} = 0$ because Y_F is closed, so, up to a constant which may depend on x :

$$P^{1F} = \left(f_i - \frac{\partial P^0}{\partial x_i}\right) y_i \tag{16}$$

and it can be proved that $\vec{V}^{1F} = 0$. Then, also in the fissures the velocity is of order ε^4.The variation with respect to y of P^{1S} is determined by a Dirichlet problem in Y_S as in section 4.2, its expression is of the form (15) with $r^i(y)$ satisfying the same differential equation in Y_S and the boundary condition

$$r^i = y_i \text{ on } \Gamma_y,$$

according to (16).

Then the variation with respect to z and y of the fields in the porous blocks is known.

To obtain the macroscopic filtration in the rock it not necessary to use explicitly the local variation of $\vec{V}^{2F}$. The use of the continuity equation

$$\mathrm{div}_y \vec{V}^{2F} = 0 \quad \text{in } Y_F$$

and the matching condition

$$\vec{V}^{2F}.\vec{n} = \tilde{\vec{V}}^{2S}.\vec{n} \quad \text{on } \Gamma_y$$

obtained by writting the masse conservation law at the first order in a control volume flattened on the fissures boundary, allow us to calculate the filtration velocity in the rock. As in section 2, it may by expressed by the volume mean value, it is at the first order $\varepsilon^4 \tilde{\tilde{\vec{V}}}^2$ with :

$$\tilde{\tilde{\vec{V}}}^2 = \frac{1}{|Y|}\left[\int_{Y_F} \vec{V}^{2F}(x,y)dy + \int_{Y_S} \tilde{\vec{V}}^{2S}(x,y)dy\right] .$$

It obeys the Darcy's law

$$\tilde{\tilde{V}}_i^2 = M_{ij}\left(f_j - \frac{\partial P^0}{\partial x_j}\right) \tag{17}$$

with

$$M_{ij} = (1 + \Pi_F)K_{ij}^Z + \frac{1}{|Y|}\int_{Y_S} K_{\ell k}^Z \frac{\partial r^i}{\partial y_\ell}\frac{\partial r^j}{\partial y_k}dy ,$$

$\Pi_F = |Y_F| / |Y|$ is the porosity of the fissures.

As K_{ij}^Z, this permeability tensor M_{ij} depends on the viscosity as μ^{-1}, it is also evidently symmetric and positive definite. The macroscopic continuity equation

$$\mathrm{div}\, \tilde{\tilde{\vec{V}}}^2 = 0 \tag{18}$$

is also valid in this case. The bulk filtration is governed by (17) and (18), as previously this determines the variation with respect to the macroscopic variable x of P^{1S}, P^{1F}, $\vec{V}^{2S}$ and $\tilde{\vec{V}}^{2S}$.

4.4. - Remark

The case of a porous medium with unidirectional fissures may be studied using 4.2 and 4.3 [6]. The global filtration satisfies too a Darcy's law, and is very much slower in the directions orthogonal to the fissures connection. Homogenization method determines the coefficients of the permeability tensor.

5. CONCLUSIONS

The use of an homogenization method gives the macroscopic filtration law and the local behaviour of the velocity and pressure. It emphasizes the importance of the microstructure geometry which determines not only the bulk coefficients but also the magnitude of the phenomena as pointed out by the different orders with respect to ε of the filtration velocity in the different studied cases.

References

[1] BENSOUSSAN A., LIONS J.L., PAPANICOLAOU G., 1978, *Asymptotic Analysis for Periodic Structures*, North-Holland.

[2] SANCHEZ-PALENCIA E., 1980, *Non-Homogeneous Media and Vibration Theory*, Lect. Notes Phys., 127, Springer-Verlag.

[3] LEVY T., 1987, Fluids in porous media and suspensions, *Lect. Notes Phys., 272*, 63-119, Springer-Verlag.

[4] LIONS J.L., 1981, *Some Methods in the Mathematical Analysis of Systems and their Control*, Science Press.

[5] LEVY T., 1990, Filtration in a porous fissured rock : influence of the fissure connexity, *Eur. J. Mech.,B/Fluids, 9*, n°4, 309-327.

[6] LEVY T., 1990, Ecoulement dans un milieu poreux avec fissures unidirectionnelles, *C.R. Acad. Sci., Paris, 310*, 685-690.

Modal Analysis of Flexible Multibody Systems

M. Pascal

**Laboratoire de Modélisation en Mécanique, associé au CNRS,
Université Pierre et Marie Curie, 4 place Jussieu,
F-75252 Paris Cedex 05, France**

Summary : We present a distributed-element method for vibration analysis of flexible multibody systems modelled by a chain of rigid and elastic bodies with tree structure. The method is based on the impedance matrix, which defines in frequency domain the linear transformation between the resultant forces and torques exerted on the boundaries of each body in the chain and the displacements of these boundaries. This impedance matrix is obtained by a spectral expansion in terms of a set of component modes.

1. INTRODUCTION

The aim of this work is the computation of natural frequencies and the associated vibrations modes of a flexible multibody system. In structural dynamics, the finite element method is very often used but for flexible structures made up from standard elements such as beams, strings and membranes, the distributed element method has been successful, leading in several cases to less computer time. This paper will discuss new research perspectives for the use of analytic representation of structural dynamics.

2. PROBLEM FORMULATION

Let the multibody systems consist of $(n + 1)$ bodies (S_i) $(i = 0, 1, ..., n)$ interconnected by n hinges ℓ_a $(a = 1, ..., n)$. The only external forces and torques are exerted on the first body which is assumed to be rigid. These external actions are represented by a force $\vec{F}_0(t)$ in the centre of mass G_0 of (S_0) and a torque $\vec{M}_0(t)$. The multibody system undergoes small vibrations around an equilibrium position in which the flexible parts are undeformed.

2.1. - Kinematics of motion of two contiguous bodies relative to one another

Let us assumed that each body (S_i) has at most two hinges attached to it (Fig. 1) and each hinge (ℓ_a) is supposed to connect only two bodies named $S_{i+(a)}$ and $S_{i-(a)}$ [1]. We assume that on one part $\gamma_{i+(a)}$ of the boundary of $S_{i+(a)}$, the elastic displacement is a rigid displacement ; the same assumption is made for one part $\gamma_{i-(a)}$ of the boundary of $S_{i-(a)}$. The relative motion of the flexible body $S_{i-(a)}$ with respect to $S_{i+(a)}$ is described by the relative motion of the rigid interface $\gamma_{i-(a)}$ with respect to the rigid interface $\gamma_{i+(a)}$: the relative position of $S_{i+(a)}$ with respect to $S_{i-(a)}$ depends on N_a

degrees of freedom, with $1 \leq N_a \leq 6$ according to the kind of articulation between $S_{i+(a)}$ and $S_{i-(a)}$.

2.2. - Equations of motion for a flexible appendage

Let us first consider the motion of one flexible body (S_i), of mass m_i and centre of mass G_i, articulated to two other flexible bodies by means of rigid links (γ_i) and (γ_i') (Fig. 2). Let (T_i) and (T_i') with origins O_i and O_i' two triads rigidly connected to the rigid interfaces (γ_i) and (γ_i'). The flexible body (S_i) undergoes small motions around an equilibrium configuration in which this body is undeformed. Let (T_i^0) be the position of the triad (T_i) in the equilibrium configuration. With repsect to (T_i^0), the position of (T_i) is given by a translation vector $\vec{r}_i$ and a rotation vector $\vec{\alpha_i}$, $\vec{\delta_i}$ is the rotation vector of (T_i') with respect to (T_i). Let us denotes by $\vec{u_i}(\bar{P},t) = \overrightarrow{\bar{P}P}$ the elastic displacement of every material point P of (S_i), $\bar{P}$ being the undeformed position of P, with $\vec{x} = \overrightarrow{O_i\bar{P}}$. It is assumed that $\vec{r_i}$, $\vec{\alpha_i}$, $\vec{\delta_i}$ are small vectors, of the same order of smallness as the displacement field $\vec{u_i}$.

The linearized equations of motion of the flexible body (S_i) are given by [2] :

$$\left.\begin{aligned} L_i\,[u_i] \;+\; \rho_i\,(\ddot{r}_i - \tilde{x}\ddot{\alpha}_i + \ddot{u}_i) &= 0 \quad \text{in } (D_i) \\ u_i &= 0 \quad \text{on } (\gamma_i) \\ u_i &= u_{0_i'} - \left(\tilde{x} - \tilde{x}_{0_i'}\right)\delta_i \quad \text{on } (\gamma_i') \\ \sigma_i(u_i)\nu_i &= 0 \quad \text{on } (\Gamma_{F_i}) = \partial D_i - (\gamma_i) \cup (\gamma_i') \end{aligned}\right\} \qquad (1)$$

Here (D_i) the domain occupied by the flexible body in its undeformed configuration, with boundary ∂D_i, $\vec{\nu_i}$ is the unitarian vector of the outward normal to ∂D_i, ρ_i is the mass density of (S_i), L_i is a linear self-adjoint differential operator with respect to the components of vector $\vec{x}$ in the (T_i) frame. $\overrightarrow{u_{0_i'}}$ and $\overrightarrow{x_{0_i'}}$ are the elastic displacement of $0_i'$ and the vector locating the undeformed position of this point in the (T_i) frame. σ_i is the stress tensor.

For every vector $\vec{X}$, X denotes the matrix column of its components (X_1, X_2, X_3) in the (T_i) frame and $\tilde{X}_i$ the skew symmetric matrix

$$\tilde{X} = \begin{pmatrix} 0 & -X_1 & X_2 \\ X_3 & 0 & -X_2 \\ -X_2 & X_1 & 0 \end{pmatrix}$$

The components in the (T_i) frame of the resultant force $\vec{F_i}$ and resultant torque $\vec{M_i}$ exerted on (γ_i) are :

$$F_i = \int_{\gamma_i} \sigma_i(u_i)\nu_i \, dS \qquad M_i = \int_{\gamma_i} \tilde{x}_i\, \sigma_i(u_i)\nu_i \, dS \qquad (2)$$

Similar expressions are obtained for the force and the torque exerted on (γ_i') :

$$F_i' = \int_{\gamma_i'} \sigma_i(u_i)\nu_i \, dS \qquad M_i' = \int_{\gamma_i'} \left(\tilde{x} - \tilde{x}_{0_i'}\right) \sigma_i(u_i)\nu_i \, dS \qquad (2)'$$

The global motion of (S_i) is deduced from (1), (2) and (2)' :

$$\left.\begin{array}{rcl} m_i\left(\ddot{r}_i - \tilde{l}_i\ddot{\alpha}_i\right) + \int_{D_i} \rho_i \ddot{u}_i dx & = & F_i + F_i' \\ m_i\tilde{l}_i\ddot{r}_i + I_i\ddot{\alpha}_i + \int_{D_i} \rho_i \tilde{x}\ddot{u}_i dx & = & M_i + M_i' + \tilde{x}_{0_i'}F_i' \end{array}\right\} \qquad (3)$$

Here $\overrightarrow{\ell_i}$ locates the undeformed position of G_i in $(\mathcal{T}_i)$ and I_i is the inertia tensor in 0_i of the body (S_i) in its undeformed configuration. By use of Fourier's transformation, we obtain from equations (1), (2), (2)', the following equations :

$$\left.\begin{array}{rcll} L_i[\bar{v}_i] - \rho_i\omega^2\bar{v}_i & = & 0 & \text{in } (D_i) \\ \bar{v}_i & = & \bar{r}_i - \tilde{x}\,\bar{\alpha}_i & \text{on } (\gamma_i) \\ \bar{v}_i & = & \bar{r}_i' - \left(\tilde{x} - x_{0_i'}\right)\bar{\alpha}_i' & \text{on } (\gamma_i') \\ \sigma_i(\bar{v}_i)\,\nu_i & = & 0 & \text{on } (\Gamma_{F_i}) \end{array}\right\} \qquad (4)$$

$$\bar{F}_i = \int_{\gamma_i} \sigma_i(\bar{v}_i)\,\nu_i \, dS \qquad \bar{M}_i = \int_{\gamma_i} \tilde{x}\sigma_i(\bar{v}_i)\,\nu_i \, dS \qquad (5)$$

$$\bar{F}_i' = \int_{\gamma_i'} \sigma_i(\bar{v}_i)\,\nu_i \, dS \qquad \bar{M}_i' = \int_{\gamma_i} \left(\tilde{x} - \tilde{x}_{0_i'}\right) \sigma_i(\bar{v}_i)\,\nu_i \, dS \qquad (5)'$$

Here $\bar{X}(\omega)$ is the Fourier's transform of $\overrightarrow{X_{(t)}}$,

$$\left.\begin{array}{rcl} \bar{v}_i & = & \bar{u}_i + \bar{r}_i - \tilde{x}\,\bar{\alpha}_i \\ \bar{r}_i' & = & \bar{u}_{0_i'} + \bar{r}_i - \tilde{x}_{0_i'}\bar{\alpha}_i \\ \bar{\alpha}_i' & = & \bar{\alpha}_i + \bar{\delta}_i \end{array}\right\}$$

2.3. - Equations of motion for the main rigid body

The linearized equations of motion of (S_0) are

$$\left.\begin{array}{rcl} m_0\ddot{R} & = & F_0 + F_0' \\ I_0\ddot{\theta} & = & M_0 + \tilde{x}_{0_i}F_0' + M_0' \end{array}\right\} \qquad (6)$$

Here m_0 and I_0 are the mass and the inertia tensor in G_0 of (S_0), R and θ are the column matrices of the $(\mathcal{T}_0)$ components of the vectors $\vec{R}$ and $\vec{\theta}$ giving the inertial position of a triad $(\mathcal{T}_0)$ rigidly connected to (S_0) with respect to an inertial reference frame. This inertial frame is chosen to have R and θ small.

(F_0, M_0) and (F'_0, M'_0) are the $(\mathcal{T}_0)$ components of $\left(\overrightarrow{F_0}, \overrightarrow{M_0}\right)$, $\left(\overrightarrow{F'_0}, \overrightarrow{M'_0}\right)$ where to two last vectors are the resultant force and the resultant torque in O_1 introduced by the link ℓ_1.

In frequency domain, we obtain :

$$\left.\begin{array}{rcl} -\,m_0\omega^2\,\bar{R} & = & \bar{F_0} + \bar{F'_0} \\ -\,I_0\omega^2\,\bar{\theta} & = & \bar{M_0} + \tilde{x}_{0_1}\,\bar{F'_0} + \bar{M'_0} \end{array}\right\} \qquad (6)'$$

3. IMPEDANCE AND INERTANCE MATRICES

The problem formulated by equations (4), (5) and (5)' being a linear problem, we can define a linear transformation depending on the frequency ω giving the resultant forces and torques $\bar{Q}_i = {}^t(\bar{F}_i, \bar{M}_i, \bar{F}'_i, \bar{M}'_i)$ exerted on the boundaries (γ_i) and (γ'_i) in terms of the displacements $\bar{q}_i = {}^t(\bar{r}_i, \bar{\alpha}_i, \bar{r}'_i, \bar{\alpha}'_i)$ of these boundaries :

$$\bar{Q} = Z_i(\omega)\bar{q}_i.$$

$Z_i(\omega)$ is a symmetrical matrix of dimensions 12 named the impedance matrix of (S_i).
The inverse transformation is given by $\bar{q}_i = \mathcal{H}_i(\omega)\bar{Q}_i$.
Here $\mathcal{H}_i(\omega) = [Z_i(\omega)]^{-1}$ is the inertance matrix of (S_i).
For the main rigid body (S_0), the impedance matrix $Z_0(\omega)$ is of dimensions 6 :

$$\left.\begin{array}{rcl} \bar{Q}_0 & = & Z_0(\omega)\bar{q}_0 \\ \bar{Q}_0 & = & {}^t(\bar{F_0} + \bar{F'_0}\,,\ \bar{M_0} + \bar{M'_0} + \tilde{x}_{0_1}\bar{F'_0}) \\ \bar{q}_0 & = & {}^t(\bar{R}, \bar{\theta}) \\ Z_0(\omega) & = & -\,\omega^2\begin{pmatrix} m_0 E & 0 \\ 0 & I_0 \end{pmatrix} \end{array}\right\}$$

E is the identity matrix of order 3.

4. COMPONENT MODES

The component modes of a flexible body (S_i) are the modes of vibrations of the body when it vibrates independently with respect to the other parts of the whole system. Two sets of component modes can be defined :

4.1. - Constrained modes

Constrained modes are vibrations modes of (S_i) with the rigid links (γ_i) and (γ'_i) fixed $(\bar{r}_i = \bar{\alpha}_i = \bar{r}'_i = \bar{\alpha}'_i = 0)$. They are defined by :

$$\left.\begin{array}{rcll} v_i & = & V_{n_i}(x)\,cos\,(\Omega_{n_i}t) & n = 1, 2, \ldots \\ L_i\,[V_{n_i}] \;-\; \rho_i\Omega^2_{n_i}\,V_{n_i} & = & 0 & \text{in } (D_i) \\ V_{n_i} & = & 0 & \text{on } (\gamma_i) \\ V_{n_i} & = & 0 & \text{on } (\gamma'_i) \\ \sigma_i\,(V_{n_i})\,\nu_i & = & 0 & \text{on } (\Gamma_{F_i}) \end{array}\right\}$$

We obtain an infinite set of constrained modes, with orthogonality property :

$$\int_{D_i} \rho_i\,{}^tV_{p_i}V_{n_i}\,dx \;=\; 0 \text{ if } \Omega^2_{n_i} \;-\; \Omega^2_{p_i} \;\neq\; 0$$

The constrained modes can be normalized by setting.

$$\int_{D_i} \rho_i \, {}^tV_{n_i} V_{n_i} \, dx = 1 \qquad n = 1, 2, \ldots$$

4.2. - Global modes

Global modes are vibrations modes of (S_i) with the rigid links (γ_i) and (γ_i') free $(\bar{F}_i = \bar{M}_i = \bar{F}_i' = \bar{M}_i' = 0)$. They are defined by :

$$\left.\begin{array}{rcll} v_i & = & v_{n_i}(x) \cos(\omega_{n_i} t) & n = 1, 2, \ldots \\ L_i[v_{n_i}] - \rho\, \omega_{n_i}^2 \, v_{n_i} & = & 0 & \text{in } (D_i) \\ \sigma_i(v_{n_i})\, \nu_i & = & 0 & \text{on } (\Gamma_{F_i}) \\ v_{n_i} & = & r_{n_i} - \tilde{x}\, \alpha_{n_i} & \text{on } (\gamma_i) \\ v_{n_i} & = & r_{n_i}' - \left(\tilde{x} - \tilde{x}_{0_i'}\right) \alpha_{n_i}' & \text{on } (\gamma_i') \end{array}\right\}$$

$$\left.\begin{array}{ll} \displaystyle\int_{\gamma_i} \sigma(v_{n_i})\, \nu_i \, dS = 0 & \displaystyle\int_{\gamma_i'} \sigma(v_{n_i})\, \nu_i \, dS = 0 \\ \displaystyle\int_{\gamma_i} \tilde{\sigma}(v_{n_i})\, \nu_i \, dS = 0 & \displaystyle\int_{\gamma_i'} \tilde{x}\sigma(v_{n_i})\, \nu_i \, dS = 0 \end{array}\right\}$$

Again, we obtain an infinite set of orthogonal vibrations modes.

5. DERIVATION OF THE IMPEDANCE/INERTANCE MATRICES IN TERMS OF COMPONENT MODES

In some particular cases like beams in axial, torsional and bending vibrations, extensible strings or membranes under uniform tension [3], it is possible to obtain an analytical representation of the matrices $Z_i(\omega)$ and $\mathcal{H}_i(\omega)$. In more general cases, for which a closed form solution for these matrices is not available, we obtain a spectral expension of $Z_i(\omega)$ and $\mathcal{H}_i(\omega)$ in terms of components modes.

5.1. - Derivation of the impedance matrix in terms of constrained modes

A spectral expansion of the impedance matrix $Z_i(\omega)$ of each flexible appendage (S_i) is obtained in the form :

$$\left.\begin{array}{rcl} Z_i(\omega) & = & Z_i(0) - \omega^2 \mathcal{M}_{0_i} - \omega^4 \displaystyle\sum_{1}^{\infty} \frac{K_{n_i}}{\Omega_{n_i}^2 - \omega^2} \\ \mathcal{M}_{O_i} & = & \displaystyle\int_{D_i} \rho_i \, {}^t[\varphi_i][\varphi_i] \, dx \\ K_{n_i} & = & \Gamma_{n_i} \, {}^t\Gamma_{n_i} \\ \Gamma_{n_i} & = & \displaystyle\int_{D_i} \rho_i \, {}^t[\varphi_i] \, V_{n_i} \, dx \end{array}\right\}$$

Here $[\varphi_i]$ is the matrix of static deformation modes of (S_i) : $[\varphi_i]$ is a rectangular matrix of dimensions (3×12) :

$$[\varphi_i] = [\{\varphi_{i1}\}, \{\varphi_{i2}\}, \{\varphi_{i3}\}, \{\varphi_{i4}\}]$$

$\{\varphi_{ip}\}$ $(p = 1,2,3,4)$ are defined by :

$$\left.\begin{array}{lcll} L_i[\{\varphi_{ip}\}] & = & 0 & \text{in } \; D_i) \\ \sigma_i[\{\varphi_{ip}\}]\,\nu_i & = & 0 & \text{on } \; (\Gamma_{Fi}) \\ \{\varphi_{ip}\} & = & E\,\delta_{p1} - \tilde{x}\,\delta_{p2} & \text{on } \; (\gamma_i) \\ \{\varphi_{ip}\} & = & E\,\delta_{p3} - \left(\tilde{x} - \tilde{x}_{O'_i}\right)\delta_{p4} & \text{on } \; (\gamma'_i) \end{array}\right\}$$

δ_{pK} $(p, K = 1, ..., 4)$ are the Kronecker's symbols.

A convergence property of the matrices K_{ni} (modal gains matrix of the nth constrained mode) holds :

$$\sum_1^\infty K_{n_i} = \mathcal{M}_{0_i}$$

This convergence property provides an upper bound of the residual modal gains when only a finite number of terms is used in the spectral expansion (7).

5.2. - Derivation of the inertance matrix in terms of global modes

A spectral expansion of the inertance matrix $\mathcal{H}_i(\omega)$ of each flexible appendage (S_i) is obtained in the form.

$$\left.\begin{array}{lcl} \mathcal{H}_i(\omega) & = & \mathcal{H}_{i0} - \dfrac{{}^tN_i\,\mathcal{M}_i^{-1}N_i}{\omega^2} + \displaystyle\sum_{n=1}^{\infty} \frac{k_{n_i}\,\omega_{n_i}^2\,\omega^2}{\omega_{n_i}^2 - \omega^2} \\ N_i & = & \begin{pmatrix} E & 0 & E & 0 \\ 0 & E & \tilde{x}_{0'_i} & E \end{pmatrix} \quad \text{rectangular matrix } (6 \times 12) \\ \mathcal{M}_i & = & \begin{pmatrix} m_iE & \tilde{l}_i \\ m_i\tilde{l}_i & I_i \end{pmatrix} \quad \begin{array}{c}\text{mass matrix in} \\ 0_i \text{ of body } (S_i) \text{ undeformed}\end{array} \\ k_{n_i} & = & \gamma_{ni}\,{}^t\gamma_{ni} \qquad \gamma_{ni} = \displaystyle\int_{D_i} \rho_i\,{}^t[P_i]\,v_{ni}\,dx \end{array}\right\} \quad (8)$$

Here $[P_i] = [\{P_{1i}\}, \{P_{2i}\}, \{P_{3i}\}, \{P_{4i}\}]$ is the matrix of attachment modes defined as static deformation modes of (S_i) subjected to unitarian forces or torques on the boundaries (γ_i) or (γ'_i) and corresponding force elements equilibrating these actions.

A convergence property of the matrix k_{ni} (modal gains matrix of the nth global mode) holds :

$$\left.\begin{array}{lcl} \displaystyle\sum_1^\infty k_{n_i} & = & \mathcal{M}^*_{0i} \\ \mathcal{M}^*_{0i} & = & \displaystyle\int_{D_i} \rho_i\,{}^t[P_i]\,[P_i]dx \end{array}\right\}$$

6. REDUCED IMPEDANCE MATRIX OF THE WHOLE SYSTEM

For each body (S_i) $(i = 0, 1, ..., n)$ we obtain an impedance matrix $Z_i(\omega)$ giving the forces and torques exerted on the boundaries in terms of the displacements of these boundaries ·

$$\overline{Q_i} = Z_i(\omega)\,\overline{q_i} \qquad i = (0, 1, ..., n)$$

It results for the whole system the following relation :

$$\left.\begin{array}{ll} \bar{Q} = Z(\omega)\bar{q} & \\ \bar{Q} = {}^t\left(\overline{Q_0}, \overline{Q_1}, \ldots, \overline{Q_n}\right) & Z_{(\omega)} = \begin{pmatrix} Z_0 & & 0 \\ & \ddots & \\ 0 & & Z_n \end{pmatrix} \\ \bar{q} = {}^t\left(\overline{q_0}, \overline{q_1}, \ldots \overline{q_n}\right) & \end{array}\right\}$$

$Z(\omega)$ is a symmetrical matrix of order $12n + 6$. The set of displacements q is not a set of independant displacements. Each link ℓ_i introduces kinetical linear constraints between the components of q.

We can write the general form of these constraints as : $q = T\,\hat{q}$. Here $\hat{q}$ is the column matrix of a set of independant displacements of the whole system and T is a constant rectangular matrix of maximum rank. It results for the whole system the relation

$$\left.\begin{array}{rcl} \bar{\hat{Q}} & = & \hat{Z}(\omega)\,\bar{\hat{q}} \\ \hat{Z}(\omega) & = & {}^tT\,Z\,T \qquad \bar{\hat{Q}} = {}^tT\bar{Q} \end{array}\right\} \tag{9}$$

$\hat{Z}(\omega)$ is the reduced impedance matrix of the whole system ; $\bar{\hat{Q}}$ has the following form : $\bar{\hat{Q}} = \mathcal{F} + K(\omega)\bar{\hat{q}}$.

Here $\mathcal{F}$ is the column matrix of the external actions for the whole system (exerted only on the main rigid body (S_0)) and $K(\omega)$ is a stiffness matrix associated with the servomotors located in the links.

The final form of the equation (9) is

$$\left.\begin{array}{rcl} H(\omega)\bar{\hat{q}} & = & \bar{\mathcal{F}} \\ H(\omega) & = & Z(\omega) - K(\omega) \end{array}\right\}$$

A modal analysis of the coupled system can be made from (9)' : the equation $det\ H(\omega) = 0$ gives the frequencies of vibrations of the whole system.

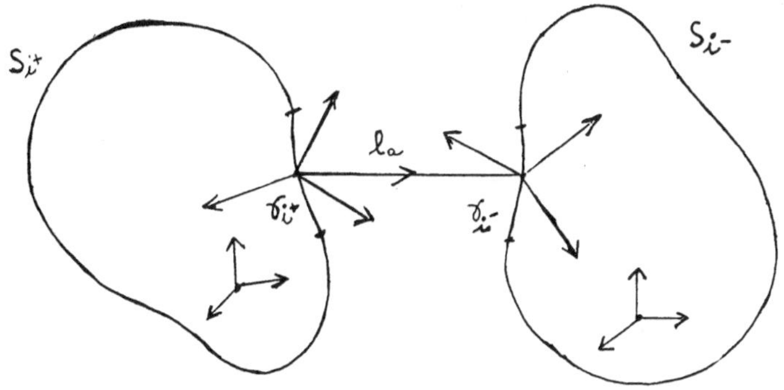

Fig. 1 : Kinematics of two contiguous bodies

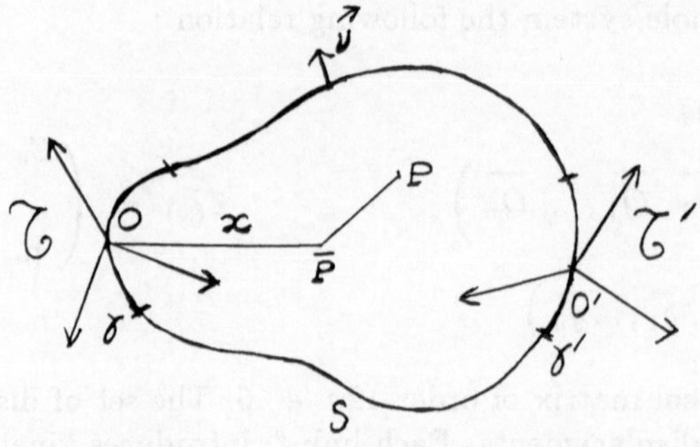

Fig. 2 : Dynamics of one flexible body

References

[1] WITTENBURG J., *Dynamics of systems of Rigid Bodies.* Stuttgart, Teubner, 1977.

[2] PASCAL M., Dynamics Analysis of a System of Hinge-connected Flexible Bodies. *Celest. Mech.*, *41*, (1988), 253-274.

[3] KOLOUSEK V., *Dynamics in Engineering Structures.* London, Butterworths, 1973.

Index of Contributors

Springer Series in Computational Physics

A Computational Method in Plasma Physics
F. Bauer, O. Betancourt, P. Garabedian

Implementation of Finite Element Methods for Navier-Stokes Equations
F. Thomasset

Finite-Difference Techniques for Vectorized Fluid Dynamics Calculations
Edited by D. Book

Unsteady Viscous Flows. D. P. Telionis

Computational Methods for Fluid Flow. R. Peyret, T. D. Taylor

Computational Methods in Bifurcation Theory and Dissipative Structures
M. Kubicek, M. Marek

Optimal Shape Design for Elliptic Systems. O. Pironneau

The Method of Differential Approximation. Yu. I. Shokin

Computational Galerkin Methods. C. A. J. Fletcher

Numerical Methods for Nonlinear Variational Problems
R. Glowinski

Numerical Methods in Fluid Dynamics. Second Edition
M. Holt

Computer Studies of Phase Transitions and Critical Phenomena
O. G. Mouritsen

Finite Element Methods in Linear Ideal Magnetohydrodynamics
R. Gruber, J. Rappaz

Numerical Simulation of Plasmas. Y. N. Dnestrovskii, D. P. Kostomarov

Computational Methods for Kinetic Models of Magnetically Confined Plasmas
J. Killeen, G. D. Kerbel, M. C. McCoy, A. A. Mirin

Spectral Methods in Fluid Dynamics. Second Edition
C. Canuto, M. Y. Hussaini, A. Quarteroni, T. A. Zang

Computational Techniques for Fluid Dynamics 1. Second Edition
Fundamental and General Techniques. C. A. J. Fletcher

Computational Techniques for Fluid Dynamics 2. Second Edition
Specific Techniques for Different Flow Categories. C. A. J. Fletcher

Methods for the Localization of Singularities in Numerical Solutions of Gas Dynamics Problems. E. V. Vorozhtsov, N. N. Yanenko

Classical Orthogonal Polynomials of a Discrete Variable
A. F. Nikiforov, S. K. Suslov, V. B. Uvarov

Flux Coordinates and Magnetic Field Structure:
A Guide to a Fundamental Tool of Plasma Theory
W. D. D'haeseleer, W. N. G. Hitchon, J. D. Callen, J. L. Shohet